Emergence of Dynamical Order

Synchronization Phenomena in Complex Systems

WORLD SCIENTIFIC LECTURE NOTES IN COMPLEX SYSTEMS

AIMS AND SCOPE

The aim of this new interdisciplinary series is to promote the exchange of information between scientists working in different fields, who are involved in the study of complex systems, and to foster education and training of young scientists entering this rapidly developing research area.

The scope of the series is broad and will include: Statistical physics of large nonequilibrium systems; problems of nonlinear pattern formation in chemistry; complex organization of intracellular processes and biochemical networks of a living cell; various aspects of cell-to-cell communication; behaviour of bacterial colonies; neural networks; functioning and organization of animal populations and large ecological systems; modeling complex social phenomena; applications of statistical mechanics to studies of economics and financial markets; multi-agent robotics and collective intelligence; the emergence and evolution of large-scale communication networks; general mathematical studies of complex cooperative behaviour in large systems.

Published
Vol. 1 Nonlinear Dynamics: From Lasers to Butterflies

World Scientific Lecture Notes
in Complex Systems– Vol. 2

Susanna C. Manrubia
Instituto Nacional de Técnica Aeroespacial, Spain

Alexander S. Mikhailov
Fritz-Haber-Institut der Max-Planck-Gesellschaft, Germany

Damián H. Zanette
Centro Atómico Bariloche, Argentina

Emergence of Dynamical Order

Synchronization Phenomena in Complex Systems

World Scientific

NEW JERSEY • LONDON • SINGAPORE • SHANGHAI • HONG KONG • TAIPEI • CHENNAI

Published by

World Scientific Publishing Co. Pte. Ltd.
5 Toh Tuck Link, Singapore 596224
USA office: Suite 202, 1060 Main Street, River Edge, NJ 07661
UK office: 57 Shelton Street, Covent Garden, London WC2H 9HE

British Library Cataloguing-in-Publication Data
A catalogue record for this book is available from the British Library.

EMERGENCE OF DYNAMICAL ORDER
Synchronization Phenomena in Complex Systems

ISBN 981-238-803-6

Printed in Singapore by World Scientific Printers (S) Pte Ltd

Preface

The last decade has brought a rapid increase of the interest in synchronization phenomena. Spontaneous synchronization is found in a broad class of systems of various origins, ranging from physics and chemistry to biology and social sciences. It is characteristic both for uniform and complexly organized populations, or networks. These phenomena include a variety of collective dynamical behaviors. Their common feature is however that they bring about dynamical order and lead to the emergence of new structural organization. In that sense, they are analogous to phase transitions and critical phenomena in physical systems.

In this book, we provide a systematic discussion of the concepts related to the emergence of collective dynamical order. Today, there are already several monographs devoted to different aspects of synchronization processes. However, a detailed exposition of recent results which involve spontaneous synchronization and dynamical clustering in large systems, requiring a statistical description, has so far been missing. Another distinguishing feature of the book is that it also includes a presentation of important applications of this theory in chemistry, cell biology, and brain science.

We hope that this book will be interesting and useful for researchers and students from different disciplines. Some basic knowledge of nonlinear dynamics and statistical mechanics is expected. This monograph can also serve as a base for graduate courses on synchronization phenomena.

Though the three authors are dispersed over the globe, we have collaborated for many years. To a large extent, this book is an outcome of our joint work and vivid conversations in Berlin, at the Fritz Haber Institute of the Max Planck Society. With respect to applications in molecular biology, we have learnt much from our extended contacts with the late Benno

Hess. The financial assistance of the Alexander von Humboldt Foundation (Germany) is gratefully acknowledged. We want to express our gratitude to H. Cerdeira, J. Hudson, K. Kaneko, and Y. Kuramoto for stimulating discussions. Many results included in this book have been presented in Berlin at the seminars and colloquia of the Joint Research Program on Complex Nonlinear Processes, and we thank its participants, particularly B. Blasius, W. Ebeling, J. Kurths, A. Pikovsky, E. Schöll, and L. Schimansky–Geier. Finally, we are pleased to acknowledge fruitful collaborations with G. Abramson, U. Bastolla, M. Bertram, M. Ipsen, H. Kori, H.–Ph. Lerch, T. Shibata, and P. Stange.

S. C. Manrubia, A. S. Mikhailov, D. H. Zanette

Contents

Preface v

1. Introduction 1

Part 1: Synchronization and Clustering of Periodic Oscillators

2. Ensembles of Identical Phase Oscillators 13
   2.1 Coupled Periodic Oscillators 13
   2.2 Global Coupling and Full Synchronization 19
   2.3 Clustering 22
   2.4 Other Interaction Models 27

3. Heterogeneous Ensembles and the Effects of Noise 35
   3.1 Transition to Frequency Synchronization 35
   3.2 Frequency Clustering 43
   3.3 Fluctuating Forces 47
   3.4 Time-Delayed Interactions 50

4. Oscillator Networks 61
   4.1 Regular Lattices with Local Interactions 62
      4.1.1 Heterogeneous ensembles 66
   4.2 Random Interaction Architectures 70
      4.2.1 Frustrated interactions 72
   4.3 Time Delays 75
      4.3.1 Periodic linear arrays 77

4.3.2 Local interactions with uniform delay 81

5. Arrays of Limit-Cycle Oscillators 83
5.1 Synchronization of Weakly Nonlinear Oscillators 83
5.1.1 Oscillation death due to time delays 91
5.2 Complex Global Coupling 94
5.3 Non-local Coupling 99

Part 2: Synchronization and Clustering in Chaotic Systems

6. Chaos and Synchronization 109
6.1 Chaos in Simple Systems 109
6.1.1 Lyapunov exponents 112
6.1.2 Phase and amplitude in chaotic systems 115
6.2 Synchronization of Two Coupled Maps 116
6.2.1 Saw-tooth maps 118
6.3 Synchronization of Two Coupled Oscillators 121
6.3.1 Phase synchronization 123
6.3.2 Lag synchronization 126
6.3.3 Synchronization in the Lorenz system 128

7. Synchronization in Populations of Chaotic Elements 131
7.1 Ensembles of Identical Oscillators 132
7.1.1 Master stability functions 137
7.1.2 Synchronizability of arbitrary connection topologies . 142
7.2 Partial Entrainment in Rössler Oscillators 146
7.2.1 Phase synchronization 152
7.3 Logistic Maps 159
7.3.1 Globally coupled logistic maps 159
7.3.2 Heterogeneous ensembles 161
7.3.3 Coupled map lattices 167

8. Clustering 171
8.1 Dynamical Phases of Globally Coupled Logistic Maps . . . 172
8.1.1 Two-cluster solutions 174
8.1.2 Clustering phase of globally coupled logistic maps . . 178
8.1.3 Turbulent phase 182
8.2 Universality Classes and Collective Behavior in Chaotic Maps 187

8.3 Randomly Coupled Logistic Maps 193
8.4 Clustering in the Rössler System 197
8.5 Local Coupling 200

9. Dynamical Glasses 203
9.1 Introduction to Spin Glasses 204
9.2 Globally Coupled Logistic Maps as Dynamical Glasses . . . 211
9.3 Replicas and Overlaps in Logistic Maps 215
9.4 The Thermodynamic Limit 217
9.5 Overlap Distributions and Ultrametricity 221

Part 3: Selected Applications

10. Chemical Systems 227
10.1 Arrays of Electrochemical Oscillators 228
10.1.1 Periodic oscillators 230
10.1.2 Chaotic oscillators 234
10.2 Catalytic Surface Reactions 245
10.2.1 Experiments with global delayed feedback 248
10.2.2 Numerical simulations 255
10.2.3 Complex Ginzburg–Landau equation with global delayed feedback 265

11. Biological Cells 273
11.1 Glycolytic Oscillations 274
11.2 Dynamical Clustering and Cell Differentiation 279
11.3 Synchronization of Molecular Machines 289

12. Neural Networks 303
12.1 Neurons 304
12.2 Synchronization in the brain 312
12.3 Cross-coupled neural networks 322

Bibliography 331

Index 345

Chapter 1

Introduction

Order is an essential property of Nature, and it is also a fundamental concept in science. Ordered patterns can easily be identified in physical, biological, and social systems. Often, order is viewed as a static aspect of structural organization. A classical example of an ordered structure is a crystal, where atoms form a perfectly periodic array. However, order can also be an important aspect of collective dynamics. In a dynamically ordered state, individual processes in different parts of a system are well coordinated and, therefore, the system is able to display coherent performance. Functioning of all living organisms is intrinsically based on dynamical order. The successful operation of social systems would also be impossible in absence of this form of order. Even in simple physical systems, coordinated action of individual elements can spontaneously develop. The emergence of collective dynamical organization is a basic problem in the theory of complex systems. One needs to investigate what kinds of collective behavior are possible. Moreover, the conditions determining the development of a particular organization form must be identified.

Dynamical order is intimately related to synchronization phenomena. Two systems are synchronized when rigid correlations between their internal dynamical states appear. Synchronization is also possible in large ensembles of interacting elements. It can be induced by the action of an external force resulting in the entrainment of the system. However, synchronization can also come up as a consequence of interaction between elements. This form of self-organization plays a fundamental role in systems of various origins.

Different kinds of synchronization phenomena are known. In the simplest case, the dynamical states of all elements in a system may become identical. Obviously, this corresponds just to a primitive type of collective organization. In more sophisticated variants, only correlations in some specific properties of individual elements develop, and the presence of dynamical order is less apparent. Moreover, some of the elements of the system may remain non-entrained.

Generally, a large ensemble of interacting elements can also exhibit the phenomenon of clustering. In this case, the population breaks down into a number of coherent groups. Inside each group, the states of all elements are close to each other or even identical. The states of elements belonging to different groups are however weakly correlated. Interactions between clusters determine the coherent behavior of the entire ensemble. Clustering is a form of self-organization: coherently operating groups spontaneously appear out of a uniform population.

The concept of order is intimately related to the notion of symmetry breaking. In physical systems at thermal equilibrium, symmetry breaking occurs through second-order phase transitions. Let us consider, for example, a system of interacting spins. In the paramagnetic state, orientations of individual spins are random and the total magnetization is zero. However, if the temperature is decreased, interactions between the spins lead to a phase transition into the ferromagnetic state with nonvanishing total magnetization. Since the system is isotropic, the direction of the magnetization remains arbitrary. However, in a particular realization a specific direction is selected, and isotropy breaks down. The order parameter of the ferromagnetic phase transition is given by the magnetization. It is zero above the critical temperature and takes finite values in the ferromagnetic state. Another example of an equilibrium second-order phase transition is provided by the phenomenon of superfluidity in liquid He^4. As the temperature is decreased, the quantum states of some of the helium atoms become identical and they form a Bose condensate. The condensate can furthermore coherently flow in a certain direction, which remains arbitrary. The fraction of the atoms belonging to the condensate can be chosen as the order parameter characterizing this transition. A similar behavior is characteristic for the superconductivity phase transition [Ginzburg and Andryushin (1994)].

Non-equilibrium systems can also exhibit phase transitions. For instance, a laser is a population of active atoms interacting through electromagnetic fields, and energetically pumped by an external source. Below the laser transition the individual activities of atoms are not correlated, and the electromagnetic field is not coherent. But as pumping is increased above a threshold, laser generation begins. In this regime, the emission events become rigidly correlated and coherent light is generated. Such coherent optical field is characterized by a certain phase. Though this phase is arbitrary, at each concrete realization it takes a particular value, and symmetry is again broken.

As demonstrated already in the pioneering studies by A. T. Winfree and Y. Kuramoto, the onset of synchronization in oscillator populations represents a phase transition [Winfree (1967); Kuramoto (1984)]. Below the transition point, the motion of individual oscillators in an ensemble is not correlated. As the interactions between them become stronger, correlations between dynamical states of oscillators in a fraction of the ensemble develop: the frequencies of these oscillators become identical. Near the transition point, the size of the coherent oscillator group is small, but the group grows as interactions are increased. Such a coherent group can be viewed as an analogue of the quantum Bose condensate and the size of this group can be again chosen as the order parameter of the synchronization transition. Note, furthermore, that the synchronization transition is accompanied by symmetry breaking. The phase of collective oscillations is arbitrary, but is fixed in a particular realization. Basic concepts of the statistical theory of critical phenomena can be used in the studies of synchronization.

Even more subtle forms of symmetry breaking are known in statistical physics. Spin glasses are systems with random interactions between individual spins. In the thermodynamical limit, when temperature is decreased, such systems undergo a special phase transition: the replica symmetry breaking. It is accompanied by the loss of ergodicity in the behavior of the system. After the transition, the state space of the glass consists of a large number of valleys, which are separated by infinite energy barriers. A trajectory starting within a certain valley cannot leave it, even in the presence of thermal fluctuations. Therefore, the average with respect to time for a particular trajectory is not equivalent to a statistical average over an ensemble of trajectories starting from all possible initial states. This means that ergodicity is broken in such systems [Mézard *et al.* (1987)].

Dynamical clustering in populations of interacting chaotic oscillators or maps exhibits a behavior which closely resembles the properties of spin glasses. Investigating ensembles of globally coupled logistic maps, K. Kaneko has found that, under certain conditions, the population of such identical chaotic elements spontaneously breaks down into a number of clusters of various sizes [Kaneko (1990a)]. The emerging cluster structure determines the collective dynamics of the ensemble. Remarkably, the same system can show a great number of different cluster partitions, depending on the initial conditions. They correspond to different attractors of the dynamical system, similar to energy valleys for spin glasses. Note that, in addition to ergodicity breakdown, the system shows yet another kind of symmetry breaking. All elements are identical but, as time goes on, they become affiliated to different clusters and, therefore, their individual dynamics also becomes different.

In our book, synchronization phenomena in complex systems are considered. What should be called a complex system? In the everyday language, "complex" is a synonym of "complicated." Therefore, any large aggregation of interacting elements would be described as being complex. The scientific concept of complexity is different. It is not enough that a system consists of many elements. A complex system must rather be able to behave as a whole, which implies a certain degree of coordination in the actions of individual parts. But such coordination, or coherence, is already a manifestation of the inherent order. All "real" complex systems are ordered! In some cases, this order is obvious and easily quantifiable. There are, however, systems where the dynamical order is deeply hidden and is not recognizable at first glance.

Because of their universality, the effects of dynamical order should play a fundamental role in the organization and functioning of many systems with various origins. The human body is full of rhythms, starting from the rapid heart beat and respiration and going to much slower circadian cycles. The rhythms are generated by cells, whose activity should be apparently synchronized, so that such macroscopic changes are generated. Moreover, the individual rhythms must be perfectly coordinated with each another, implying interactions between different cyclic subsystems.

Information processing and control of body functions in the brain are performed by a very large population of neural cells. The operation of the brain is based on coherent patterns of electrical activity and provides an extreme example of organization and dynamical order. Essentially, it is a giant dynamical system with billions of coupled individual elements. Through the collective dynamics of neurons, the brain can efficiently emulate, or *model*, the processes in the outside world. It becomes increasingly evident that, to a large extent, various brain functions involve synchronization and clustering in neural populations.

A society is a form of organization of a large population of active agents. The agents are organized into groups. The behaviors of the groups of agents are coordinated and coherent collective action thus becomes possible. Through a joint effort, a society achieves goals that are out of reach for its individual agents. Depending on the composition of the groups, their mutual interactions and the degree of synchronization, different collective tasks can be exercised by a socially organized population. Thus, synchronization should play an essential role in social phenomena.

Turning attention to the processes at very small scales, inside individual biological cells, one notices that they are also characterized by a high degree of organization and dynamical order. A living cell is a tiny chemical reactor where tens of thousands of chemical reactions can simultaneously go on. These reactions proceed in a regular and predictable manner, despite thermal fluctuations and variations in the environmental conditions. The biochemical activity of a cell can be compared with the operation of a large industrial factory, where certain parts are produced by a system of machines. Products of one machine are then used by other machines for manufacturing of their products or for regulation of their functions. In a synchronous operation mode ("just-in-time production"), the intermediate products, required for a certain operation step in a given machine, are released by other machines and become available at the moment when they are needed. The role of machines in a cell is played by individual proteins and their complexes, which operate with single molecules. The phenomena of mutual synchronization and dynamical clustering are crucial for this operation mode.

Dynamical order is found in systems with different properties and various structures. Synchronization is possible both for periodic and chaotic oscillators. Identical periodic oscillators can synchronize at any interaction strength, whereas mutual synchronization of chaotic elements becomes possible only when their interactions are sufficiently strong. In real systems the elements are however only rarely identical. Usually, some heterogeneity in the individual properties of the elements, periodic or chaotic, is present. Heterogeneous ensembles can still synchronize, though often only a fraction of elements becomes entrained.

The simplest structure of an interacting ensemble corresponds to global coupling, where each element is connected in the same way and with the same strength with any other element. A different simple form of structural organization is represented by regular arrays, where each element interacts only with its immediate neighbors. Generally, an ensemble is characterized by a network of connections with complex topology. The architecture of the network determines the synchronization and clustering behavior in the system.

Another significant aspect of all real systems is that, typically, they are subject to noise resulting from the irregular action of the environment. Noise has a pronounced effect on synchronization and clustering. For instance, already for relatively weak noise levels, synchronous clusters become fuzzy and elements can occasionally switch from one cluster to another. Very strong noise can destroy synchronization. Still, it should be said that the effects of dynamical order are robust with respect to fluctuations.

Studies of synchronization phenomena represent an important part of modern nonlinear science. Today, many groups worldwide are actively working on these problems. There is a large literature devoted to such phenomena, and a number of monographs has already been published. An early introduction into the collective behavior of biological oscillators was given by A. T. Winfree [Winfree (2001)]. Many important concepts in the theory of synchronization were formulated in the classical text by Y. Kuramoto [Kuramoto (1984)]. A good textbook on nonlinear science, including synchronization phenomena, has been written by S. H. Strogatz [Strogatz (1994)] (see also his recent popular book on synchronization [Strogatz (2003)]). A systematic approach and many examples can be found in the extensive monograph by A. Pikovsky, M. Rosenblum, and J. Kurths [Pikovsky *et al.* (2001)]. Large populations of interacting chaotic elements are considered by K. Kaneko and I. Tsuda [Kaneko and Tsuda (2000)]. Selected topics in biological synchronization phenomena have also been discussed [Mosekilde *et al.* (2002)]. Some aspects related to dynamical order are also considered in a previous book by one of the present authors [Mikhailov and Calenbuhr (2002)].

Synchronization and clustering are viewed by us from the perspective of statistical physics, for large populations of interacting elements. There are also interesting problems related to synchronization of two coupled chaotic oscillators or to the entrainment of a single chaotic oscillator by an external force. These problems are of much importance in the application to secure communication and chaos control [Pecora (1998a); Boccaletti *et al.* (2000)]. However, they are outside of the scope of our book.

Here, the attention is focused on the spontaneous emergence of dynamical order as a consequence of interactions between elements in a large system. The concepts relevant in our discussion are those of nonequilibrium phase transitions, fluctuations, order parameters and other statistical properties. We also discuss how the internal static organization, or architecture, of a complex system is affecting its dynamical order. A systematic presentation of these topics is given. We have also selected some applications, which are used to illustrate the practical importance of these results.

The book is divided into three parts. Part I is devoted to the analysis of synchronization phenomena in ensembles formed by interacting periodic oscillators. In Chapter 2, after formulating a general model for coupling between dynamical systems, we introduce phase oscillators as the simplest representation of periodic motion. Then, the joint dynamics of a pair of coupled phase oscillators is studied. Collective behavior in large ensembles of globally coupled phase oscillators, where all elements interact with the same strength, is discussed in Chapter 3. We begin by considering ensembles of identical phase oscillators, which exhibit full synchronization for attractive coupling. Then, we study the transition to frequency synchronization in ensembles where the natural frequencies of individual oscillators are not identical. We consider interaction models which induce clustering of identical oscillators, where the ensemble splits into internally synchronized groups. Finally, we analyze the effects of noise and time delays in the synchronization properties of globally coupled ensembles.

More complicated interaction architectures are considered in Chapter 4, where we study networks of coupled phase oscillators. First, we characterize the spatial structures emerging in regular arrays of both identical and non-identical elements with local interactions. Such structures include static patterns as well as propagating waves. Then, we turn the attention to random networks with disordered interaction strengths. These systems exhibit collective dynamical properties similar to spin glasses, such as frustration and slow relaxation to equilibrium. Time-delayed dynamics is also considered in oscillator networks. The last chapter of Part I is devoted to the study of collective behavior in ensembles of periodic oscillators whose individual dynamics is characterized by the presence of a stable limit cycle. We pay special attention to those forms of behavior which are not observed for phase oscillators, in particular, to oscillation death and to collective chaos. We also study the effects of non-local coupling of limit-cycle oscillators, whose interaction strength depends on their mutual distance.

Part II deals with synchronization phenomena in ensembles of elements with chaotic dynamics. Chapter 6 reviews some notions related to chaotic behavior, including the definition of Lyapunov exponents and phase in chaotic systems. Several systems formed by two coupled chaotic elements are used to present different forms of coherent behavior observed in such populations. Large ensembles are introduced in Chapter 7. In the first part of this chapter, we discuss some analytical procedures to determine the stability of the fully synchronous state in chaotic ensembles of identical elements under different coupling schemes. Then, we show that the fully synchronous state is also present in systems of heterogeneous elements, and discuss how the transition to such a regime proceeds.

Chapter 8 is devoted to clustering phenomena. The different dynamical phases present in ensembles of globally coupled logistic maps are explored in detail. Some weaker forms of coherent behavior, such as hidden order, are subsequently discussed. Later, we derive some relationships between the universality class of individual maps and the collective behavior of globally coupled ensembles of identical elements. This chapter is closed with a brief discussion of coherent evolution in coupled map lattices. The last chapter of Part II explores the deep analogy existing between globally coupled logistic maps and glassy systems. We first introduce the phenomenology of spin glasses and its thermodynamical description. We continue with a discussion of the macroscopic behavior of globally coupled logistic maps and show that the clustering phase has a clear counterpart in the phenomenon of replica symmetry breaking observed in spin glasses. Finally, we show that replicas and overlaps can be suitably defined in dynamical systems, this eventually leading to the formal introduction of dynamical glasses.

Part III is devoted to the applications of the synchronization theory. Here, we do not aim to review all available literature, but rather focus our attention on several selected fields. In Chapter 10, synchronization and clustering in chemical systems are discussed. We show that investigations of arrays of electrochemical oscillators provide clear experimental evidence of synchronization transitions and cluster formation both for periodic and chaotic elements. Later in the same chapter, synchronization phenomena in catalytic surface reactions are considered. A special aspect of this experimental system is that, in addition to local diffusive coupling between chemical oscillators, global delayed coupling between them can be easily introduced and controlled. As a result of such feedback, chemical turbulence can be suppressed and various spatiotemporal patterns can be induced.

Synchronization in systems of biological cells and at the intracellular level is the subject of Chapter 11. It begins with a discussion of the experiments demonstrating synchronization in populations of yeast cells. Then, theoretical studies based on abstract models, where cells are described by randomly generated networks of catalytic reactions, are considered. We show that evolution in such abstract cell populations leads to spontaneous differentiation of cells, proceeding through synchronization and dynamical clustering. At the end of this chapter, biochemical processes inside individual biological cells are analyzed. We point out that many biological macromolecules, such as enzymes, effectively represent cyclic molecular machines and discuss the possibility of synchronization in molecular networks.

Neural networks and brain operation are considered in the last Chapter 12. Out of the vast volume of related research, several topics are chosen here. First, the problems of modeling of neural networks are discussed. We show that the models of integrate-and-fire neurons can be derived, based on very general considerations, as a canonical form of oscillators in the vicinity of a special bifurcation, i.e. the saddle-node bifurcation on a limit cycle. Subsequently, the available experimental evidence of synchronization phenomena in the brain is briefly presented. We emphasize that such phenomena are often accompanied not by synchronization of states of all neurons in their population, but by the development of correlations and cross-synchronization between different neural networks. A simple theoretical model, displaying synchronization and dynamical clustering in populations of cross-coupled neural networks, is finally presented.

The book includes an extensive bibliography, intending to cover the majority of contributions in this discipline. To a large extent, the knowledge of synchronization phenomena in complex systems is based on numerical studies. When presenting numerical results of other authors, we have usually repeated all relevant numerical simulations. Most of the graphical illustrations in Parts I and II are not simple copies of the figures from the original articles. They have been plotted anew using our own simulation data. Sometimes, such plots are made for parameter values or in intervals of variables which are different from the original work. When the new plots coincide with the previously published ones (up to a difference in labeling or notations), they are described as "adapted" from the respective publications. Many illustrations in the last part of the book have however been directly copied from the original articles, as indicated in the figure captions.

PART 1

Synchronization and Clustering of Periodic Oscillators

Chapter 2

Ensembles of Identical Phase Oscillators

Dynamical systems with oscillatory motion are a basic ingredient in the mathematical modeling of a broad class of physical, physicochemical, and biological phenomena. Ensembles of interacting elements with periodic dynamics are used to represent natural systems with collective rhythmic behavior. A simplified model for the periodic evolution of each individual element is given by a single variable with cyclic uniform motion, like an elementary clock. This simple dynamical system is called phase oscillator. Coupled phase oscillators provide a phenomenological description of complex systems whose collective evolution is driven by synchronization processes. They reproduce the main features of the emergence of coherent behavior found in more elaborate models of interacting oscillators.

We begin this chapter by introducing the equations of motion of coupled periodic oscillators, and the phase oscillator model. After discussing the synchronization properties of a system of two phase oscillators, we focus the attention on large ensembles of identical oscillators subject to global coupling, where interactions are uniform for all oscillator pairs. We characterize the state of full synchronization induced by attractive interactions. Then, the regimes of clustering and incoherent behavior for more complex interaction models are analyzed.

2.1 Coupled Periodic Oscillators

Macroscopic oscillations may emerge from the mutual synchronization of a large number of more elementary, individual oscillatory processes [Wiener (1948)]. The mechanisms governing the spontaneous organization of such cyclic elements are intricate, and may be considerably dissimilar for different systems. However, all these systems can be phenomenologically represented as ensembles of interacting dynamical elements with cyclic individual evolution. We assume that the internal state of each element i is mathematically described by a set of time-dependent variables $\mathbf{r}_i(t) = (x_i(t), y_i(t), z_i(t), \dots)$, whose evolution is governed, in the absence of interactions, by

$$\dot{\mathbf{r}}_i = \mathbf{f}_i(\mathbf{r}_i). \tag{2.1}$$

In the specific class of models we have in mind, each non-interacting element behaves as a periodic oscillator. Therefore, the function $\mathbf{f}_i$ is such that the solutions to Eq. (2.1) are periodic or, more generally, approach a periodic limit cycle for asymptotically large times. Coupling between N periodic oscillators described by Eq. (2.1) is introduced by means of pairwise interactions, given by interaction functions $\mathbf{U}_{ij}(\mathbf{r}_i, \mathbf{r}_j)$, as

$$\dot{\mathbf{r}}_i = \mathbf{f}_i(\mathbf{r}_i) + \sum_{j=1}^{N} \mathbf{U}_{ij}(\mathbf{r}_i, \mathbf{r}_j), \qquad i = 1, \dots, N. \tag{2.2}$$

Perhaps the simplest representation of periodic motion is given by a single *phase* variable $\phi(t)$ which, as time elapses, varies as $\phi(t) = \omega t + \phi(0)$ [Winfree (2001)]. Conventionally, the phase is defined on the interval $[0, 2\pi)$. When ϕ reaches the limiting value 2π, it is reset to $\phi = 0$. The equation of motion for the phase is

$$\dot{\phi} = \omega. \tag{2.3}$$

This one-dimensional dynamical system is called *phase oscillator*. It performs uniform periodic motion of *natural frequency* ω.

In analogy with Eqs. (2.2), the evolution of an ensemble of interacting phase oscillators is governed by the equations

$$\dot{\phi}_i = \omega_i + \sum_{j=1}^{N} F_{ij}(\phi_i, \phi_j), \qquad i = 1, \dots, N, \tag{2.4}$$

where ω_i is the natural frequency of oscillator i, and the functions F_{ij} describe interactions. Since the phase variables ϕ_i are defined on the interval $[0, 2\pi)$, the interaction functions $F_{ij}(\phi_i, \phi_j)$ must be 2π-periodic with respect to their two variables, namely, $F_{ij}(\phi_i + 2\pi n_i, \phi_j + 2\pi n_j) = F_{ij}(\phi_i, \phi_j)$ for any integers n_i and n_j.

If the interaction functions F_{ij} depend on the phase differences only, $F_{ij}(\phi_i, \phi_j) = F_{ij}(\phi_i - \phi_j)$, Eqs. (2.4) are invariant under the transformation

$$\begin{aligned} \phi_i &\to \phi_i + \phi_0(t), \\ \omega_i &\to \omega_i + \dot{\phi}_0(t), \end{aligned} \tag{2.5}$$

for all $i = 1, \ldots, N$. Here, $\phi_0(t)$ is an arbitrary function of time. The transformation represents a time-dependent phase shift, and is equivalent to the change to a reference system rotating with frequency $-\dot{\phi}_0(t)$. In the special case $\phi_0(t) = \omega_0 t$, this invariance implies that the natural frequencies ω_i are defined up to an arbitrary additive constant ω_0.

Before analyzing the emergence of order in large ensembles of interacting phase oscillators, it is illustrative to study the simpler case of just two oscillators [Sakaguchi *et al.* (1987)]. We consider symmetric interaction functions

$$F_{12}(\phi_1, \phi_2) = -F_{21}(\phi_2, \phi_1) = \frac{K}{2} \sin(\phi_2 - \phi_1), \tag{2.6}$$

where K is the *coupling intensity.* When $K > 0$, the interaction is attractive. The sign of the force acting on oscillator 1 is opposite to the phase displacement of this oscillator with respect to oscillator 2. For $K < 0$, the interaction is repulsive. The equations for the phases $\phi_1(t)$ and $\phi_2(t)$ read

$$\begin{aligned} \dot{\phi}_1 &= \omega_1 + \tfrac{K}{2} \sin(\phi_2 - \phi_1), \\ \dot{\phi}_2 &= \omega_2 + \tfrac{K}{2} \sin(\phi_1 - \phi_2). \end{aligned} \tag{2.7}$$

These equations take a more convenient form if they are written for the variables $\rho(t) = \phi_1(t) + \phi_2(t)$ and $\Delta\phi(t) = \phi_2(t) - \phi_1(t)$:

$$\begin{aligned} \dot{\rho} &= \omega_1 + \omega_2, \\ \Delta\dot{\phi} &= \Delta\omega - K \sin \Delta\phi, \end{aligned} \tag{2.8}$$

with $\Delta\omega = \omega_2 - \omega_1$. The first of them implies that the sum of the two phases performs uniform motion with frequency $\omega_1 + \omega_2$:

$$\rho(t) = \rho(0) + (\omega_1 + \omega_2)t. \tag{2.9}$$

Figure 2.1 shows the time derivative of the phase difference, $\Delta\dot{\phi}$, as a function of $\Delta\phi$ for different values of $\Delta\omega$ and $K > 0$. When the natural frequencies of the two oscillators are identical, $\Delta\omega = 0$, the system has fixed points at $\Delta\phi = 0 (= 2\pi)$ and π. The fixed point $\Delta\phi = 0$ is stable,

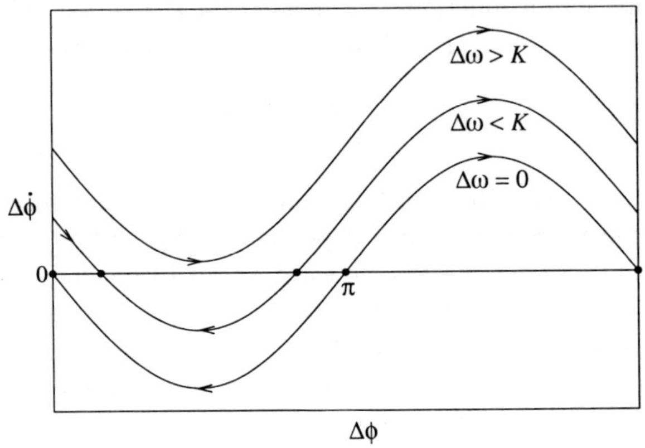

Fig. 2.1 Time derivative of the phase difference of two coupled oscillators, according to the second of Eqs. (2.8), for different values of the frequency difference $\Delta\omega$ and coupling constant $K > 0$. Full dots on the horizontal axis stand for stable and unstable fixed points. Arrows indicate the direction of motion.

while $\Delta\phi = \pi$ is unstable. At long times the two oscillators asymptotically reach a state of *full synchronization*. In this state,

$$\phi_1(t) = \phi_2(t) = \Omega t, \tag{2.10}$$

where

$$\Omega = \omega_1 = \omega_2. \tag{2.11}$$

In the case of repulsive interaction, $K < 0$, the stability of the fixed points changes. The equilibrium $\Delta\phi = 0$ becomes unstable, while $\Delta\phi = \pi$ is now stable. In this situation, the asymptotic motion of the oscillators is

$$\begin{aligned} \phi_1(t) &= \Omega t, \\ \phi_2(t) &= \Omega t + \pi, \end{aligned} \tag{2.12}$$

with $\Omega = \omega_1 = \omega_2$. The oscillators have the same frequency, but their phases are opposite. Repulsive interaction, therefore, gives rise to a different kind of coherent evolution for the two oscillators, which we call *anti-phase synchronization*.

Coming back to the case of attractive coupling, if the natural frequencies of the two oscillators are different but $\Delta\omega < K$, there are still two fixed points. They are given by the two solutions of the equation

$\sin \Delta\phi = \Delta\omega/K$. Again, one of them is stable, while the other is unstable. In these conditions, the phase difference asymptotically approaches a fixed value. The two oscillators do not reach full synchronization but they move uniformly with the same frequency Ω. This is a state of *frequency synchronization.* The common frequency is given by the average of the natural frequencies of the two oscillators,

$$\Omega = \frac{\omega_1 + \omega_2}{2}. \tag{2.13}$$

The asymptotic motion of the oscillators is

$$\begin{aligned} \phi_1(t) &= \Omega t, \\ \phi_2(t) &= \Omega t + \arcsin(\Delta\omega/K). \end{aligned} \tag{2.14}$$

Finally, if $\Delta\omega > K$, the natural frequencies are too disparate to allow for any kind of synchronization. The motion of the two oscillators remains *incoherent.* As a function of time, the phase difference is then given by

$$\Delta\phi(t) = 2\arctan\left\{\frac{K}{\Delta\omega} + \sqrt{1 - \frac{K^2}{\Delta\omega^2}} \tan\left[\frac{\sqrt{\Delta\omega^2 - K^2}}{2}(t - t_0)\right]\right\}, \tag{2.15}$$

with

$$t_0 = \frac{2}{\sqrt{\Delta\omega^2 - K^2}} \arctan\left\{\frac{K - \Delta\omega \tan[\Delta\phi(0)/2]}{\sqrt{\Delta\omega^2 - K^2}}\right\}. \tag{2.16}$$

The phase difference (2.15) can be written in the form

$$\Delta\phi(t) = \sqrt{\Delta\omega^2 - K^2}\, t + \tilde{\phi}(t), \tag{2.17}$$

where $\tilde{\phi}(t)$ is a periodic function of time. On the average, $\Delta\phi(t)$ grows linearly with time, indicating that the two oscillators fail to be entrained. Their motion is now described by

$$\begin{aligned} \phi_1(t) &= \tfrac{1}{2}(\omega_1 + \omega_2 + \tfrac{1}{2}\sqrt{\Delta\omega^2 - K^2})t + \tfrac{1}{2}\tilde{\phi}_0(t) + \phi_1(0), \\ \phi_2(t) &= \tfrac{1}{2}(\omega_1 + \omega_2 - \tfrac{1}{2}\sqrt{\Delta\omega^2 - K^2})t - \tfrac{1}{2}\tilde{\phi}_0(t) + \phi_2(0). \end{aligned} \tag{2.18}$$

with $\tilde{\phi}_0(t) = \tilde{\phi}(t) - \tilde{\phi}(0)$. In this case, coupling is not strong enough to synchronize the oscillators.

Figure 2.2 shows the evolution of $\phi_1(t)$ and $\phi_2(t)$ in the three regimes discussed above. In summary, under the action of attractive coupling, $K > 0$, two phase oscillators with identical natural frequencies reach a state of full

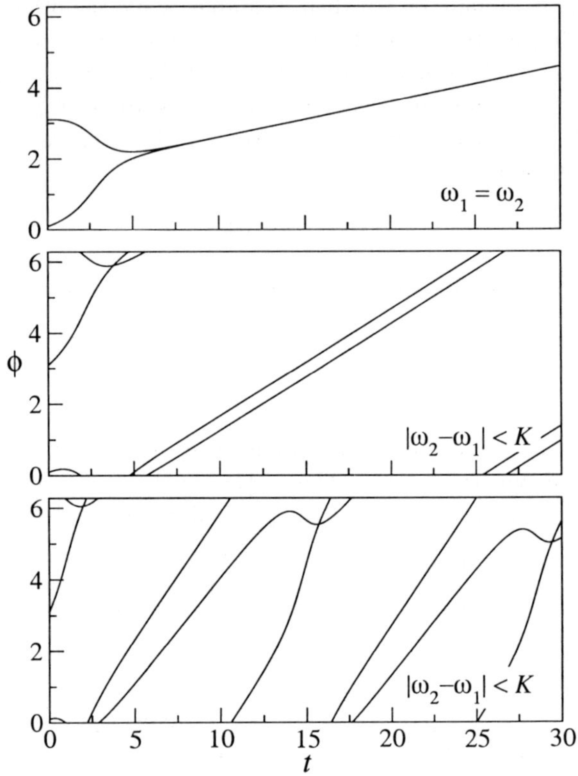

Fig. 2.2 Evolution of the phases ϕ_1 and ϕ_2 of two coupled oscillators governed by Eqs. (2.7), with $K = 1$ and $\omega_1 = 0.1$. In the upper plot, $\omega_2 = \omega_1$, and the oscillators become fully synchronized. In the middle plot, $\omega_2 = 0.5$, and the oscillators synchronize only in frequency. In the lower plot, $\omega_2 = 1.2$, and the oscillators do not synchronize.

synchronization, where their phases are exactly the same and move at the natural oscillator frequency. When the natural frequencies are different but coupling is strong enough, the oscillators become synchronized in frequency, and move uniformly with a constant phase difference. If, on the other hand, the difference of natural frequencies is too large, the two oscillators do not synchronize.

These different synchronization regimes are shown in parameter space in Fig. 2.3. The shaded region where frequency synchronization is stable is known as the Arnol'd tongue. We show below how these results are generalized to the case of ensembles of many phase oscillators.

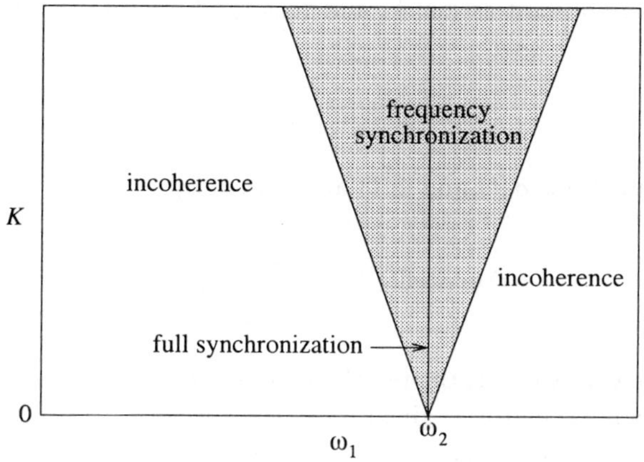

Fig. 2.3 Synchronization regimes for two phase oscillators in the (ω_1, K)-plane. Synchronization occurs in the shaded triangular zone (Arnol'd tongue) with vertex at $\omega_1 = \omega_2$ and $K = 0$. Full synchronization is found on the vertical line $\omega_1 = \omega_2$.

2.2 Global Coupling and Full Synchronization

In the rest of this chapter and in Chapter 3, we deal with ensembles of phase oscillators where the interaction function is the same for all pairs, $F_{ij}(\phi_i, \phi_j) = F(\phi_i, \phi_j)$ for all i and j. This kind of uniform interaction, whose range encompasses the whole ensemble of oscillators, is called *global coupling*. We pay particular attention to the case where the interaction function is given by $F(\phi_i, \phi_j) = k \sin(\phi_j - \phi_i)$ for all oscillator pairs, as in the system of two oscillators considered above. This interaction function represents an attractive force for $k > 0$ and a repulsive force for $k < 0$. The interaction constant k is usually written as $k = K/N$, where K specifies the coupling intensity. With this choice, Eq. (2.4) reads

$$\dot{\phi}_i = \omega_i + \frac{K}{N} \sum_{j=1}^{N} \sin(\phi_j - \phi_i). \tag{2.19}$$

Expanding $\sin(\phi_j - \phi_i)$, this equation can be cast in the form

$$\dot{\phi}_i = \omega_i + K(\langle \sin \phi \rangle \cos \phi_i - \langle \cos \phi \rangle \sin \phi_i), \tag{2.20}$$

with $\langle \sin \phi \rangle = N^{-1} \sum_j \sin \phi_j$ and $\langle \cos \phi \rangle = N^{-1} \sum_j \cos \phi_j$. The interaction of each oscillator with the ensemble occurs effectively through the global average quantities $\langle \sin\phi \rangle$ and $\langle \cos\phi \rangle$. Equation (2.20) can in turn be rewritten as

$$\dot{\phi}_i = \omega_i + K\sigma \sin(\Phi - \phi_i), \tag{2.21}$$

where the functions $\sigma(t)$ and $\Phi(t)$ are defined by

$$\sigma(t)\exp[i\Phi(t)] = \frac{1}{N}\sum_{j=1}^{N} \exp[i\phi_j(t)]. \tag{2.22}$$

From a formal viewpoint, the problem reduces to the solution of Eq. (2.21) for each single oscillator i, given its initial phase $\phi_i(0)$, and for arbitrary forms of $\sigma(t)$ and $\Phi(t)$. Then, these two functions must be calculated self-consistently from their definition (2.22).

First, we study the case of identical oscillators, where all the natural frequencies coincide. As a consequence of the invariance of the system under transformations (2.5), we can fix $\omega_i = 0$ for all oscillators. Equations (2.19) become

$$\dot{\phi}_i = \frac{K}{N}\sum_{j=1}^{N} \sin(\phi_j - \phi_i). \tag{2.23}$$

The discrete-time version of these equations has also been considered [Kaneko (1991a)]. Note that the coupling intensity K fixes the time scale of evolution. Its absolute value can be chosen arbitrarily by redefining time units. Equations (2.23) have a stationary solution where all phases are equal: $\phi_i = \phi^*$ for all i, where ϕ^* is an arbitrary constant. Since $F_{ij}(\phi^*, \phi^*) = 0$, interactions play no role in this state. As natural frequencies are all zero, oscillations cease. Such stationary situation corresponds to full synchronization of the ensemble, because the individual states of all oscillators coincide at all times.

The state of full synchronization will actually be reached if it is stable. To analyze its linear stability, we consider the ensemble in a state close to full synchronization, where each oscillator deviates from the stationary phase ϕ^* by a small quantity:

$$\phi_i(t) = \phi^* + \delta\phi_i(t). \tag{2.24}$$

Assuming $\delta\phi_i(t) \ll 1$ for all oscillators, and neglecting higher-order terms in Eqs. (2.23), the evolution equations for the deviations $\delta\phi_i(t)$ read

$$\delta\dot{\phi}_i = -K\delta\phi_i + \frac{K}{N}\sum_{j=1}^{N}\delta\phi_j. \tag{2.25}$$

The state of full synchronization is linearly stable if these equations have only decaying solutions. This requires that all the eigenvalues of the $N \times N$ matrix $\mathbf{S} = \{s_{ij}\}$ with elements

$$s_{ij} = \frac{K}{N}(1 - N\delta_{ij}), \tag{2.26}$$

are negative or have negative real parts. Here, δ_{ij} is the Kronecker delta symbol.

The eigenvalue problem for the matrix $\mathbf{S}$ can be completely solved for any N. The eigenvalue $\lambda_1 = 0$, associated with the eigenvector $\mathbf{e}_1 = (1, 1, \ldots, 1)$, is directly related to the invariance of the system under a rotation of all the phases by a fixed angle. This rotation corresponds to a shift along the vector $\mathbf{e}_1$ in the space of phase deviations. The eigenvalue $\lambda_1 = 0$ is called *longitudinal* eigenvalue. All the remaining eigenvalues are called *transversal*.

For the matrix whose elements are given in Eq. (2.26), all the $N-1$ transversal eigenvalues are equal, $\lambda_2 = \cdots = \lambda_N = -K$. Therefore, full synchronization is linearly stable for any positive value of the coupling intensity, and unstable if $K < 0$. The numerical solution of Eqs. (2.23) shows that, in the case of attractive coupling, full synchronization is globally stable. Figure 2.4 illustrates the evolution towards full synchronization in an ensemble of 100 identical phase oscillators, with coupling intensity $K = 1$. Initially, their phases are uniformly distributed over $[0, 2\pi)$. At $t = 10$, synchronization is almost complete.

In addition to the state of full synchronization, Eqs. (2.23) have many other stationary states, given by the solutions of the equations

$$\sum_{j=1}^{N}\sin(\phi_j - \phi_i) = 0, \qquad i = 1, \ldots, N. \tag{2.27}$$

For $K > 0$, all of them are unstable. These other states may become, however, stable for repulsive interactions, $K < 0$, where the fully synchronized state is unstable. For $K < 0$, the phases asymptotically approach fixed values, uniformly scattered over the interval $[0, 2\pi)$. While phases do not

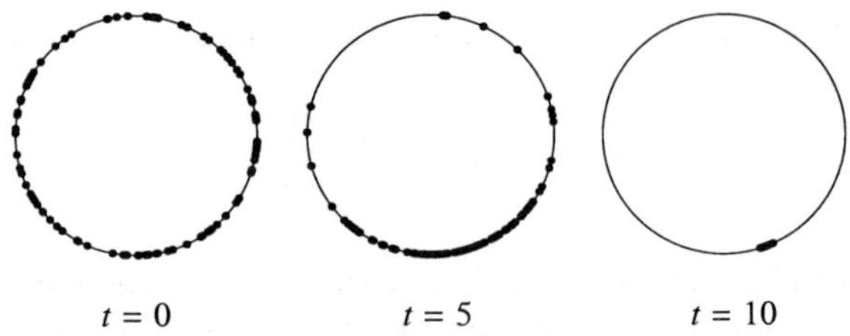

Fig. 2.4 Three snapshots of the distribution of phases, plotted on the unit circle, in a system of 100 oscillators with identical natural frequencies, $\omega_i = 0$, and coupling intensity $K = 1$. The initial phases $\phi_i(0)$ have a homogeneous distribution on $[0, 2\pi)$.

become identical, frequencies converge to the natural frequency common to all oscillators.

2.3 Clustering

Clustering is a regime of collective evolution where an ensemble of interacting dynamical elements spontaneously splits into two or more groups. While each of these clusters follows its own orbit, all the elements within a given cluster are mutually synchronized and their individual orbits coincide. Later in this book, we show that clustering occurs in ensembles of globally coupled chaotic dynamical systems for coupling intensities just below the threshold at which full synchronization becomes stable. However, clustered states are also possible in ensembles of interacting phase oscillators. Clustering of identical phase oscillators is found when the interaction functions in Eqs. (2.4) are more complex than those studied so far, $F_{ij}(\phi_i, \phi_j) \propto \sin(\phi_j - \phi_i)$ [Hansel *et al.* (1993); Okuda (1993)].

We consider interactions of the form $F_{ij}(\phi_i, \phi_j) = N^{-1}F(\phi_i - \phi_j)$, so that

$$\dot{\phi}_i = \frac{1}{N}\sum_{j=1}^{N} F(\phi_i - \phi_j). \tag{2.28}$$

Taking advantage of the symmetry (2.5), the natural frequency of all oscillators has been chosen equal to zero. The function $F(\phi)$ is required to be 2π-periodic, $F(\phi) = F(\phi + 2\pi)$ for all ϕ. Moreover, in contrast with the case considered in previous sections, it may now contain harmonic contributions of any order. In other words, $F(\phi)$ can be written as a Fourier series:

$$F(\phi) = \sum_{n=1}^{\infty}(A_n \sin n\phi + B_n \cos n\phi). \tag{2.29}$$

The previously considered systems correspond to the choice $A_1 = -K$, $B_1 = 0$, and $A_n = B_n = 0$ for all $n > 1$.

The stability analysis of M-cluster states, where the system splits into M groups, can be explicitly carried out in the case where the clusters have identical sizes N/M [Okuda (1993)]. First of all, we note that Eqs. (2.28) have a solution where the oscillators are segregated into groups of identical sizes, if all of them move with the same collective frequency and if their phases are equally spaced in $[0, 2\pi)$. Denoting as Φ_m the phase of cluster m ($m = 1, \ldots, M$), we find that

$$\Phi_m = \Omega t + \frac{2\pi m}{M} \tag{2.30}$$

is a solution of Eqs. (2.28) if the synchronization frequency Ω satisfies

$$\Omega = \frac{1}{M}\sum_{m=1}^{M} F\left[\frac{2\pi}{M}(m-1)\right]. \tag{2.31}$$

In terms of the Fourier coefficients of Eq. (2.29), the synchronization frequency reads

$$\Omega = \sum_{n=1}^{\infty} B_{nM}. \tag{2.32}$$

Here, nM denotes the product of n times M. Note that this expression for Ω involves only the coefficients of even M-order harmonics.

As it has been done for the state of full synchronization in Sec. 2.2, the stability of the M-cluster solution is analyzed in the linear approximation by assuming small deviations from the stationary state. For an oscillator i in cluster m, we introduce the phase deviation $\delta\phi_i(t)$ as

$$\phi_i(t) = \Phi_m + \delta\phi_i(t). \tag{2.33}$$

Equations (2.28) can now be linearized around the stationary state by expanding the interaction function up to the first order in the phase deviations. The M-cluster state is stable if the solutions of the linearized equations vanish asymptotically. This requires that all the eigenvalues of the $N \times N$ matrix

$$\mathbf{S} = \begin{pmatrix} \alpha\mathbf{I} - \beta_0\mathbf{U} & -\beta_1\mathbf{U} & -\beta_2\mathbf{U} & \dots & -\beta_{M-1}\mathbf{U} \\ -\beta_{M-1}\mathbf{U} & \alpha\mathbf{I} - \beta_0\mathbf{U} & -\beta_1\mathbf{U} & \dots & -\beta_{M-2}\mathbf{U} \\ -\beta_{M-2} & -\beta_{M-1} & \alpha\mathbf{I} - \beta_0\mathbf{U} & \dots & -\beta_1\mathbf{U} \\ \vdots & \vdots & \vdots & \ddots & \vdots \\ -\beta_1\mathbf{U} & -\beta_2\mathbf{U} & -\beta_3\mathbf{U} & \dots & \alpha\mathbf{I} - \beta_0\mathbf{U} \end{pmatrix}. \tag{2.34}$$

are negative or have negative real parts. In Eq. (2.34), the matrix $\mathbf{S}$ has been expressed as an array of $M \times M$ blocks, each of them consisting of an $\frac{N}{M} \times \frac{N}{M}$ matrix. There, $\mathbf{I}$ is the $\frac{N}{M} \times \frac{N}{M}$ identity matrix, and $\mathbf{U}$ is a matrix of the same dimensions whose elements are all equal to unity. Moreover,

$$\alpha = \frac{1}{M} \sum_{m=1}^{M} F' \left[\frac{2\pi}{M}(m-1) \right] \tag{2.35}$$

and

$$\beta_k = \frac{1}{M} F' \left(-\frac{2\pi k}{M} \right), \qquad k = 0, \dots, M, \tag{2.36}$$

where $F'(\phi)$ is the first derivative of the interaction function.

The eigenvector problem for matrix $\mathbf{S}$ can be worked out explicitly, yielding M non-degenerate eigenvalues

$$\lambda_p = \frac{1}{M} \sum_{k=0}^{M-1} F' \left(\frac{2\pi k}{M} \right) [1 - \exp(-2\pi i k p / M)], \qquad p = 0, \dots, M-1, \tag{2.37}$$

and an eigenvalue with multiplicity $N - M$,

$$\lambda_M = \frac{1}{M} \sum_{k=0}^{M-1} F' \left(\frac{2\pi k}{M} \right). \tag{2.38}$$

Note that $\lambda_0 = 0$ is the longitudinal eigenvalue discussed in Sec. 2.2. For $M = 1$, these results collapse to those obtained for the state of full synchronization in that section. The eigenvalues of matrix $\mathbf{S}$ are given in terms of the Fourier coefficients of Eq. (2.29) as

$$\lambda_p = \sum_{n=1}^{\infty} \left[B'_{nM} - \frac{B'_{nM+p-M} + B'_{nM-p}}{2} + i \frac{A'_{nM+p-M} - A'_{nM-p}}{2} \right] \tag{2.39}$$

and

$$\lambda_M = \sum_{n=1}^{\infty} B'_{nM}, \tag{2.40}$$

where $A'_n = -nB_n$ and $B'_n = nA_n$ are the Fourier coefficients of the derivative $F'(\phi)$. Since linear stability depends just on the sign of the real part of the eigenvalues, only the coefficients B'_n determine whether M-cluster states are stable. In other words, the stability condition is completely given by the odd part of the interaction function $F(\phi)$ or, equivalently, the even part of its derivative. This is a direct consequence of the fact that we are restricting the analysis to the case of identical size clusters. The stability of less symmetric states involves the even part of $F(\phi)$ as well [Okuda (1993)].

Equations (2.37) to (2.40) make it possible to calculate the eigenvalues λ_p and λ_M and, thus, to determine the linear stability of the M-cluster state for any value of M and any interaction function $F(\phi)$. As an example, let us consider the case

$$F(\phi) = -\sin\phi + a_2 \sin 2\phi + a_3 \sin 3\phi. \tag{2.41}$$

This choice corresponds to the situation analyzed in Sec. 2.2 ($K = 1$), given by the first term, with the addition of two higher-order harmonics. Table 2.1 displays the transversal eigenvalues of matrix $\mathbf{S}$ for the first few values of M. For $M > 3$, one or several transversal eigenvalues (or their real parts) are equal to zero, which implies that the corresponding clustered states are not stable.

Table 2.1 Transversal eigenvalues for M-cluster states with the interaction function (2.41).

eigenvalue	$M = 1$	$M = 2$	$M = 3$
λ_1	$-1 + 2a_2 + 3a_3$	$1 + 2a_2 - 3a_3$	$1/2 - a_2 + 3a_3$
λ_2	–	$2a_2$	$1/2 - a_2 + 3a_3$
λ_3	–	–	$3a_3$

Figure 2.5 shows the stability regions of M-cluster states for $M = 1$, 2, and 3 in the parameter space (a_2, a_3). Note that the origin, $a_2 = a_3 = 0$, is excluded from the stability regions of two and three clusters, but belongs to that of one cluster. This implies that full synchronization ($M = 1$) is the only stable M-cluster state when higher harmonics are absent. Similarly, the 3-cluster state is not stable on the axis $a_3 = 0$, indicating that third-

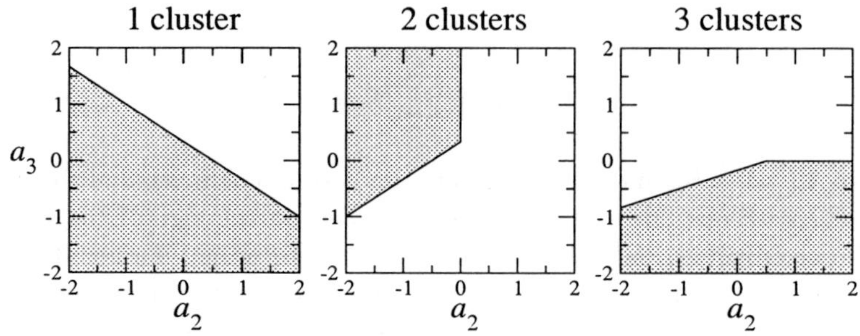

Fig. 2.5 Stability regions (shaded) of 1, 2, and 3-cluster states in the parameter space (a_2, a_3), for the interaction function of Eq. (2.41).

order harmonics are necessary to make such a state stable. As a rule, M-cluster states are not stable unless M-order or higher harmonics are present in the interaction function $F(\phi)$ [Okuda (1993)]. On the other hand, stability regions overlap in several zones of parameter space. There, M-cluster states are simultaneously stable for two or more values of M, and the system is multistable. In these zones, the asymptotic state of the ensemble depends on the initial condition.

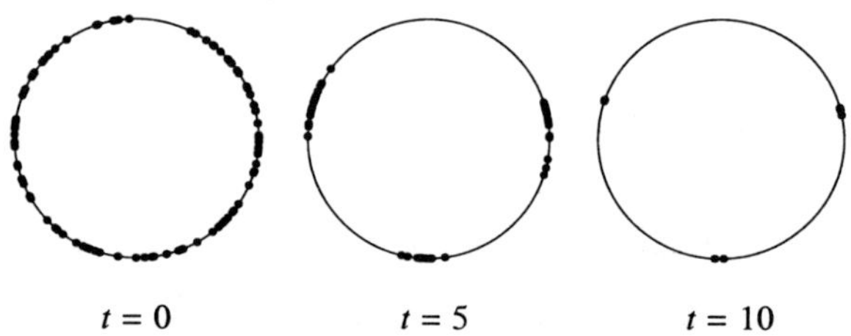

Fig. 2.6 Three snapshots of the distribution of phases, plotted on the unit circle, in an ensemble of 100 oscillators with the interaction function of Eq. (2.41), for $a_2 = 1.5$ and $a_3 = -0.5$. Initially, phases are homogeneously distributed in $[0, 2\pi)$.

It is important to stress that the evolution of the present model from an arbitrary initial condition will not necessarily lead to one of the clustered states analyzed above, where the ensemble is evenly segregated into identical clusters. Configurations with several clusters of different sizes may also be stable [Tass (1997)]. For example, Fig. 2.6 shows the results of numerical integration of the equations of motion for a 100-oscillator ensemble with the interaction function of Eq. (2.41), for $a_2 = 1.5$ and $a_3 = -0.5$, and with a random initial condition. The parameters correspond to the region where the 3-cluster state is stable, whereas full synchronization and 2-cluster states are unstable. We find that, indeed, the system splits into three clusters, but they are not equally spaced in phase. This is due to the fact that the clusters are not equal in size. The ensemble has segregated into three groups of 45, 28, and 27 elements. This result illustrates a characteristic feature of the regime of clustering in large ensembles of interacting dynamical elements. The ensemble may have a large number of asymptotic states, with many different partitions into clusters, which in turn lead to many different phase configurations. The system thus exhibits a large degree of multistability, and the asymptotic state is highly dependent on the initial condition.

In connection with the description of neural systems as ensembles of interacting phase oscillators, it has been conjectured that the presence of a multitude of stable states in populations of neurons may be exploited to encode and classify information. The configuration of a given clustered state could play the role of a code for the attributes of sensory signals in the brain. Storage of memory and activity patterns associated with motility functions may also take advantage of this form of coding [Abarbanel *et al.* (1996); Tass (1997)].

2.4 Other Interaction Models

In some applications of the phase-oscillator model, the interaction functions do not depend on the difference of phases, as we have assumed so far. For example, a different form of the interaction function is necessary to describe arrays of Josephson junctions [Wiesenfeld and Hadley (1989); Tsang *et al.* (1991); Domínguez and Cerdeira (1993)]. In this case, the dynamics of phases is approximately described by Eqs. (2.4) with

$$F_{ij}(\phi_i, \phi_j) = \frac{1}{N}[f_1(\phi_i) - f_2(\phi_j)], \tag{2.42}$$

where both $f_1(\phi)$ and $f_2(\phi)$ are proportional to $\sin\phi$. For an ensemble of identical elements, we have the equations

$$\dot{\phi}_i = \omega + f_1(\phi_i) - \frac{1}{N}\sum_{j=1}^{N} f_2(\phi_j). \tag{2.43}$$

Since the interaction function F_{ij} does not depend on ϕ_i and ϕ_j through their difference, these equations are not invariant under transformations (2.5). This implies that, even for oscillators with identical natural frequencies, the first term in the right-hand side of Eqs. (2.43) cannot be eliminated by a uniform shift in the value of the natural frequencies. Without loss of generality, we can at most choose $\omega = 1$, by rescaling time and the functions f_1 and f_2.

Equations (2.43) have been studied for arbitrary forms of $f_1(\phi)$ and $f_2(\phi)$, with the only requirement that they are 2π-periodic functions [Golomb *et al.* (1992)]. In spite of the substantial differences in the interaction functions, it has been found that the forms of collective motion and synchronization in this system are qualitatively very similar to those occurring for the interactions considered in preceding sections. The same is found to happen with other interaction models [Kuramoto (1991); Daido (1996); Ariaratnam and Strogatz (2001)].

First of all, Eqs. (2.43) may have a fixed-point solution, where all the phases are equal and do not depend on time, $\phi_i = \phi^*$ for all i. In contrast with the case analyzed in Sec. 2.2, however, the value of ϕ^* is not arbitrary, but satisfies the equation

$$\omega + f_1(\phi^*) - f_2(\phi^*) = 0. \tag{2.44}$$

This fully synchronized fixed-phase state exists if Eq. (2.44) has at least one solution. Linear stability analysis of this state, carried out along the same lines as in Sects. 2.2 and 2.3, shows that the eigenvalues

$$\lambda_1 = f_1'(\phi^*) - f_2'(\phi^*) \tag{2.45}$$

and

$$\lambda_2 = \cdots = \lambda_N = f_1'(\phi^*) \tag{2.46}$$

must be negative to have a stable state. Here, λ_1 is the longitudinal eigenvalue. It corresponds to an eigenvector that represents rigid displacements of the ensemble along the oscillator orbit. Due to the lack of rotation symmetry of the interaction functions in Eqs. (2.43), this longitudinal eigenvalue is generally different from zero. The remaining eigenvalues correspond to transversal deviations from the fully synchronized state.

When Eq. (2.44) has no solution, the oscillators may be entrained in a state of full synchronization where the phases are identical and evolve with time, $\phi_i(t) = \phi^*(t)$ for all i. The motion of this collective phase is given by

the equation

$$\dot{\phi}^* = \omega + f_1(\phi^*) - f_2(\phi^*). \tag{2.47}$$

Due to the periodicity of functions f_1 and f_2, the collective phase ϕ^* moves periodically, with period

$$T = \left| \int_0^{2\pi} \frac{d\phi}{\omega + f_1(\phi) - f_2(\phi)} \right|. \tag{2.48}$$

To carry out the stability analysis for this solution, we must bear in mind that the reference orbit to be perturbed with small deviations depends on time. Writing $\phi_i(t) = \phi^*(t) + \delta\phi_i(t)$ and linearizing Eqs. (2.43) yields

$$\delta\dot{\phi}_i = f_1'(\phi^*)\delta\phi_i - \frac{1}{N} f_2'(\phi^*) \sum_{j=1}^{N} \delta\phi_j. \tag{2.49}$$

The coefficients of these linear equations depend on time through the function $\phi^*(t)$. Summation of Eqs. (2.49) over the index i gives an equation for the quantity $\Delta(t) = \sum_i \delta\phi_i(t)$, whose solution reads

$$\sum_{i=1}^{N} \delta\phi_i(t) = \Delta(0) \exp \int_0^t [f_1'(\phi^*) - f_2'(\phi^*)] dt'. \tag{2.50}$$

This quantity can now be substituted into the last term of the right-hand side of Eq. (2.49), resulting in an equation for $\delta\phi_i$ only. Its solution is

$$\delta\phi_i(t) = \left[\delta\phi_i(0) - \frac{\Delta(0)}{N} \right] \exp \int_0^t f_1'(\phi^*) dt' + \frac{\Delta(0)}{N} \exp \int_0^t [f_1'(\phi^*) - f_2'(\phi^*)] dt'. \tag{2.51}$$

The fully synchronized periodic state $\phi^*(t)$ is stable if, at long times, $\delta\phi_i(t)$ asymptotically vanishes for all i. This requires that the functions of time in both terms of the right-hand side of Eq. (2.51) tend to zero. Due to the periodicity of the motion $\phi^*(t)$ and of the functions $f_1(\phi)$ and $f_2(\phi)$, the first term tends to zero if the inequality

$$\int_0^T f_1'(\phi^*) dt = \int_0^{2\pi} \frac{f_1'(\phi) d\phi}{\omega + f_1(\phi) - f_2(\phi)} < 0 \tag{2.52}$$

holds. As for the time dependence of the second term, the integral calculated over a whole period is

$$\int_0^T [f_1'(\phi^*) - f_2'(\phi^*)]dt = \int_0^{2\pi} \frac{[f_1'(\phi) - f_2'(\phi)]}{\omega + f_1(\phi) - f_2(\phi)} d\phi = 0, \tag{2.53}$$

because the integrand in the second integral represents an exact differential of a 2π-periodic function of ϕ, $d \ln |\omega + f_1(\phi) - f_2(\phi)|$. In other words, the second term in the right-hand side of Eq. (2.51) does not tend to zero, but is instead a 2π-periodic function of time. The presence of this non-vanishing contribution to the deviation $\delta\phi_i(t)$ is related to the symmetry of Eqs. (2.43) and (2.47) under constant shifts in the time scale, $t \to t + t_0$. If the deviations from $\phi^*(t)$ include a longitudinal component, this part of the perturbation will not fade out. A net shift along the orbit is equivalent to a change in the phase of the synchronized motion. Note, in fact, that the non-vanishing contribution is proportional to the mean deviation $\Delta(0)/N$. If, on the other hand, the deviations average to zero, no effective longitudinal shift is being applied and such contribution is not present. The situation is therefore similar to that encountered for an ensemble of identical oscillators with interaction functions $F_{ij}(\phi_i, \phi_j) \propto \sin(\phi_j - \phi_i)$. The non-vanishing deviations in Eq. (2.51) are equivalent to the longitudinal eigenvectors with vanishing eigenvalues of the time-independent linearized problem analyzed in Sec. 2.2. Hence, the stability of the fully synchronized state $\phi^*(t)$ is associated with the asymptotic disappearance of transversal perturbations, ensured if inequality (2.52) holds.

Equations (2.43) also have solutions representing clustered states, where the ensemble splits into M internally synchronized groups with phases Φ_m. The equations of motion for these clusters are

$$\dot{\Phi}_m = \omega + f_1(\Phi_m) - \frac{1}{N} \sum_{k=1}^{M} n_k f_2(\Phi_k), \tag{2.54}$$

where n_m is the number of oscillators in cluster m. Under phase perturbations $\delta\phi_i$ that do not affect the average phase of the clusters, namely when within each cluster $\sum_i \delta\phi_i = 0$, the stability condition coincides with inequality (2.52) [Golomb *et al.* (1992)]. If, on the other hand, clusters are not broken up by perturbations but their relative positions change, the linear stability analysis can be performed directly for Eqs. (2.54). Note that if the clusters are identical in size, $n_m = N/M$ for all m, these equations coincide with those of an ensemble of M coupled oscillators. It is interesting to point out that clustered states have been observed for Eqs. (2.43) only when the function $f_1(\phi)$ includes higher-order harmonics, in analogy with the situation studied in Sec. 2.3.

Numerical analysis of Eqs. (2.43) shows that, for certain choices of the functions $f_1(\phi)$ and $f_2(\phi)$, the ensemble may approach a stationary state where phases are distributed over the whole interval $[0, 2\pi)$. This situation is analogous to that of the phase-distributed stationary state discussed at the end of Sec. 2.2. For Eqs. (2.43), however, such stationary phase distribution is not uniform and depends on ϕ. To show this, it is useful to consider the limit $N \to \infty$, where the ensemble is statistically described by the phase density

$$n(\phi, t) = \frac{1}{N} \sum_{i=1}^{N} \delta[\phi - \phi_i(t)]. \tag{2.55}$$

The product $n(\phi, t)d\phi$ represents the fraction of oscillators in the infinitesimal interval $d\phi$ of $[0, 2\pi)$ at time t, and can be interpreted as the probability of finding an oscillator in $d\phi$ at that time. Note that

$$\int_0^{2\pi} n(\phi, t)d\phi = 1 \tag{2.56}$$

for all t. For the present interaction model, the phase density satisfies the equation

$$\frac{\partial n}{\partial t} + \frac{\partial}{\partial \phi}\{[w(t) + f_1(\phi)]n\} = 0, \tag{2.57}$$

where

$$w(t) = \omega - \int_0^{2\pi} f_2(\phi)n(\phi, t)d\phi. \tag{2.58}$$

Stationary solutions to the equation for the phase density have the form

$$n_s(\phi) = \frac{\nu}{w_s + f_1(\phi)}, \tag{2.59}$$

where ν is a normalization constant. The stationary value w_s results from the self-consistency relation

$$w_s = \omega - \nu \int_0^{2\pi} \frac{f_2(\phi)}{w_s + f_1(\phi)} d\phi. \tag{2.60}$$

Thus, the stationary phase distribution depends on ϕ. Note that such solution describes a possible state of the system if $n_s(\phi)$ is positive for any

ϕ. Its linear stability can be studied by considering the evolution of small perturbations to $n_s(\phi)$ determined by Eq. (2.57).

In the stationary distributed state, the equation of motion is

$$\dot{\phi}_i = w_s + f_1(\phi_i). \tag{2.61}$$

Since $w_s + f_1(\phi) \neq 0$ for any ϕ and $f_1(\phi)$ is 2π-periodic, the phases perform periodic motion. They are closer to each other near the maxima of $n_s(\phi)$, where they move more slowly, and become more separated where $n_s(\phi)$ is smaller and $\dot{\phi}$ is larger.

Equations (2.43) also exhibit regimes of non-periodic incoherent motion, either quasiperiodic or aperiodic, where oscillators are not entrained in synchronous evolution. Quasiperiodic and aperiodic motions may coexist, and are then selected by the initial conditions [Golomb *et al.* (1992)].

As a specific realization of the present system let us consider Eqs. (2.43) for $f_1(\phi) = A\sin\phi$ and $f_2(\phi) = B\cos\phi$. Without loosing generality, we fix $\omega = 1$ by rescaling time. Moreover, we can take $A > 0$. The fully synchronized fixed-phase state ϕ^* exists for $A^2 + B^2 > 1$. According to Eqs. (2.45) and (2.46) it is stable if $A > 1$ or $B < 0$. In the region where the fixed-phase synchronized state does not exist, we find a fully synchronized periodic orbit $\phi^*(t)$. Calculation of the integral in Eq. (2.52) shows that this state is stable if $B < 0$.

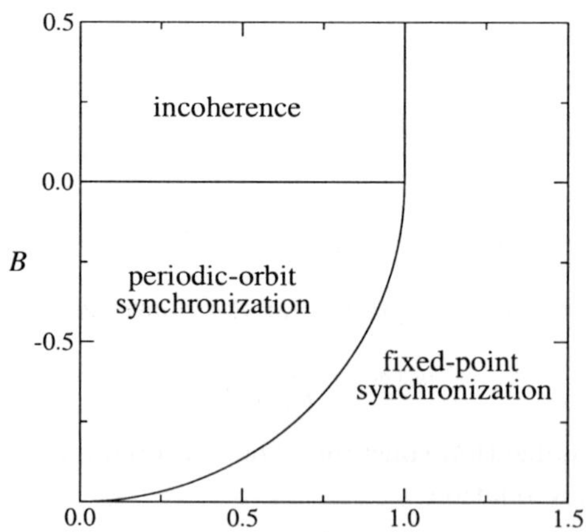

Fig. 2.7 Phase diagram of system (2.43) with $f_1(\phi) = A\sin\phi$, $f_2(\phi) = B\cos\phi$, and $\omega = 1$. Labels indicate the kind of collective evolution that is stable in each region (adapted from [Golomb *et al.* (1992)]).

For this choice of f_1 and f_2, clustered states and stationary distributed states are unstable or marginally stable, because the eigenvalues in the corresponding stability analysis turn out to have negative or vanishing real parts. In the parameter region where both the fixed-phase and the periodic synchronized states are unstable, numerical results show that the ensemble does not approach any of the trajectories discussed above. Instead, the asymptotic trajectory is strongly dependent on the initial condition and is typically characterized by a continuous distribution of phases over $[0, 2\pi)$. For some initial conditions, this distribution varies periodically. Other initial conditions lead to partially synchronized states, with coexistence of a cluster of synchronized oscillators and a background of non-entrained elements. Figure 2.7 shows the stability regions in parameter space for this form of incoherent collective motion and for synchronized states.

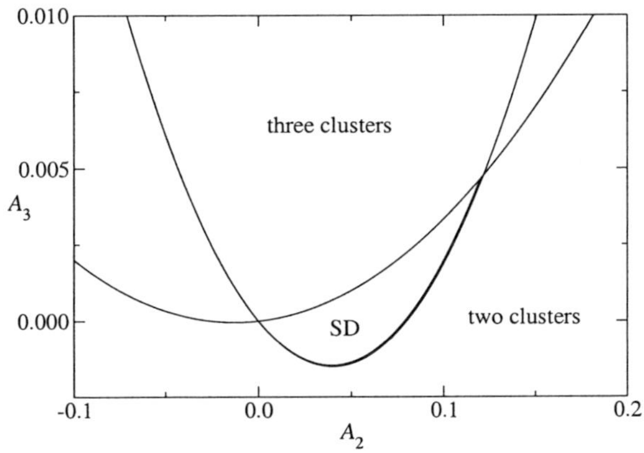

Fig. 2.8 Phase diagram of system (2.43) with $f_1(\phi) = A\sin\phi + A_2\sin 2\phi + A_3\sin 2\phi$, $f_2(\phi) = B\cos\phi$, $A = B = 0.5$ and $\omega = 1$. In the zones between two- and three-cluster regimes, both clustered configurations are stable. The stationary distribution of Eq. (2.59) is stable in the region marked SD. Incoherent collective motion is found in the narrow band separating the SD region and the two-cluster region (adapted from [Golomb *et al.* (1992)]).

When the functions $f_1(\phi)$ and $f_2(\phi)$ have more complex shapes, the incoherent collective behavior observed in the region with $B > 0$ and $A < 1$ becomes restricted to small zones of parameter space. Moreover, if a few higher harmonics are added to $f_1(\phi)$, stable clustered states or a stable distribution of phases become possible. In the case with

$$f_1(\phi) = A \sin \phi + A_2 \sin 2\phi + A_3 \sin 3\phi, \tag{2.62}$$

even small values of A_2 and A_3 are enough to stabilize two- and three-cluster configurations. Figure 2.8 show the stability regions for two- and three-cluster states, and for the stationary distribution of Eq. (2.59), on the plane (A_2, A_3). The other two parameters, $A = B = 0.5$, are chosen in such a way that, in the absence of higher harmonics, collective motion is incoherent. Incoherent behavior persists only in a narrow band of parameter space, separating the regions where two clusters and the stationary phase distribution are stable.

Chapter 3

Heterogeneous Ensembles and the Effects of Noise

Heterogeneities and fluctuations are always present in macroscopic natural systems. All real populations of coupled dynamical elements are characterized by a certain degree of diversity both in the individual properties of their components and in their interaction. At the same time, they are subject to external forces originating in the environment. These forces act at many different time scales, and can be represented as randomly varying contributions to the dynamics of each single element. Disorder and noise are therefore important ingredients in any realistic model of a complex system. They compete with the mechanisms that induce the emergence of order and, thus, can drastically modify the properties of collective evolution.

This chapter is devoted to the analysis of the effects of individual heterogeneities and of fluctuations on globally coupled phase oscillators. First, we discuss the phenomenon of frequency synchronization in an ensemble of oscillators with different natural frequencies. We study the emergence and mutual interaction of many frequency-synchronized clusters when natural frequencies are distributed in groups. Fluctuating forces are then introduced as additive noise in the individual dynamics of identical oscillators. We show that they may lead to the desynchronization of the ensemble. Finally, we consider time-delayed interactions, both in uniform and in heterogeneous ensembles, which give rise to many coexisting synchronized states.

3.1 Transition to Frequency Synchronization

We begin our study of heterogeneous systems by analyzing the emergence of collective evolution in an ensemble of interacting non-identical periodic oscillators. As discussed below, the ensemble may become entrained in a form of partial synchronization where the frequencies of a group of oscillators coincide [Winfree (1967)]. This phenomenon can be analytically studied for a large ensemble of globally coupled phase oscillators whose individual evolution is given by

$$\dot{\phi}_i = \omega_i + \frac{K}{N}\sum_{j=1}^{N} \sin(\phi_j - \phi_i), \tag{3.1}$$

where ω_i is the natural frequency of oscillator i [Kuramoto (1984)]. The natural frequencies are chosen at random from a distribution $g(\omega)$.

As a result of interactions, the phase of any individual oscillator displays complicated evolution. Its motion is typically chaotic. In general, the frequency $\dot{\phi}_i$ of each oscillator differs from its natural frequency ω_i. It is useful to define the *effective frequency* ω_i' as the average of $\dot{\phi}_i$ over long times,

$$\omega_i' = \lim_{T\to\infty} \frac{1}{T}\int_0^T \dot{\phi}_i(t)dt. \tag{3.2}$$

A cluster formed by a fraction of oscillators with identical effective frequencies appears at some critical value of the coupling strength. For any two oscillators i and j in the cluster, we have $\omega_i' = \omega_j' = \Omega$, where Ω is the *synchronization frequency*. The phases of these oscillators, however, are not identical. The number of elements inside the cluster increases as K grows beyond the critical coupling K_c. We show below that the values of Ω and K_c depend on the distribution of natural frequencies $g(\omega)$. The onset of frequency synchronization in the system described by Eqs. (3.1) for $N \to \infty$ corresponds to a bifurcation, and has the properties of a critical phenomenon.

A statistical description of the solutions to Eqs. (3.1) in the limit $N \to \infty$ is constructed in terms of the phase density $n(\phi, t)$ introduced in Eq. (2.55). Using the density $n(\phi, t)$ to replace the summation over the oscillator ensemble in Eqs. (3.1) by an integral,

$$\frac{1}{N}\sum_j \sin(\phi_j - \phi_i) \to \int_0^{2\pi} n(\phi)\sin(\phi - \phi_i)d\phi, \tag{3.3}$$

the evolution equation for ϕ_i takes the form of Eq. (2.21) where

$$\sigma(t)\exp[i\Phi(t)] = \int_0^{2\pi} n(\phi, t)\exp(i\phi)\, d\phi. \tag{3.4}$$

In the simplest stationary state, corresponding to a uniform distribution of oscillators in $[0, 2\pi)$, the phase density is a constant, $n = (2\pi)^{-1}$. In this state, $\sigma(t) = 0$ and, according to Eq. (2.21), the effect of coupling vanishes and each oscillator moves with its natural frequency.

The simplest form of collective motion, on the other hand, corresponds to rigid rotations of the ensemble at a certain frequency Ω. In this case, $n(\phi, t) = n_0(\phi - \Omega t)$, while

$$\Phi(t) = \Phi_0 + \Omega t, \tag{3.5}$$

and σ turns out to be independent of time:

$$\sigma \exp(i\Phi_0) = \int_0^{2\pi} n_0(\phi) \exp(i\phi) d\phi. \tag{3.6}$$

Introducing now the relative phase Φ_i of oscillator i with respect to the average phase $\Phi(t)$, $\Phi_i = \phi_i - \Phi_0 - \Omega t$, Eq. (2.21) becomes

$$\dot{\Phi}_i = \omega_i - \Omega - K\sigma \sin \Phi_i. \tag{3.7}$$

The quantities Ω and σ must be determined self-consistently.

Equation (3.7) is formally identical to the second of Eqs. (2.8). It has a stable fixed point when the natural frequency ω_i is sufficiently close to the synchronization frequency Ω, i.e. when $|\omega_i - \Omega| \leq K\sigma$. In this case, the phase of oscillator i evolves with time as

$$\phi_i(t) = \Omega t + \psi_i, \tag{3.8}$$

with $\psi_i = \Phi_0 + \arcsin[(\omega_i - \Omega)/K\sigma]$. Due to the interaction with the ensemble, the frequency of the oscillator has shifted from its natural frequency ω_i to $\omega_i' = \Omega$.

On the other hand, if $|\omega_i - \Omega| > K\sigma$, the solution to Eq. (3.7) is

$$\phi_i(t) = \omega_i' t + \psi[(\omega_i - \Omega)t], \tag{3.9}$$

where $\psi(t)$ is a 2π-periodic function of t and

$$\omega_i' = \Omega + (\omega_i - \Omega)\sqrt{1 - \left(\frac{K\sigma}{\omega_i - \Omega}\right)^2} \tag{3.10}$$

is the effective frequency of oscillator i. Now, ω_i' depends on ω_i. For $|\omega_i - \Omega| \gg K\sigma$, we find $\omega_i' \approx \omega_i$, whereas $\omega_i' \approx \Omega$ if $|\omega_i - \Omega| \approx K\sigma$. Therefore, these oscillators do not become entrained in the periodic collective motion with frequency Ω.

For a given distribution of natural frequencies and a fixed value of the coupling intensity K, the ensemble of phase oscillators governed by Eqs. (3.1) splits into two groups. Oscillators with $|\omega_i - \Omega| \leq K\sigma$ are collectively entrained in periodic motion with frequency Ω, while the remaining population moves incoherently. The size of these two groups is determined by the values of the synchronization frequency Ω and the amplitude σ which, as already pointed out, must be found self-consistently.

According to Eq. (3.6), σ determines the size of the synchronous cluster. The phase density $n_0(\phi)$ corresponding to this cluster can be calculated from the distribution of natural frequencies $g(\omega)$, taking into account the identity

$$n_0(\phi)d\phi = g(\omega)d\omega, \tag{3.11}$$

and the relation (3.8) between the phase ϕ and the natural frequency ω of each oscillator. This yields

$$n_0(\phi) = K\sigma g[\Omega + K\sigma \sin(\phi - \Phi_0)] \cos(\phi - \Phi_0) \tag{3.12}$$

for $|\phi - \Phi_0| \leq \pi/2$, and $n_0(\phi) = 0$ otherwise. Replacing this result into Eq. (3.6), we obtain the following self-consistency equation for σ:

$$\sigma = K\sigma \int_{-\pi/2}^{\pi/2} g(\Omega + K\sigma \sin\phi) \cos\phi \exp(i\phi) d\phi. \tag{3.13}$$

This equation has a trivial solution $\sigma = 0$ for any set of values of the relevant parameters. Assuming the existence of additional solutions, $\sigma \neq 0$, we separate real and imaginary parts to get two coupled equations for σ and Ω as functions of $g(\omega)$ and K, namely,

$$1 = K \int_{-\pi/2}^{\pi/2} g(\Omega + K\sigma \sin\phi) \cos^2\phi \; d\phi \tag{3.14}$$

and

$$0 = \int_{-\pi/2}^{\pi/2} g(\Omega + K\sigma \sin\phi) \cos\phi \sin\phi \; d\phi. \tag{3.15}$$

If the distribution $g(\omega)$ is symmetric around a frequency ω_0, $g(\omega_0 + \omega) = g(\omega_0 - \omega)$, $\Omega = \omega_0$ is a solution to Eq. (3.14). A nonzero solution for σ, found from Eq. (3.15), exists above a certain critical coupling intensity K_c. Just above this threshold σ increases rapidly, and then saturates to $\sigma = 1$ for large K. In other words, a synchronous cluster moving with collective frequency $\Omega = \omega_0$ appears at K_c, and grows in size as coupling becomes stronger. Figure 3.1 shows the solution of Eq. (3.15) for σ as a function of K, with $g(\omega) = \exp(-\omega^2/2)/\sqrt{2\pi}$. Since σ measures the size of the cluster, it plays the role of an order parameter for the transition to frequency synchronization.

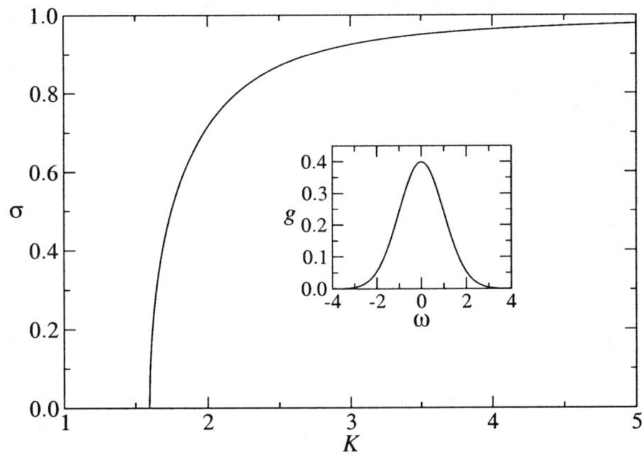

Fig. 3.1 The order parameter σ as a function of the coupling intensity K for an ensemble of phase oscillators with Gaussian distribution of natural frequencies. The inset shows the frequency distribution $g(\omega) = \exp(-\omega^2/2)/\sqrt{2\pi}$. The transition to frequency synchronization occurs at $K_c \approx 1.596$.

We can obtain an approximate expression for σ as a function of K near the transition by examining Eq. (3.15) for $\Omega = \omega_0$ and $\sigma \approx 0$. Expanding $g(\omega_0 + K\sigma \sin x)$ up to second order around $\sigma = 0$, Eq. (3.15) becomes

$$\frac{\pi}{2} K g(\omega_0) + \frac{\pi}{16} K^3 g''(\omega_0) \sigma^2 = 1. \tag{3.16}$$

Note that $g''(\omega_0) < 0$, since $g(\omega)$ reaches a maximum at ω_0. The polynomial equation (3.16) has nonzero roots for $K > K_c$, with

$$K_c = \frac{2}{\pi g(\omega_0)}. \tag{3.17}$$

The solution reads

$$\sigma = \sqrt{\frac{8 g(\omega_0)}{K_c^3 |g''(\omega_0)|} (K - K_c)}. \tag{3.18}$$

At the transition, the order parameter behaves as $\sigma \propto (K - K_c)^{1/2}$. This result holds for any symmetric distribution $g(\omega)$ with a smooth maximum at ω_0, as long as $g''(\omega_0) \neq 0$. The critical exponent 1/2 is characteristic of second order phase transitions in the mean field approximation.

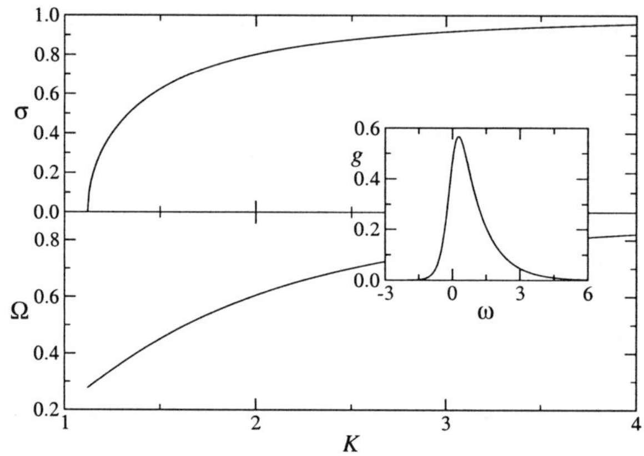

Fig. 3.2 The order parameter σ as a function of the coupling intensity K for an ensemble of phase oscillators with an asymmetric distribution of natural frequencies. The inset shows the frequency distribution $g(\omega) \propto [\exp(-4\omega) + \exp(\omega)]^{-1}$. The transition to frequency synchronization occurs at $K_c \approx 1.122$.

The situation is qualitatively similar for asymmetric frequency distributions with a single maximum. Figure 3.2 shows the solutions to Eqs. (3.14) and (3.15) as functions of K, for $g(\omega) \propto [\exp(-4\omega) + \exp(\omega)]^{-1}$. This frequency distribution has a maximum at $\omega_0 = \frac{2}{5} \ln 2 \approx 0.277$. The threshold of the transition to frequency synchronization has the same form as in Eq. (3.17), and the critical behavior of σ as a function of K is given by Eq. (3.18). Now, however, the collective frequency Ω varies with the coupling intensity. It coincides with ω_0 for $K = K_c$ and shifts to larger values as K grows.

As discussed above, interactions modify the distribution of frequencies in the ensemble. Entrained oscillators, all of which have the same effective frequency Ω, are represented by a distribution

$$G_0(\omega') = r\delta(\omega' - \Omega), \tag{3.19}$$

where

$$r = \int_{\Omega - K\sigma}^{\Omega + K\sigma} g(\omega) d\omega \tag{3.20}$$

is the entrained fraction of the population. The frequency distribution $G(\omega')$ of non-entrained oscillators, on the other hand, can be found taking into account the relation (3.10) between the natural frequency ω and the effective frequency ω', through the identity

$$G(\omega')d\omega' = g(\omega)d\omega. \tag{3.21}$$

For a symmetric distribution of natural frequencies, this yields

$$G(\omega') = g[\Omega + \sqrt{(\omega' - \Omega)^2 + K^2\sigma^2}] \frac{|\omega' - \Omega|}{\sqrt{(\omega' - \Omega)^2 + K^2\sigma^2}}. \tag{3.22}$$

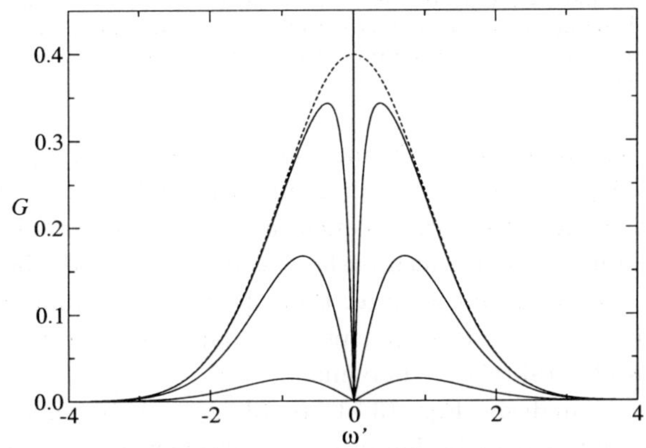

Fig. 3.3 The distribution of effective frequencies $G(\omega')$ for the case of a Gaussian distribution of natural frequencies, and three values of the coupling intensity (from top to bottom: $K = 1.6$, $K = 1.7$, and $K = 2.2$; cf. Fig. 3.1). The dotted curve represents the distribution of natural frequencies $g(\omega) = \exp[-(\omega - \omega_0)^2/2]/\sqrt{2\pi}$ with $\omega_0 = 0$. The vertical line stands for the distribution of entrained oscillators, Eq. (3.19).

Figure 3.3 illustrates this result for $g(\omega) = \exp[-(\omega - \omega_0)^2/2]/\sqrt{2\pi}$, and some values of K. We find that the frequency distribution becomes depleted around $\omega' = \Omega = \omega_0$, as a consequence of the entrainment of oscillators from that region. Figure 3.4 shows the evolution of the distribution of frequencies ω_i' in a system of 10^4 phase oscillators with a Gaussian distribution of natural frequencies centered a $\omega = 0$, with K larger than

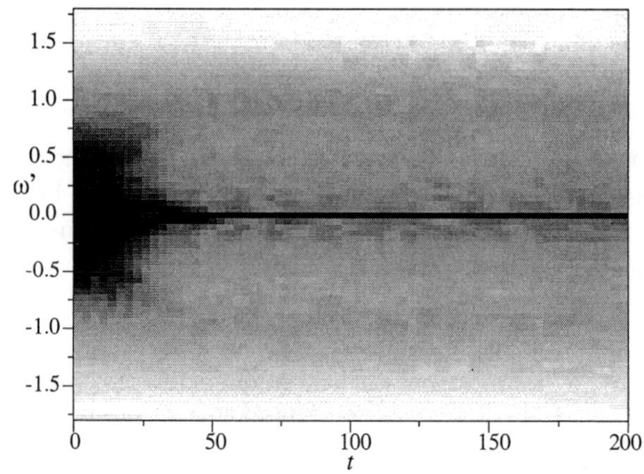

Fig. 3.4 Density plot of the histograms of frequencies ω_i', as a function of time, for an ensemble of 10^4 coupled phase oscillators with Gaussian distribution of natural frequencies, obtained from the numerical solution of Eqs. (3.1). Darker shading corresponds to larger concentrations.

the critical coupling intensity. Numerically, the effective frequencies ω_i' are calculated taking averages of $\dot{\phi}_i$ over a finite time T [cf. Eq. (3.2)]. In order to reveal the change of natural frequencies in time, the averaging interval T is fixed to a finite value which is short as compared with the time scales of frequency evolution, but is larger than the typical oscillation periods in the ensemble. Note the development of a sharp concentration at $\omega' = 0$, and of the two lateral relative maxima.

The present analysis, Eqs. (3.11) to (3.22), can be explicitly worked out for a Lorentzian distribution of natural frequencies, $g(\omega) \propto [\gamma^2 + (\omega - \omega_0)^2]^{-1}$ [Kuramoto (1984)]. We study below another case where explicit results can be obtained, namely

$$g(\omega) = \frac{1+\alpha}{2\alpha}(1 - |\omega - \omega_0|^\alpha) \tag{3.23}$$

for $|\omega - \omega_0| < 1$, and $g(\omega) = 0$ otherwise. Here, $\alpha > 0$. Since $g(\omega)$ is symmetric around ω_0, Eq. (3.14) is satisfied if $\Omega = \omega_0$. Limiting the analysis to the region where $K\sigma < 1$, which includes in particular the entrainment transition, we find that the solution to Eq. (3.15) for $K > K_c$ is

$$\sigma = \frac{1}{K}\left[\frac{2\Gamma\left(2+\frac{\alpha}{2}\right)}{\Gamma\left(\frac{1}{2}+\frac{\alpha}{2}\right)}\left(\frac{\pi}{2} - \frac{2\alpha}{1+\alpha}\frac{1}{K}\right)\right]^{1/\alpha}, \tag{3.24}$$

with

$$K_c = \frac{4\alpha}{\pi(1+\alpha)}. \tag{3.25}$$

Consequently, near the entrainment transition, the order parameter behaves as $\sigma \propto (K - K_c)^{1/\alpha}$. The critical exponent $1/\alpha$ is in general different from that found from Eq. (3.16) and in the case of a Lorentzian distribution of natural frequencies [Kuramoto (1984)]. The exponent is determined by the shape of $g(\omega)$ at its maximum, and equals 1/2 only for quadratic profiles.

The transition to frequency synchronization has also been studied in a time-discrete version of Eqs. (3.1) [Daido (1986)], and when the interaction function includes a constant phase shift, $F(\phi_i, \phi_j) \propto \sin(\phi_j - \phi_i + \alpha)$ [Sakaguchi and Kuramoto (1986)]. Dynamical properties in the non-entrained regime, $K < K_c$, have been analyzed as well [Strogatz *et al.* (1992); Strogatz (2000)].

3.2 Frequency Clustering

The theory of frequency synchronization presented in Sec. 3.1 assumes that, as elements become entrained and move at the same frequency, only one cluster of synchronized oscillators is present in the ensemble. This is the case when the distribution of natural frequencies $g(\omega)$ has a single maximum. As we have shown, the cluster forms out of this maximum as the coupling intensity is increased. When $g(\omega)$ has more than one maximum, on the other hand, it may happen that several clusters appear during the synchronization process [Kuramoto (1984)]. From our study of frequency distributions with a single maximum, we expect that a distribution with several well-separated maxima develops a cluster of entrained oscillators at each maximum, if the number of oscillators there is large enough to trigger mutual synchronization. We find that, once formed, these clusters behave as interacting individual oscillators, each of them affecting the motion of the others.

Figure 3.5 shows the temporal evolution of the distribution of effective frequencies for an ensemble of 10^4 phase oscillators whose natural frequencies are grouped into two peaks centered at $\omega = \pm 1.5$. The peaks have slightly different populations. The averaging interval T used to calculate ω_i' has been chosen to encompass a few oscillation periods of a typical element in the peaks, in order to reveal the time evolution of the effective frequencies. The coupling intensity is such that, as time elapses, two clusters build

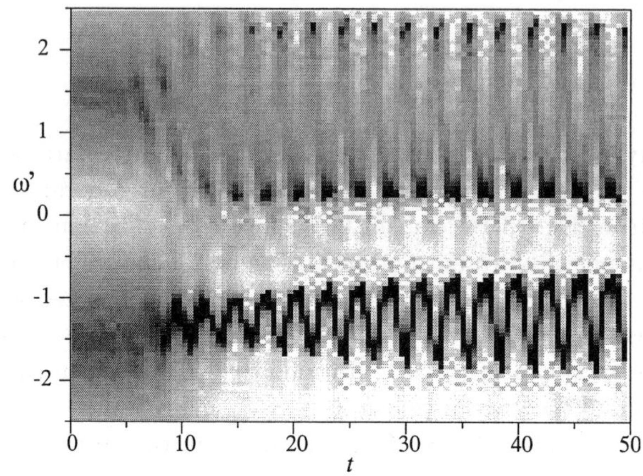

Fig. 3.5 Density plot of the histograms of effective frequencies ω_i', as a function of time, for an ensemble of 10^4 coupled phase oscillators. The distribution of natural frequencies has two peaks at $\omega = -1.5$ and $\omega = 1.5$, with 55% of the population in the first peak, and 45% in the second. Darker tones corresponds to larger concentrations.

up at the peaks, while a part of the ensemble remains non-entrained. The evolution of the two clusters is similar to that of two coupled oscillators, studied in Sec. 2.1. Each cluster maintains its integrity, and its effective frequency oscillates as a consequence of its interaction with the other cluster. Note that, since the populations of the clusters are different, their interaction is not symmetric. The frequency oscillations of the smaller cluster have larger amplitude.

When the distribution of natural frequencies has several overlapping maxima, the gradual emergence of frequency synchronization as coupling becomes stronger is an intricate collective process. Several clusters form at different coupling intensities, and new oscillators keep joining these clusters as K is increased. In turn, clusters approach each other and successively collapse. During this process, the distribution of effective frequencies changes steadily, due to the mutual interaction of clusters and non-entrained oscillators. Eventually, the whole ensemble becomes synchronized and all oscillators move with the same effective frequency. The complex hierarchical aggregation of oscillators and clusters is illustrated in Fig. 3.6, which shows the distribution of effective frequencies as a function of the coupling intensity for an ensemble of 10^3 oscillators. Their natural frequencies are distributed in six groups of different sizes and widths. The histogram in the

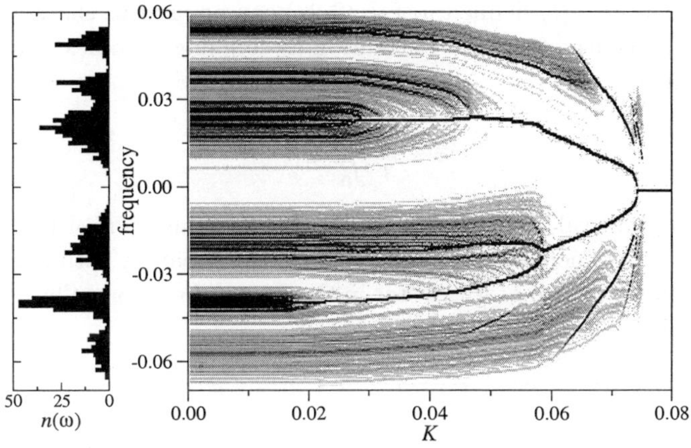

Fig. 3.6 Density plot of the histograms of effective frequencies ω_i', as a function of the coupling intensity, for an ensemble of 10^3 coupled phase oscillators. Darker tones correspond to larger concentrations. The left panel shows a histogram of the distribution of natural frequencies.

left panel shows the number of oscillators $n(\omega)$ as a function of their natural frequency. Averaging times in the calculation of effective frequencies are long as compared with their oscillation periods so that, for a fixed value of the coupling intensity, ω_i' has a well-defined constant value for each oscillator. In Fig. 3.7 we have plotted the effective frequencies against the natural frequencies for the same ensemble, and for three values of the coupling intensity. In this kind of plot, clusters of frequency-synchronized elements are revealed by the plateaus of constant ω'. These plateaus become broader as K grows.

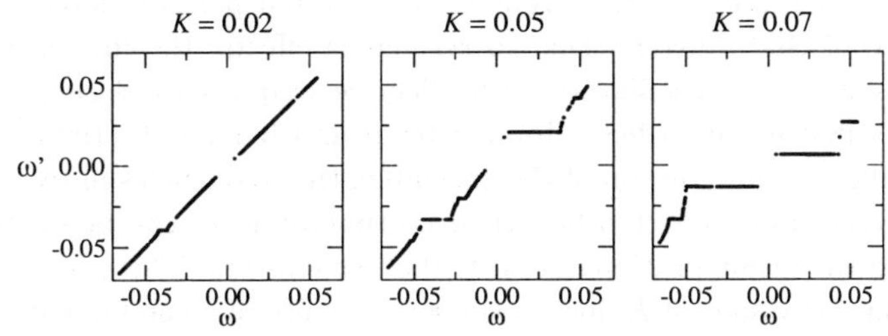

Fig. 3.7 Effective frequency ω' as a function of the natural frequency ω for the oscillators of the ensemble of Fig. (3.6), and three values of the coupling intensity K.

The quantity $\sigma(t)$ defined in Eq. (2.22) can be used to characterize the gradual emergence of coherent evolution as coupling becomes stronger. Its time average

$$\bar{\sigma} = \frac{1}{T}\int_0^T \sigma(t)dt \tag{3.26}$$

over a long interval T is plotted in Fig. 3.8 as a function of the coupling intensity. As K increases, $\bar{\sigma}$ grows steadily. Comparing with Fig. 3.6, we find that the variation of $\bar{\sigma}$ is faster in the zones where the collapse of large clusters takes place.

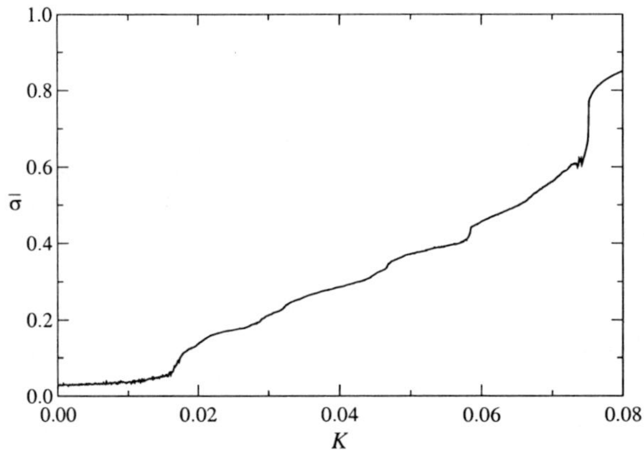

Fig. 3.8 The time average $\bar{\sigma}$ of the quantity $\sigma(t)$ as a function of the coupling intensity K, for the same ensemble as in Fig. 3.6.

The action of a cluster on the non-entrained oscillators and on the other clusters is qualitatively equivalent to an external periodic force with the frequency of that cluster. Due to resonance effects, the influence of the cluster is larger on oscillators with effective frequencies close to its own frequency [Sakaguchi (1988); Hoppensteadt and Izhikevich (1998)]. If, due to the collective interaction of the ensemble, the frequencies of two clusters become very close to each other, their mutual influence can be so strong as to lead to the disintegration of one or the two clusters. This process is seen in Fig. 3.6 for values of K just below the collapse of some big clusters.

Moreover, when most of the ensemble is entrained in a few clusters which dominate the collective dynamics of the system, non-linearities in the interaction function induce resonance effects at frequencies which do not coincide with those of the clusters, but which are given by linear combinations of them. These non-linear resonance effects have also been discussed for ensembles of phase oscillators at intermediate stages of frequency synchronization, under the action of external periodic forcing [Sakaguchi (1988)]. They can result in the formation of frequency-synchronized clusters at higher-harmonic frequencies. For instance, in the frequency distributions of Fig. 3.6 this phenomenon is seen for $K = 0.07$ where, apart from some non-entrained elements, the ensemble is divided into four clusters (see also Fig. 3.7). Clearly, the two large clusters with the central frequencies result from the successive aggregation of smaller groups. The other two, on the other hand, appear rather suddenly, at $K \approx 0.06$, out of the groups of non-entrained elements with the most lateral natural frequencies. Numerical results show that, along the whole interval of coupling intensities where these lateral clusters exist, their frequencies are $\omega'_C = 2\omega'_A - \omega'_B$ and $\omega'_D = 2\omega'_B - \omega'_A$, where ω'_A and ω'_B are the frequencies of the central clusters. This is an indication that the lateral clusters are induced by higher-harmonic resonance of the other two.

3.3 Fluctuating Forces

In the preceding sections of this chapter, we have studied the emergence of collective order in heterogeneous oscillator ensembles, where elements have different natural frequencies. Heterogeneities can also be introduced as dynamical disorder, in the form of fluctuating forces acting on each individual oscillator. In this section, we analyze the effect of noise on the synchronization phenomena studied so far. We show that low levels of noise allow for partially synchronized states. Sufficiently strong fluctuations, however, lead to a transition to incoherent collective dynamics [Kuramoto (1984); Shinomoto and Kuramoto (1986a); Shinomoto and Kuramoto (1986b); Shinomoto and Kuramoto (1988)].

Consider an ensemble of identical phase oscillators subject to the action of independent random forces $\xi_i(t)$,

$$\dot{\phi}_i = \frac{K}{N} \sum_{j=1}^{N} \sin(\phi_j - \phi_i) + \xi_i(t), \tag{3.27}$$

such that

$$\langle \xi_i(t) \rangle = 0,$$
$$\langle \xi_i(t)\xi_j(t') \rangle = 2S\delta_{ij}\delta(t-t'). \tag{3.28}$$

Introducing the function $\sigma(t)$ and $\Phi(t)$ as in Eq. (2.22), the equation of motion for the phase under the action of noise becomes

$$\dot{\phi}_i = K\sigma \sin(\Phi - \phi_i) + \xi_i(t). \tag{3.29}$$

For sufficiently long times, in the absence of fluctuations, the ensemble approaches a state of full synchronization if $K > 0$. When noise is acting, the condensate of synchronized oscillators breaks down but, if the noise level is not too high, the oscillators form a well-localized "cloud" around their average position ϕ^*. In the limit of an infinitely large ensemble, $N \to \infty$, the phase distribution asymptotically reaches a stationary profile $n(\phi)$, peaked around ϕ^*. Its width is determined by the intensity of noise.

The Fokker–Planck equation for the time-dependent phase distribution $n(\phi, t)$ reads

$$\frac{\partial n}{\partial t} = S\frac{\partial^2 n}{\partial \phi^2} + K\sigma\frac{\partial}{\partial \phi}[\sin(\phi - \Phi)n]. \tag{3.30}$$

Then, the stationary distribution $n(\phi)$ satisfies

$$\frac{1}{n}\frac{dn}{d\phi} = -\frac{K\sigma}{S}\sin(\phi - \Phi). \tag{3.31}$$

The solution of this equation is

$$n(\phi) = \frac{1}{2\pi I_0(K\sigma/S)} \exp\left[\frac{K\sigma}{S}\cos(\phi - \Phi)\right], \tag{3.32}$$

where $I_0(x)$ is the modified Bessel function of the first kind. Here, σ and Φ are constants, related to $n(\phi)$ as in Eq. (3.4). The above solution for $n(\phi)$ can be replaced into Eq. (3.4) to obtain a self-consistency equation for σ [Mikhailov and Calenbuhr (2002)]:

$$\sigma = \frac{I_1(K\sigma/S)}{I_0(K\sigma/S)}. \tag{3.33}$$

We recall from Sec. 3.1 that σ acts as an order parameter for the synchronization transition. Due to the identity $I_1(0) = 0$, the trivial solution $\sigma = 0$ of Eq. (3.33) exists for any values of S and K. This order parameter corresponds to a flat phase distribution, with the ensemble in a completely incoherent state. The nontrivial solution is shown in Fig. 3.9 as a function of the noise intensity S. In the absence of noise, $S = 0$, the order parameter reaches its maximum value $\sigma = 1$, corresponding to full synchronization. As S grows, the value of σ decreases, showing that the degree of coherence in the synchronized state becomes lower. At the critical point $S_c = K/2$ the order parameter drops to zero, and from then on the only solution to Eq. (3.33) is $\sigma = 0$. The incoherent state $n(\phi) = (2\pi)^{-1}$ is stable for $S > S_c$ and unstable otherwise [Kuramoto (1984)]. Just below the transition, the order parameter behaves as

$$\sigma \approx \frac{2}{\sqrt{K}}(S_c - S)^{1/2}. \tag{3.34}$$

Compare this behavior with the transition to frequency synchronization as a function of coupling intensity in ensembles of non-identical phase oscillators, Eq. (3.18). In the present case, the transition is induced by time-dependent fluctuations. For the ensemble of non-identical oscillators the transition between incoherence and synchronization takes place when the degree of disorder, given by the distribution of natural frequencies, varies.

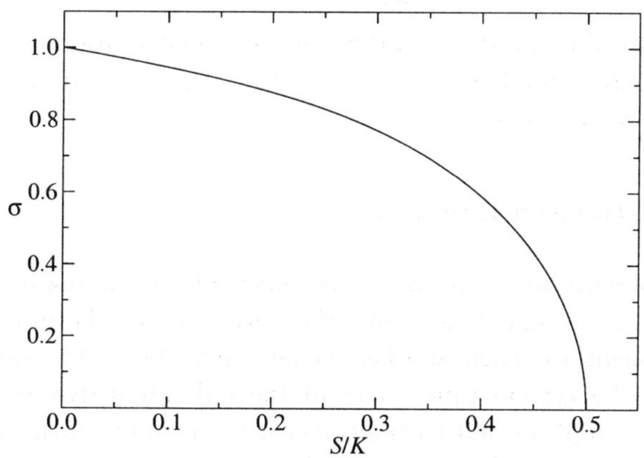

Fig. 3.9 The synchronization order parameter σ as a function of the ratio S/K between the intensity of noise and the coupling constant for an ensemble of identical phase oscillators, obtained from the numerical solution of Eq. (3.33).

The effect of random fluctuations on interaction models of the type studied in Sec. 2.4 has also been analyzed [Golomb and Rinzel (1994)]. It has been found that noise stabilizes the stationary solution given in

Eq. (2.59), where phases are distributed on the interval $[0, 2\pi)$ with a profile determined by the interaction function. In the parameter regions where this stationary solution is unstable for $S = 0$, increasing the level of noise induces a transition and it becomes stable. If in the absence of noise the ensemble is in the incoherent regime (see Fig. 2.7), the transition takes place at $S = 0$, and the stationary distribution is stable for any noise intensity $S > 0$. It becomes a global attractor of the system. On the other hand, when fully synchronized or clustered states are stable for $S = 0$, the transition takes place at a finite noise intensity S_c. For clustered states the quantities σ and Φ, defined as in Eq. (3.4), depend on time even at asymptotically large times. A time-independent order parameter $\bar{\sigma}$ can be defined as the temporal average

$$\bar{\sigma} = \left[\frac{1}{T} \int_0^T \left| \bar{z} - \sigma(t) \exp[i\Phi(t)] \right|^2 dt \right]^{1/2} \tag{3.35}$$

over a sufficiently long interval T, where

$$\bar{z} = \frac{1}{T} \int_0^T \sigma(t) \exp[i\Phi(t)] dt. \tag{3.36}$$

Near the critical point at which the stationary distribution becomes stable, this order parameter behaves as $\bar{\sigma} \propto (S_c - S)^{1/2}$, as for the interaction model considered above.

3.4 Time-Delayed Interactions

In many potential applications of ensembles of interacting oscillators, the time needed for a signal carrying information about the internal state of a given element to reach another element may be of the same order or larger than the typical time scales of the individual dynamics. In such cases, the assumption that each element acts instantaneously on any other element of the ensemble, implicit in Eqs. (2.2) and (2.4), does not hold. The role of this kind of delay in the collective behavior of interacting elements has been emphasized, particularly, for biological systems. In neural tissues, the propagation of electrochemical perturbations along axons occurs at relatively slow rates [Abarbanel *et al.* (1996)]. In populations of interacting organisms, communication involves visual, acoustic or chemical signals that must travel through air, water or soil [Buck and Buck (1976); Walker (1969); Sismondo (1990)]. It is therefore interesting to analyze the dynamics of coupled elements when time delays are introduced in the interaction functions.

In the model of globally coupled phase oscillators of Eqs. (2.4), time delays can be introduced as

$$\dot{\phi}_i(t) = \omega_i + \sum_{j=1}^{N} F_{ij}[\phi_i(t), \phi_j(t-\tau_{ij})], \qquad i = 1, \ldots, N. \tag{3.37}$$

The delay $\tau_{ij} > 0$ represents the time needed for the signal carrying information about the state of oscillator j to travel from j to i. For the globally coupled ensembles considered in this chapter, the interaction functions do not depend on the specific pair of interacting oscillators. Therefore, we focus the attention on the case of uniform time delays, $\tau_{ij} = \tau$ for all $i \neq j$, and assume that $\tau_{ii} = 0$ for all elements.

To gain insight on the effect of time delays in the collective dynamics of coupled phase oscillators it is useful to begin studying the case of two oscillators [Schuster and Wagner (1989)]. We consider the pair of equations of motion

$$\begin{aligned} \dot{\phi}_1(t) &= \omega_1 + \tfrac{K}{2} \sin[\phi_2(t-\tau) - \phi_1(t)], \\ \dot{\phi}_2(t) &= \omega_2 + \tfrac{K}{2} \sin[\phi_1(t-\tau) - \phi_2(t)]. \end{aligned} \tag{3.38}$$

These equations have solutions of the form

$$\phi_{1,2}(t) = \Omega t \pm \frac{\alpha}{2}, \tag{3.39}$$

where the two oscillators are synchronized in frequency but not in phase. Their common frequency is Ω, and their phases differ by α. The solutions (3.39) satisfy Eqs. (3.38) if the following identities hold:

$$\begin{aligned} \omega_1 - \omega_2 &= K \cos \Omega\tau \sin \alpha, \\ \omega_1 + \omega_2 &= 2\Omega + K \sin \Omega\tau \cos \alpha. \end{aligned} \tag{3.40}$$

These equations give the synchronization frequency and the phase difference as functions of the natural frequencies, the coupling intensity, and the delay. Eliminating α, we get an equation for Ω,

$$\omega_1 + \omega_2 - 2\Omega - K\sqrt{\cos^2 \Omega\tau - \left(\frac{\omega_1 - \omega_2}{K}\right)^2} \tan \Omega\tau = 0, \tag{3.41}$$

which must be solved numerically. When Ω is known, the phase difference is calculated as

$$\alpha = \arcsin\left(\frac{\omega_1 - \omega_2}{K\cos\Omega\tau}\right). \tag{3.42}$$

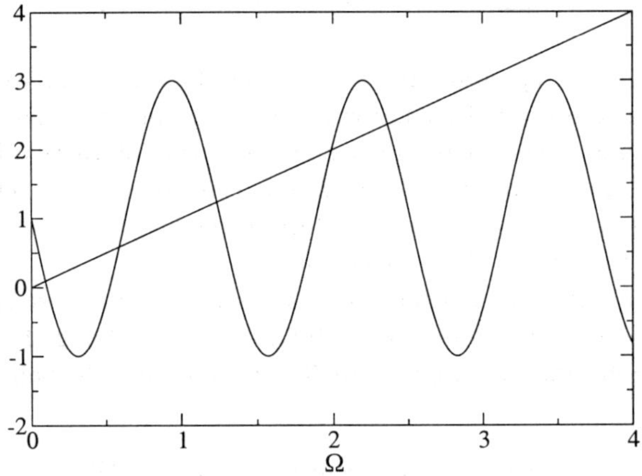

Fig. 3.10 Graphical solution of Eq. (3.43), for $\omega = 1$, $K = 4$, and $\tau = 5$.

Even in the case of identical natural frequencies, $\omega_1 = \omega_2 = \omega$, where Eq. (3.41) reduces to the simpler form

$$\Omega = \omega - \frac{K}{2}\sin\Omega\tau, \tag{3.43}$$

there will typically be many solutions for the coherent motion of the two oscillators. For small K and τ, Eq. (3.43) has only one solution, $\Omega \approx \omega$. As the coupling intensity and the time delay grow, however, new solutions appear both at $\Omega < \omega$ and $\Omega > \omega$. Figure 3.10 shows the left-hand and right-hand sides of Eq. (3.43) as functions of the synchronization frequency Ω, for $\omega = 1$, $K = 4$, and $\tau = 5$. The intersections give the frequencies of the possible synchronized states. For this case of identical natural frequencies, the phase difference α is always zero, irrespectively of the value of Ω.

An important consequence of the presence of time delays in the interaction of two phase oscillators is, therefore, that more than one synchronization frequency may exist for a given set of parameters. It is now necessary to analyze whether one or more of these coherent states are stable. Linear stability analysis of delay equations is carried out following the same lines as for ordinary differential equations. However, the corresponding eigenvalue problem leads typically to a transcendental equation, instead of the polynomial equation of the standard problem [Kuang (1993)]. For Eqs. (3.38), a solution with synchronization frequency Ω and phase difference α is stable, if all the roots λ of

$$\lambda^2 - 2\lambda \cos \Omega\tau \cos \alpha + K^2[1 - \exp(2\lambda\tau)] \cos(\Omega\tau + \alpha) \cos(\Omega\tau - \alpha) = 0 \quad (3.44)$$

are negative or have negative real parts.

This problem has been studied numerically, as a function of the coupling intensity [Schuster and Wagner (1989)]. It is found that, as K grows, new solutions to Eq. (3.41) appear in pairs. For large values of K, the total number of solutions is of order $K\tau$. For the case $\omega_1 = \omega_2$, the appearance of these pairs of solutions can be immediately inferred from Eq. (3.43) and Fig. 3.10. In each pair, one of the solutions is stable, while the other is unstable. New stable solutions have increasingly large synchronization frequencies. Each new stable solution is "more stable" than the pre-existing states, in the sense that the (negative) real part of the dominant eigenvalue associated with the new solution is larger in modulus than for the previous solutions. Overall, the real parts of the eigenvalues decrease in modulus as K grows and the number of solutions increases, which implies that all solutions become "less stable." Meanwhile, the phase differences α of successively new stable solutions alternate between $\alpha \approx 0$ and $\alpha \approx \pi$. Synchronized stable solutions are therefore almost in-phase or anti-phase states.

As discussed for clustering in Sec. 2.3, in connection with the application of oscillator models to neural activity, the simultaneous existence of many stable synchronized states in small groups of interacting neurons subject to time delays could be used by the brain to encode sensory information and functional patterns [Schuster and Wagner (1989); Abarbanel *et al.* (1996)].

We now consider the effect of time-delayed interactions on the collective dynamics of large ensembles of phase oscillators. Let us first analyze the case of identical natural frequencies,

$$\dot{\phi}_i(t) = \omega + \frac{K}{N} \sum_{j \neq i} \sin[\phi_j(t - \tau) - \phi_i(t)]. \quad (3.45)$$

In this case, changing $\phi_i \to \phi_i + \omega t$ eliminates the first term in the right-hand side, but introduces an additional term $-\omega\tau$ in the interaction function.

This time, therefore, we do not apply the symmetry transformation and work with Eqs. (3.45) in their standard form.

Equations (3.45) have a fully synchronized solution representing uniform rotations, $\phi_i(t) = \Omega t$ for all i, if the synchronization frequency Ω satisfies

$$\Omega = \omega - \beta K \sin \Omega\tau, \tag{3.46}$$

with $\beta = 1 - N^{-1}$ [cf. Eq. (3.43)]. We see that, as in the case of two identical oscillators, many fully synchronized states may exist simultaneously. Linear stability analysis shows that full synchronization is stable if all the solutions λ of the transcendental equation $\det \mathbf{S}(\lambda) = 0$ are negative or have negative real parts. Here, the $N \times N$ matrix $\mathbf{S} = \{s_{ij}\}$ has elements

$$\begin{aligned} s_{ij} &= \beta \frac{K}{N} \exp(-\lambda\tau) \cos \Omega\tau \\ &\quad + \left[\lambda + \beta K \cos \Omega\tau - \beta \frac{K}{N} \exp(-\lambda\tau) \cos \Omega\tau \right] \delta_{ij}, \end{aligned} \tag{3.47}$$

where δ_{ij} is the Kronecker delta symbol. The general analysis of such equation is difficult, but it can be shown that in the limit $N \to \infty$ there are $N - 1$ identical solutions

$$\lambda = -K \cos \Omega\tau, \tag{3.48}$$

while the remaining solution satisfies

$$\lambda = -K \cos \Omega\tau [1 - \exp(-\lambda\tau)]. \tag{3.49}$$

The stability condition applied to solution (3.48) requires

$$K \cos \Omega\tau > 0. \tag{3.50}$$

If this inequality is satisfied, the only real solution of Eq. (3.49) is $\lambda = 0$, which corresponds to the longitudinal eigenvalue discussed in Sec. 2.2. Therefore, condition (3.50) is necessary and sufficient for the stability of the fully synchronized state.

The conditions for existence and stability of at least one fully synchronized state are equivalent to the following inequalities [Yeung and Strogatz (1999)]:

$$K < \frac{\omega}{2(2m-1)}, \qquad \frac{(4m-3)\pi}{2\omega - 2K} < \tau < \frac{(4m-1)\pi}{2\omega + 2K}, \tag{3.51}$$

where m is an arbitrary positive integer. The non-shaded zone in Fig. 3.11 shows the parameter region where there exists at least one stable fully

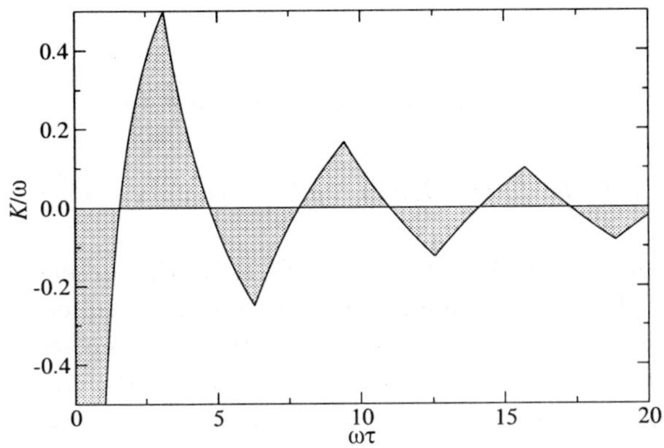

Fig. 3.11 The non-shaded zone corresponds to the region where at least one fully synchronized state exists and is stable for an ensemble of identical phase oscillators of natural frequency ω subject to time-delayed interactions, in the plane of rescaled parameters $(\omega\tau, K/\omega)$.

synchronized solution [Earl and Strogatz (2003)]. Generally, several states of full synchronization with different frequencies are simultaneously stable at each point. Note that, except for small delays, full synchronization can also be stable for $K < 0$.

For heterogeneous oscillator ensembles, where natural frequencies are not identical, the theory discussed in Sec. 3.1 can be generalized to the case of time-delayed coupling. As in the absence of delays, a cluster of frequency-synchronized oscillators develops if interactions are strong enough [Choi *et al.* (2000)]. In the present case, however, several synchronization frequencies Ω are simultaneously possible. In particular, even if the distribution of natural frequencies $g(\omega)$ is symmetric around a certain frequency ω_0, it is not possible to ensure that the only synchronization frequency will be $\Omega = \omega_0$. For a symmetric distribution of natural frequencies centered at $\omega = 0$, the self-consistency equations for the order parameter σ and the synchronization frequency Ω read

$$\cos \Omega\tau = K \int_{-\pi/2}^{\pi/2} g(\Omega + K\sigma \sin\phi) \cos^2\phi \, d\phi \tag{3.52}$$

and

$$\sin \Omega\tau = K \int_{-\pi/2}^{\pi/2} g(\Omega + K\sigma \sin\phi) \cos\phi \sin\phi \, d\phi$$
$$+ K \int_{1}^{\infty} [g(\Omega + K\sigma x) - g(\Omega - K\sigma x)](x - \sqrt{x^2 - 1})dx. \quad (3.53)$$

These equations replace Eqs. (3.14) and (3.15) in the presence of a uniform time delay τ. If $g(\omega)$ is symmetric around a nonzero frequency ω_0, the self-consistency equations are the same, except that Ω is replaced by $\Omega + \omega_0$.

Equations (3.52) and (3.53) have no solutions if the coupling intensity K and the delay τ are sufficiently small. Increasing K, the first solution appears at

$$K_c = \frac{2}{\pi g(0)} \quad (3.54)$$

[cf. Eq. (3.17)]. The synchronization frequency for this solution is $\Omega = 0$. Therefore, the corresponding order parameter does not depend on the delay τ. This fully synchronized state is equivalent to that found in Sec. 3.1 just above the critical coupling intensity K_c. It corresponds to condensation in frequencies around the maximum of $g(\omega)$, with the condensate moving coherently in a background of non-entrained oscillators. As K grows further, however, new fully synchronized states appear in pairs. They have non-vanishing synchronization frequency Ω, and one of them is stable while the other is unstable. For these new solutions, the order parameter at the transition is different from zero. Their frequencies are practically constant as functions of the coupling intensity. Meanwhile, for fixed K, Ω decreases monotonically as the delay grows.

The effects of noise on a heterogeneous ensemble of coupled phase oscillators have also been analyzed with time-delayed interactions, considering the equations

$$\dot{\phi}_i(t) = \omega_i + \frac{K}{N} \sum_{j \neq i} \sin[\phi_j(t - \tau) - \phi_i(t)] + \xi_i(t) \quad (3.55)$$

in the limit $N \to \infty$, with $\langle \xi_i(t) \rangle = 0$ and $\langle \xi_i(t)\xi_j(t') \rangle = 2S\delta_{ij}\delta(t - t')$ [Yeung and Strogatz (1999)]. In the presence of noise, it becomes important to study the stability of the incoherent state with stationary phase density $n(\phi) = (2\pi)^{-1}$, as discussed in Sec. 3.3. Using the Fokker–Planck equation for $n(\phi, t)$, linear stability analysis of the incoherent state shows that part

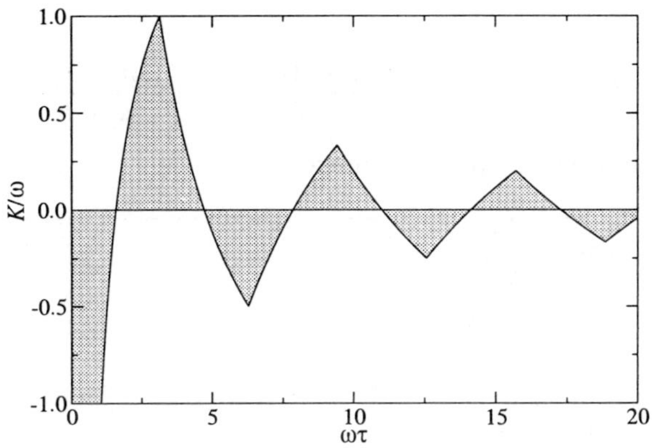

Fig. 3.12 Stability region of the incoherent state (shaded) for an infinitely large ensemble of identical phase oscillators with natural frequency ω, in the limit of vanishingly small noise intensity. The incoherent state is unstable in the remaining of the parameter space.

of the eigenvalue spectrum is continuous. The corresponding eigenvalues are

$$\lambda = -S - i\omega, \tag{3.56}$$

where ω takes the values of all the natural frequencies for which the distribution $g(\omega)$ is different from zero. The discrete eigenvalues, on the other hand, satisfy:

$$\int_{-\infty}^{\infty} \frac{g(\omega)}{\lambda + S + i\omega} d\omega = \frac{2}{K} \exp(\lambda\tau). \tag{3.57}$$

Equation (3.56) shows that the incoherent state is unstable (or, more specifically, marginally stable) in the absence of noise, since the real part of λ vanishes for $S = 0$. As for the discrete spectrum, Eq. (3.57) is considerably simplified if all the oscillators have the same frequency ω. In this case, it reduces to

$$(\lambda + S + i\omega) \exp(\lambda\tau) = \frac{K}{2}. \tag{3.58}$$

It is instructive to consider the solutions to this equation in the limit of vanishingly small noise, $S \to 0$, keeping however $S > 0$. In this case, the incoherent state is stable if the following conditions are simultaneously fulfilled:

$$K < \frac{\omega}{2m-1}, \qquad \frac{(4m-3)\pi}{2\omega - K} < \tau < \frac{(4m-1)\pi}{2\omega + K} \tag{3.59}$$

where m is an arbitrary positive integer.

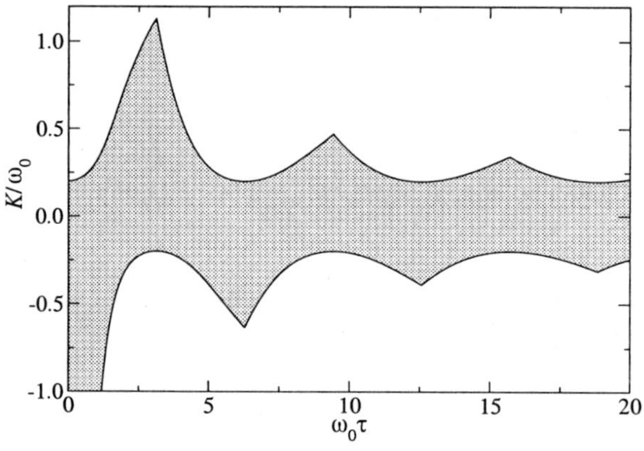

Fig. 3.13 Stability region for the incoherent state with uniform phase distribution (shaded) in an infinitely large ensemble of identical phase oscillators with Lorentzian distribution of frequencies, Eq. (3.60), subject to noise of intensity S, with $(\gamma+S)/\omega_0 = 0.1$.

Figure 3.12 shows the zones of parameter space where the incoherent state is stable or unstable, in the limit of $S \to 0$. Comparison with Fig. 3.11 shows that the zone where full synchronization is unstable is completely included in the region where the incoherent state is stable. Similarly, the zone where the incoherent state is unstable is fully contained in the region where at least one state of full synchronization is stable. Such zones, however, do not coincide, and there is an intermediate region where both full synchronization and the incoherent state are simultaneously stable.

As in the case of heterogeneous ensembles without time delays [Kuramoto (1984)], a distribution of natural frequencies that allows for explicit analysis is a Lorentzian,

$$g(\omega) = \frac{\gamma}{\pi} \frac{1}{\gamma^2 + (\omega - \omega_0)^2}, \tag{3.60}$$

where γ measures the width of the distribution around the central frequency ω_0. In this case, the eigenvalue equation (3.57) reduces to [Yeung and

Strogatz (1999)]

$$(\lambda + \gamma + S + i\omega)\exp(\lambda\tau) = \frac{K}{2}. \tag{3.61}$$

According to this equation, the boundary of stability for the incoherent state is given by

$$K = 2\frac{\gamma + S}{\cos \chi\tau} \tag{3.62}$$

where χ is a solution to

$$\chi = \omega_0 - (\gamma + S)\tan \chi\tau. \tag{3.63}$$

Figure 3.13 shows the regions of parameter space where the incoherent state is stable or unstable, for $(\gamma + S)/\omega_0 = 0.1$. Comparing with Fig. 3.12, we see that the main effect of having a distribution of natural frequencies is a smoothing of the boundaries between those regions.

Note that Eqs. (3.61) to (3.63) depend on the width of the frequency distribution and on the intensity of noise just through their sum, $\gamma + S$. This points out that heterogeneity in the natural frequencies and noise play similar roles in the dynamics of coupled oscillators.

Chapter 4

Oscillator Networks

Natural systems formed by ensembles of coupled elements are characterized by complex interaction architectures. The interaction strength between any pair of elements may strongly vary from pair to pair. While some pairs may be connected by coupling, others may not interact at all. This situation contrasts with globally coupled ensembles, considered in the preceding chapters, where connections are uniform and all interacting pairs have identical coupling intensities. A well-known class of models with more complex interaction architectures is that of neural networks. They can perform elaborate collective tasks, reminiscent of the function of biological organisms. In the same spirit as neural networks, coupled elements with heterogeneous interactions can be thought of as occupying the nodes of a network where links are present between those elements that may potentially interact. In turn, each link can be weighted by a different coupling intensity. The collection of sites where the elements are located and the links between them constitute the graph of the network. This underlying structure drives and controls the emergence of coherent evolution.

In this chapter, we analyze synchronization phenomena in networks of phase oscillators. We begin with regular arrays of identical oscillators, which support propagating structures whose complexity increases as the coupling intensity grows. When natural frequencies are different, on the other hand, the system splits into spatially localized domains of frequency synchronization. We then consider fully connected ensembles where connections are weighted by randomly distributed coefficients. This form of quenched disorder induces a dynamical regime which resembles the glassy phase of disordered spin systems. Finally, we analyze propagation phenomena in regular arrays where the distance between the elements determines time delays in their interaction. Time delays are also studied in locally connected networks.

4.1 Regular Lattices with Local Interactions

We first consider the case where oscillators occupy the nodes of a regular lattice, so that all sites are equivalent. Links starting at a given site reach only a certain neighborhood of that site. Interactions are uniform over the whole ensemble, but take place between connected sites only. For the case of interaction functions of the form $F_{ij}(\phi_i, \phi_j) \propto F(\phi_j - \phi_i)$, the equations of motion for the oscillator phases are

$$\dot{\phi}_i = \omega_i + \frac{K}{z} \sum_{j \in \mathcal{N}_i} F(\phi_j - \phi_i), \tag{4.1}$$

where $\mathcal{N}_i$ indicates the neighborhood of oscillator i, and z is the number of neighbors of each site. Periodic boundary conditions are assumed.

When all the natural frequencies are identical, $\omega_i = \omega$ for all i, we can apply the symmetry transformation (2.5) and fix $\omega = 0$. In this case, Eqs. (4.1) have a fully synchronized solution, with synchronization frequency $\Omega = KF(0)$. The fully synchronized state is stable if $KF'(0) > 0$, where $F'(\phi)$ is the derivative of the interaction function. Besides the state of full synchronization, there is a class of solutions that represents propagating structures. We study these solutions for a one-dimensional array, a ring, where the evolution of phases is governed by the equations

$$\dot{\phi}_i = \frac{K}{2}[F(\phi_{i-1} - \phi_i) + F(\phi_{i+1} - \phi_i)], \tag{4.2}$$

for $i = 1, \ldots, N$. Periodic boundary conditions imply $\phi_{N+1} \equiv \phi_1$ and $\phi_0 \equiv \phi_N$. Propagating solutions to Eqs. (4.2) have the form

$$\phi_i(t) = \Omega t + i\alpha, \tag{4.3}$$

with

$$\Omega = \frac{K}{2}[F(\alpha) + F(-\alpha)]. \tag{4.4}$$

The boundary conditions fix the possible values of α, given by $\alpha = 2\pi m/N$ with $m = 0, 1, \ldots N - 1$. For $m \neq 0$, Eq. (4.3) represents a linear profile of phases which propagates around the ring at velocity $V = -\Omega L/2\pi m$, where L is the length of the array. The index m gives the number of times that the phase varies over the interval $[0, 2\pi)$ in a whole rotation around the ring. For $m = 0$, we have the fully synchronized state.

Linear stability analysis of these propagating solutions shows that they are stable if all the eigenvalues of the tridiagonal matrix $\mathbf{S} = \{s_{ij}\}$ with elements

$$s_{ij} = -K[F'(\alpha) + F'(-\alpha)]\delta_{ij} + KF'(\alpha)\delta_{i,j+1} + KF'(-\alpha)\delta_{i,j-1}, \quad (4.5)$$

are negative or have negative real parts. A necessary and sufficient condition for stability is

$$K[F'(\alpha) + F'(-\alpha)] > 0. \quad (4.6)$$

The emergence of spatiotemporal structures in the solutions of Eqs. (4.2) has been studied with the interaction function [Daido (1997)]

$$F(\phi) = \sin\phi + A\cos 2\phi. \quad (4.7)$$

It has been found that, besides the traveling waves of Eq. (4.3), other kinds of propagating solutions can exist. They are found for values of A above a certain critical level and, as A grows, they change from temporally periodic but spatially disordered waves, to completely disordered patterns. Figure 4.1 shows the long-time evolution of spatiotemporal structures for several values of A, starting from initial conditions where phases are distributed at random over $[0, 2\pi)$. The absolute value of the coupling constant fixes the evolution time scale and thus can be chosen arbitrarily. The numerical results shown in Fig. 4.1 correspond to $K = 1$. For $A = 2$, the system has reached a state corresponding to a traveling wave of the type of Eq. (4.3) in the mode with $m = 2$. As A becomes larger we find two kinds of structures. In the first one, the system is divided into spatial domains of different sizes ($A = 5$, upper-right plot of Fig. 4.1). The boundaries of these domains are fixed, and irregularly distributed in space. Within each domain, a traveling wave with linear profile, like the waves of Eq. (4.3), develops. The associated wavelength is the same in all the domains, but in contiguous domains the structures propagate in opposite directions. In these structures, the ensemble is synchronized in frequency, $\phi_i(t) = \Omega t + \psi_i$, but the phase shifts ψ_i have a complex dependence with position. They vary linearly within each spatial domain, and are discontinuous at the boundaries.

In the second kind of pattern ($A = 5$, lower-left plot of Fig. 4.1) there are no fixed domains. The traveling wave moves through the whole system, with occasional changes in velocity and direction. The two kinds of patterns

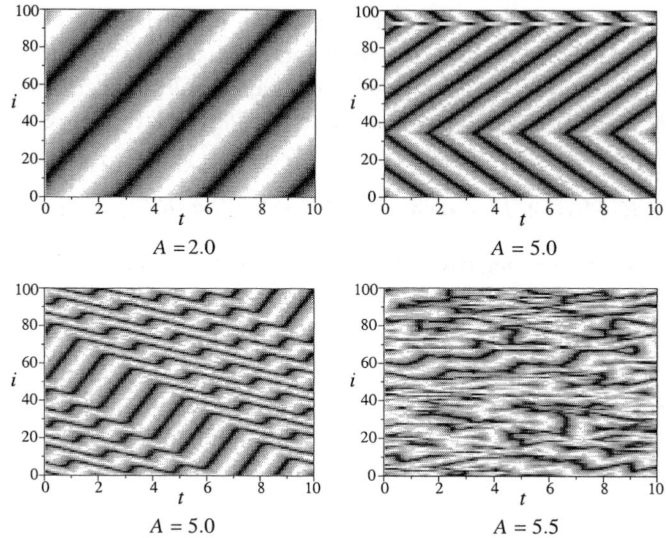

Fig. 4.1 Temporal evolution of an array of 100 identical phase oscillators with nearest-neighbor coupling given by Eq. (4.7), for different values of A. The natural frequency of all oscillators is $\omega = 0$, and the coupling constant is $K = 1$. Each plot displays phases in gray scale, darker for $\phi \approx 0$ or 2π and lighter for $\phi \approx \pi$. Horizontal and vertical axes correspond to time and position, respectively. Evolution is shown along 10 time units in the long-time regime.

found for $A = 5$ are simultaneously stable. They are reached from different initial conditions. In the same range of parameters, moreover, many of the traveling modes of Eq. (4.3) are still stable. As A increases further, most initial conditions lead to a highly disordered state, where a rather strong correlation of phases persists at short ranges, but where highly irregular spatiotemporal structures form. They are shown in Fig. 4.1 for $A = 5.5$. Phase correlations decrease as A grows and, correspondingly, patterns become more complex, with smaller typical length scales.

The appearance of the frequency-synchronized states with irregular spatial distribution illustrated in the upper-right plot of Fig. 4.1 can be predicted analytically. Assuming that the ensemble has reached a state of frequency synchronization, $\phi_i(t) = \Omega t + \psi_i$, the equations of motion for the oscillator phases reduce to an iterative mapping for the difference of phase shifts, $x_i = \psi_{i+1} - \psi_i$, given by

$$\Omega = \frac{K}{2}[F(x_{i-1}) + F(x_i)], \tag{4.8}$$

From this equation, $F(x_i)$ can be obtained as a function of x_{i-1}. For the interaction function of Eq. (4.7), the map takes the two-branch form

$$w_{i+1} = \frac{1}{\sqrt{\frac{1}{2} - \frac{4A\Omega}{K} + 4A^2}} \pm \sqrt{1 - w_i^2}, \tag{4.9}$$

for the variable

$$w_i = \frac{1 + 4A \sin x_i}{2\sqrt{\frac{1}{2} - \frac{4A\Omega}{K} + 4A^2}}. \tag{4.10}$$

The sign in the second term of the right-hand side of Eq. (4.9) is determined, at each iteration step, by the value of x_i. Since the relation (4.10) between w_i and x_i is not one-to-one, it is not possible to calculate the variation of w_i with map (4.9) independently of that of x_i. Nevertheless, this approximate representation of the problem is useful to analyze the distribution of phases in frequency-synchronized states [Daido (1997)].

Frequency-synchronized solutions where the ensemble splits into spatial domains of fixed phase difference exist when the map has two fixed points, $w_\pm$, one of them stable and the other unstable. This condition requires that $\sqrt{1/2 - 4A\Omega/K + 4A^2} > 1$. The fixed points are given by the solutions of

$$w_\pm = \frac{1}{\sqrt{\frac{1}{2} - \frac{4A\Omega}{K} + 4A^2}} \pm \sqrt{1 - w_\pm^2}. \tag{4.11}$$

The corresponding values of x, given by Eq. (4.10), satisfy $x_+ = 2\pi - x_-$. In terms of these values, the synchronization frequency is given by

$$\Omega = KA \cos 2x_+ = KA \cos 2x_-. \tag{4.12}$$

Within one of the domains, phases are such that the variable w_i is very close to one of the fixed points. If w_i is close to the stable point, it will remain there until the opposite branch of the mapping acts. If, on the other hand, w_i is in the neighborhood of the unstable point, it will be either repelled by that point, or just taken away when the branch changes. In the first case, a transition from the stable state to the unstable state will occur if both points are connected by an orbit of the mapping. It can be shown that such an orbit exists if the coefficient A in the higher-harmonic term of the interaction function (4.7) is above a certain critical value, $A > 9/4 = 2.25$ [Daido (1997)]. However, the number of iterations needed to connect the two points may be very large, giving rise to a broad interface between the domains. Narrow interfaces, up to a minimum of three sites, are possible if

$$A \geq \frac{5+\sqrt{17}}{4} \approx 2.28. \tag{4.13}$$

Numerical results show that the probability of getting one of these solutions from randomly chosen initial conditions jumps abruptly from zero to a finite value at this second threshold.

4.1.1 *Heterogeneous ensembles*

Let us now turn back the attention to Eqs. (4.1) in the case where the natural frequencies ω_i are chosen at random from a distribution $g(\omega)$. We consider the interaction function $F(\phi) = \sin\phi$.

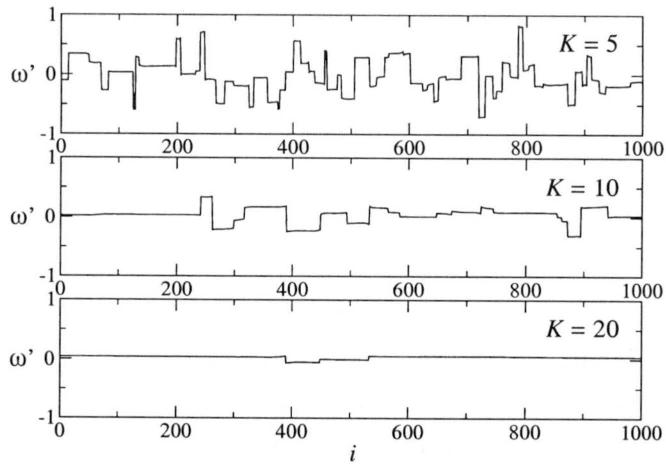

Fig. 4.2 Asymptotic distribution of effective frequencies in a linear array of 10^3 phase oscillators with nearest-neighbor interactions, for three values of the coupling intensity K. The distribution of natural frequencies is $g(\omega) = \exp(-\omega^2/2)/\sqrt{2\pi}$.

Figure 4.2 shows numerical results for the asymptotic distribution of effective frequencies ω_i', defined as in Eq. (3.2), in a linear array of 10^3 oscillators with nearest-neighbor connections and periodic boundary conditions. The distribution of natural frequencies is given by a Gaussian,

$g(\omega) = \exp(-\omega^2/2)/\sqrt{2\pi}$. The effective frequency is plotted as a function of the oscillator index i, which coincides with the position on the lattice. Each plot corresponds to a different coupling intensity K. The ensemble segregates into spatial domains where oscillators are almost synchronized in frequency, separated by abrupt jumps in the synchronization frequency [Ermentrout and Kopell (1984); Sakaguchi *et al.* (1987)]. The size of these domains grows with K, and the frequency differences become smaller. The synchronization frequency in all domains approaches $\Omega = 0$. As discussed later, however, it can be argued that in the limit $N \to \infty$, the ensemble will not collapse into a single synchronized domain as long as K remains finite.

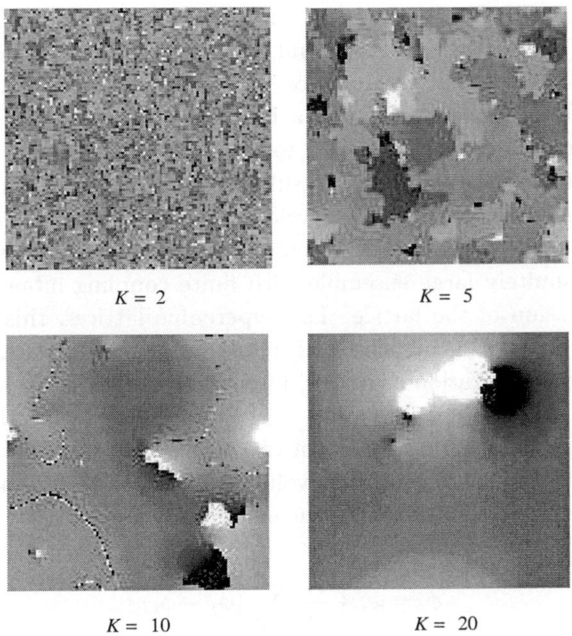

Fig. 4.3 Asymptotic distribution of effective frequencies in a square array of 100 × 100 phase oscillators with nearest-neighbor interactions, for four values of the coupling intensity K. The distribution of natural frequencies is $g(\omega) = \exp(-\omega^2/2)/\sqrt{2\pi}$. The gray scale varies from black for the lowest (negative) frequencies to white for the highest (positive) frequencies.

Figure 4.3 illustrates frequency-synchronization patterns in two-dimensional square 100×100-site lattices with nearest-neighbor connections and periodic boundary conditions. Again, frequency-synchronized domains develop, and their size increases as the coupling intensity grows, while their synchronization frequencies collapse to $\Omega = 0$. Even for large values of K, however, zones of non-entrained oscillators with sharp frequency differences persist for long times. The process of aggregation of synchronized domains is characterized by the order parameter

$$r(K) = \lim_{N \to \infty} \frac{N_s(K)}{N}, \tag{4.14}$$

where $N_s(K)$ is the average number of oscillators in the largest domain, for a given value of K [Sakaguchi *et al.* (1987)]. Numerical results for different values of N ($N = 32^2$ to 128^2) and a Gaussian distribution of natural frequencies, suggest that the function $r(K)$ has a well-defined profile, but does not exhibit any transition as a function of K. Rather, $r(K)$ grows smoothly and monotonously from 0 to 1, as K is increased.

On the other hand, numerical results for a three-dimensional cubic lattice show that, as N grows, $r(K)$ develops a step-like profile, with a jump at $K \approx 3$. Thus, the possibility of reaching a state of frequency synchronization in an infinitely large ensemble with finite coupling intensity depends on the dimension of the lattice. For hypercubic lattices, this fact can be explained as follows [Sakaguchi *et al.* (1987)]. We note first that a state of frequency synchronization coincides, for sufficiently large K, with full synchronization. In fact, if $\phi_i(t) = \Omega t + \psi_i$, Eqs. (4.1) imply that the difference $|\psi_i - \psi_j|$ between the phase shifts of two interacting oscillators is of order K^{-1}. Under these conditions, the evolution of the phases is approximately given by the linear equations [Niebur *et al.* (1991)]

$$\dot{\phi}_i = \omega_i + \frac{K}{z} \sum_{j \in \mathcal{N}_i} (\phi_j - \phi_i). \tag{4.15}$$

Using a discrete Fourier representation of these equations, it is possible to show that for any two oscillators i and j at positions $\mathbf{r}_i$ and $\mathbf{r}_j$, the average square difference of their phases is

$$\langle |\psi_i - \psi_j|^2 \rangle = \frac{1}{N} \sum_q |K_q|^{-2} [1 - \cos(\mathbf{q} \cdot \mathbf{r}_{ij})], \tag{4.16}$$

with $\mathbf{r}_{ij} = \mathbf{r}_j - \mathbf{r}_i$. Here,

$$K_q = \frac{K}{z} \sum_{j \in \mathcal{N}_0} [1 - \exp(i\mathbf{q} \cdot \mathbf{e}_j)], \tag{4.17}$$

where $\mathbf{e}_j$ are the nearest-neighbor positions relative to the origin. The vector $\mathbf{q}$ has components $q_\alpha = 2\pi n_\alpha / L$, where n_α runs over the integers, and L is the linear size of the lattice. The average in Eq. (4.16) is performed over the distribution of natural frequencies. This distribution is such that $\langle \omega_i \rangle = 0$ and $\langle \omega_i \omega_j \rangle = \delta_{ij}$, where δ_{ij} is the Kronecker delta symbol.

The dimension of the lattice determines the set of values of q in the summation of Eq. (4.16). For a fixed distance $|\mathbf{r}_{ij}|$, the average $\langle |\psi_i - \psi_j|^2 \rangle$ may or may not diverge. Large contributions to the summation come from small values of $q = |\mathbf{q}|$ in the limit $L \to \infty$. For small q, we have $K_q \sim K a^2 q^2$, with $a = |\mathbf{e}_j|$. The contribution from $q < q_r = |\mathbf{r}_{ij}|^{-1}$ is approximately given by

$$\frac{1}{q_r^2 K^2} \int_{q_0}^{q_r} q^{d-3} dq, \tag{4.18}$$

where $q_0 \sim L^{-1}$ and d is the dimension of the hypercubic lattice. For $L \to \infty$, this contribution diverges if $d \leq 2$. This implies that at least a finite fraction of oscillator pairs do not remain synchronized even for very large K. The divergence of the average square phase difference shows that also frequency synchronization is impossible. For $2 < d \leq 4$, the contribution becomes finite even for $L \to \infty$, but still diverges for $|\mathbf{r}_{ij}| \to 0$, where the upper limit in the integral (4.18) goes to infinity. This fact indicates that the order parameter of Eq. (4.14) may still not attain the value $r = 1$ for finite K and $N \to \infty$, if the distance between the oscillators tends to zero. For $d > 4$, on the other hand, frequency synchronization of the whole ensemble is ensured at a finite, sufficiently high coupling intensity.

By means of renormalization-group analysis, it is possible to give a more rigorous proof of the existence of a lower critical dimension d_c for frequency synchronization of non-identical oscillators [Daido (1988)]. The result suggests that d_c is sensitive to the decay of the distribution of natural frequencies $g(\omega)$ for large values of $|\omega|$. In particular, for distributions with a power-law tail, $g(\omega) \sim |\omega|^{-a-1}$, the critical dimension satisfies the inequality

$$d_c \geq \frac{a}{a-1} \tag{4.19}$$

for $1 < a \leq 2$. For $0 < a \leq 1$, on the other hand, frequency synchronization is not possible at any finite dimension.

4.2 Random Interaction Architectures

Let us now consider an ensemble of non-identical phase oscillators where coupling involves all pairs of elements, as in globally coupled systems, but such that the interaction intensity is different for each pair. The equations of motion for the phases are

$$\dot{\phi}_i = \omega_i + \frac{K}{N} \sum_{j=1}^{N} J_{ij} \sin(\phi_j - \phi_i), \tag{4.20}$$

where natural frequencies are chosen at random from a distribution $g(\omega)$. The coefficient J_{ij} weights the interaction between oscillators i and j. This problem has been studied for several choices of the interaction coefficients [Daido (1987); Daido (1992); Daido (2000)]. The effects of noise on Eqs. (4.20) have also been analyzed [Stiller and Radons (1998)].

First, we assume that the interaction coefficients can be factored as $J_{ij} = s_i s_j$, where s_i is chosen independently for each oscillator i from a distribution function $P(s)$. Remarkably, in the case where this distribution has the form

$$P(s) = p\delta(s-1) + (1-p)\delta(s+1), \tag{4.21}$$

where $\delta(x)$ is the Dirac delta function and $0 \leq p \leq 1$, the randomness given by the choice of s_i can be "removed" [Daido (1987)]. In fact, the transformed phases $\phi'_i = \phi_i - \pi s_i/2$ satisfy Eqs. (3.1), as in a heterogeneous ensemble of globally coupled oscillators. From the analysis of Sec. 3.1, we know that in the limit $N \to \infty$ the system undergoes a transition at a critical value of the coupling intensity K, above which a cluster of frequency-synchronized oscillators appears. Averaging over realizations of the random choice of s_i, we find a relation between the order parameter σ defined by Eq. (2.22), and $\sigma' = N^{-1}|\sum_j \exp(i\phi'_j)|$, namely,

$$\sigma = |2p-1|\sigma'. \tag{4.22}$$

Since $\sigma \leq \sigma'$, we conclude that for Eqs. (4.20) the distribution of phases in the frequency-synchronized cluster is more symmetric than in the case of global coupling, Eqs. (3.1). Note, in particular, the case $p = 1/2$, for which $\sigma = 0$ even at coupling intensities above the critical point. It can be seen that, in this case, the phases of entrained oscillators are distributed in two symmetric broad groups with a phase difference of π. These oscillators rotate at the synchronization frequency Ω in the uniform background of non-entrained elements. The symmetry of the phase distribution in the cluster implies that no macroscopic organization emerges, at least as measured by the order parameter σ, though order is present in the form of frequency synchronization.

This new form of coherent behavior is revealed by a different order parameter, defined as

$$\rho = \frac{1}{N}\left|\sum_{j=1}^{N} s_j \exp(i\phi_j)\right|. \tag{4.23}$$

The analysis of Sec. 3.1 can be adapted to derive self-consistency equations for the order parameters σ and ρ, for arbitrary distributions of the coefficients s_i [Daido (1987)]. Assuming, for simplicity, that the distribution of natural frequencies $g(\omega)$ is symmetric around zero, which implies that the synchronization frequency is $\Omega = 0$, we get

$$\rho = K\rho \int_{-\infty}^{\infty} P(s)s^2\, ds \int_{-\pi/2}^{\pi/2} g(K\rho|s|\sin\phi)\cos^2\phi\, d\phi, \tag{4.24}$$

and

$$\sigma = K\rho \left|\int_{-\infty}^{\infty} P(s)s\, ds \int_{-\pi/2}^{\pi/2} g(K\rho|s|\sin\phi)\cos^2\phi\, d\phi\right|. \tag{4.25}$$

The first equation makes it possible to find ρ as a function of the coupling intensity K. As expected, it reduces to Eq. (3.14) for $P(s) = \delta(s-1)$ and $\rho \neq 0$. The second equation shows that, independently of the values of K and ρ, the order parameter σ vanishes if $P(s)$ is an even function, as in the case of Eq. (4.21) for $p = 1/2$.

Equations (4.24) and (4.25) can be solved explicitly for $P(s)$ as in Eq. (4.21), with arbitrary p, and for the Lorentzian distribution of Eq. (3.60) centered at $\omega_0 = 0$. In this case, we get

$$\rho = \sqrt{1 - \frac{2\gamma}{K}} \tag{4.26}$$

and

$$\sigma = |2p-1|\sqrt{1 - \frac{2\gamma}{K}}. \tag{4.27}$$

Thus, the critical coupling intensity is $K_c = 2\gamma$. Generally, both ρ and σ are zero for $K < K_c$ and positive for $K > K_c$. For $p = 1/2$, however, $\sigma = 0$ for all K. Qualitatively similar results have been obtained numerically with Gaussian distributions for the coefficients s_i [Daido (1987)].

The distribution $Q(\psi)$ of the phase shifts ψ_i of entrained oscillators, whose phases are given by $\phi_i(t) = \Omega t + \psi_i$, reads

$$Q(\psi) = \frac{|\cos\psi|}{Z} \int_0^\infty sP[h(\psi)s]\, g(\Omega + K\rho s \sin\psi) ds, \qquad (4.28)$$

where $h(\psi) = \text{sign}(\cos\psi)$ and Z is a normalization constant. Figure 4.4 is a plot of $Q(\psi)$ for $P(s)$ as in Eq. (4.21), with a Lorentzian distribution of natural frequencies. Note that the symmetry between the two anti-phase groups is broken for $p \neq 1/2$. Numerical results for other forms of symmetric and asymmetric $P(s)$ show the same feature.

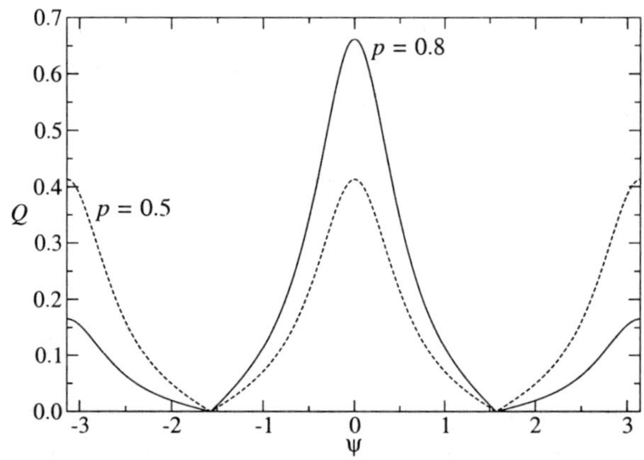

Fig. 4.4 Distribution of phase shifts in the frequency-synchronized cluster for a Lorentzian distribution of natural frequencies, Eq. (3.60) with $\gamma = 1$ and $\omega_0 = 0$, and coupling intensity $K = 1$. The interaction coefficients s_i are distributed according to Eq. (4.21), for two values of p.

4.2.1 *Frustrated interactions*

With the choice $J_{ij} = s_i s_j$, mutual entrainment of many oscillator pairs can occur simultaneously. Thus, frequency synchronization is possible in system (4.20). This is not the general case, though. For other forms of J_{ij} it typically happens that, while the conditions for synchronization may hold between, say, oscillators i and j, and i and k, they do not hold between j and k. As a consequence, synchronization does not take place, and interactions are "frustrated." By analogy with spin systems [Mézard *et al.* (1987); Marinari *et al.* (1994)], where the interactions are qualitatively similar to those of Eqs. (4.20), this regime has been identified as glass-like behavior.

Frustration is found, for instance, when the coefficients J_{ij} are chosen independently from a distribution $P(J)$ which allows for both positive and negative values. Let us consider, specifically, a Gaussian distribution

$$P(J) = \frac{1}{\sqrt{2N\pi\kappa^2}} \exp\left(-\frac{J^2}{2N\kappa^2}\right), \tag{4.29}$$

and take $J_{ij} = J_{ji}$, so that interactions are symmetric for all oscillator pairs. The natural frequencies are distributed according to $g(\omega) = \exp(-\omega^2/2\gamma^2)/\sqrt{2\pi\gamma^2}$, and $\gamma = 2\pi$ [Daido (1992)].

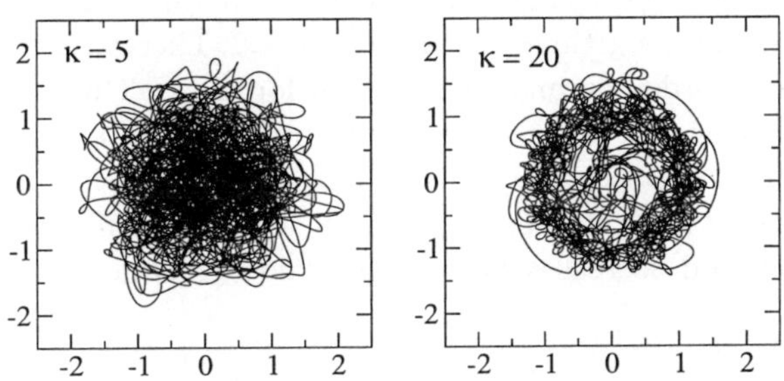

Fig. 4.5 Trajectories of the complex order parameter $\rho_i \exp(i\Theta_i)$, Eq. (4.30), for an oscillator chosen at random from an ensemble of $N = 100$ elements governed by Eqs. (4.20). The interaction coefficients J_{ij} are distributed according to Eq. (4.29), for two values of the dispersion κ. The distribution of natural frequencies is a Gaussian of dispersion $\gamma = 2\pi$, centered at $\omega = 0$. Each trajectory was obtained from numerical integration over 200 time units.

To introduce an order parameter similar to ρ, Eq. (4.23), we define the complex "local field" of each oscillator i as

$$\rho_i(t) \exp[i\Theta_i(t)] = \frac{1}{\kappa N} \sum_{j=1}^{N} J_{ij} \exp[i\phi_j(t)]. \tag{4.30}$$

While this local field is different for each oscillator, its statistical properties are uniform all over the ensemble. Figure 4.5 shows the trajectory of $\rho_i \exp(i\Theta_i)$ in the complex plane, for an oscillator chosen at random from an ensemble of $N = 100$ elements, and for two values of the mean square dispersion κ. For small values of the dispersion κ of interaction coefficients, numerical results show that local fields are distributed in time with a maximum at the origin and a Gaussian-like profile in their modulus. For $\kappa \to 0$ it is possible to prove that the distribution is proportional to $\exp(-\rho_i^2)$. As κ grows, however, a qualitative change takes place. Above a certain threshold $\kappa_c \approx 8$, the local field is distributed with a maximum at a finite value of its modulus [Daido (1992)].

This transition reflects the onset of a form of mutual entrainment where the distribution of the effective frequencies ω_i' develops a sharp peak at the average value of natural frequencies, as shown in Fig. 4.6. In contrast with the case of frequency synchronization studied so far, however, this peak is not isolated from the rest of the distribution (cf. Fig. 3.3). As a consequence, phase differences between entrained oscillators are not bounded and grow diffusively, $\langle |\phi_i(t) - \phi_j(t)|^2 \rangle \propto t$. In this form of entrainment, and for $N \to \infty$, the average $\sigma(t) \exp[i\Phi(t)]$ defined in Eq. (2.22) tends to zero for asymptotically long times, even for $\kappa > \kappa_c$. Therefore, the associated synchronization order parameter is $\sigma = 0$ for all κ. On the other hand, the time decay of $\sigma(t)$ is sensitive to the transition at κ_c. Specifically, it is observed that for $\kappa < \kappa_c$, $\sigma(t)$ decays exponentially with time, while for $\kappa > \kappa_c$ the decay is algebraic, $\sigma(t) \propto T^{-a}$. The exponent is $a \approx 2$ close to the transition, and decreases as κ grows [Daido (2000)].

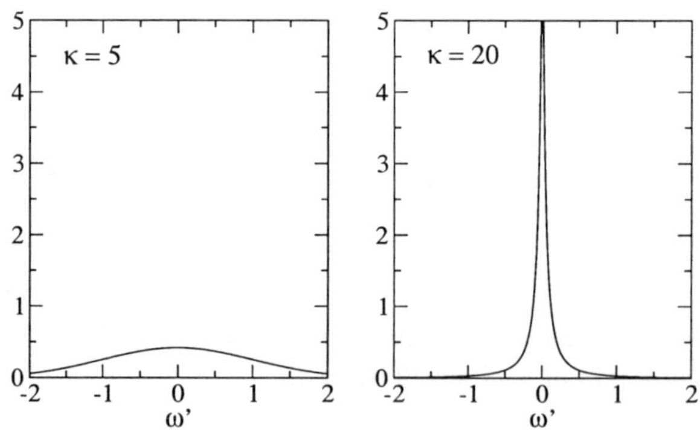

Fig. 4.6 Distribution of effective frequencies ω' for an ensemble of coupled oscillators with random interaction coefficients taken from the distribution of Eq. (4.29), for two values of the dispersion κ. The distribution of natural frequencies is Gaussian (adapted from [Daido (1992)]).

While the asymptotic value of σ does not depend on the dispersion κ, the order parameter defined through the local field, Eq. (4.30), detects the transition at κ_c. This suggests, as discussed above, a kind of glass-like behavior due to the quenched disorder of coupling and to the symmetry of the distribution of interaction weights around zero. This analogy has been explored in detail for the case where disorder is introduced through random phase shifts α_{ij} in the interaction function [Park *et al.* (1998)], as

$$\dot{\phi}_i = \omega_i + \frac{K}{N}\sum_{j=1}^{N}\sin(\phi_j - \phi_i + \alpha_{ij}). \tag{4.31}$$

Adding randomly fluctuating forces, the replica method [Mézard *et al.* (1987)] makes it possible to obtain self-consistency equations for suitably defined order parameters, which disclose critical transitions between phases of incoherent, glass-like, and synchronized evolution.

4.3 Time Delays

As discussed in Sec. 3.4, time delays in the interaction of coupled dynamical elements become important when the time needed for transferring information between elements is similar to or longer than the time scales associated with the individual dynamics. When the elements occupy the nodes of a network distributed in space, their mutual distances imply different time delays if the information propagates at a relatively small velocity.

In order to study the effect of time delays on the dynamics of an ensemble of phase oscillators distributed over a network, we first consider the following model [Zanette (2000)]:

$$\dot{\phi}_i(t) = \omega + \frac{K}{N}\sum_{j=1}^{N}\sin[\phi_j(t-\tau_{ij}) - \phi_i(t)]. \tag{4.32}$$

Here, all the oscillators have identical natural frequencies and the intensity of their interaction does not depend on the relative position of the elements over the network. However, a different time delay τ_{ij} is in general assigned to each pair of oscillators. This time delay is identified with the time required by the interaction signal to travel from element j to element i at velocity v. In other words, $\tau_{ij} = d_{ij}/v$, where d_{ij} is the distance between i and j. Equations (4.32) are fully specified once all the pair distances d_{ij} and the velocity v are given. As in the case of globally coupled identical oscillators with time delays, studied in Sec. 3.4, Eqs. (4.32) are not invariant under the transformation $\phi_i \to \phi_i + \omega t$ and, therefore, the natural frequency cannot be arbitrarily chosen.

The numerical solution to Eqs. (4.32) shows that, when the time delays are small enough, $\tau_{ij} \ll \omega^{-1}, K^{-1}$ for all i and j, the ensemble becomes synchronized in frequency, but the oscillators have different phases. Phase differences, however, turn out to be of the same order as the time delays. Therefore, we can assume $|\phi_j(t-\tau_{ij}) - \phi_i(t)| \ll 1$ for all i, j and t. This makes it possible, in the linear approximation, to simplify Eqs. (4.32) as

$$\dot{\phi}_i(t) = \omega - K\phi_i(t) + \frac{K}{N}\sum_{j=1}^{N} \phi_j(t-\tau_{ij}). \tag{4.33}$$

It is convenient to consider first the evolution of the individual frequencies $\dot{\phi}_i \equiv \Omega_i$. Approximating $\Omega_i(t-\tau_{ij}) \approx \Omega_i(t) - \tau_{ij}\dot{\Omega}_i(t)$, we have

$$\dot{\Omega}_i = -K\Omega_i + \frac{K}{N}\sum_{j=1}^{N}(\Omega_j - \tau_{ij}\dot{\Omega}_j), \tag{4.34}$$

where all the functions are now evaluated at the same time. These equations can be seen as a perturbation to the problem without delays,

$$\dot{\Omega}_i^0 = -K\Omega_i^0 + \frac{K}{N}\sum_{j=1}^{N}\Omega_j^0, \tag{4.35}$$

whose solution reads

$$\Omega_i^0(t) = \Omega_i^0(0)\exp(-Kt) + \langle\Omega^0\rangle[1 - \exp(-Kt)]. \tag{4.36}$$

Here, $\langle\Omega^0\rangle = N^{-1}\sum_j \Omega_j$ is a constant of motion, associated with the invariance of the unperturbed problem under uniform shifts in frequency. Equations (4.34) can now be approximately solved by putting $\Omega_i(t) = \Omega_i^0(t) + \Xi_i(t)$, where $\Xi_i(t)$ should be of the same order as the time delays. Expanding the equation of motion for Ω_i up to the first order in the time delays, we find an equation for $\Xi_i(t)$, whose solution reads

$$\Xi_i(t) = -\frac{K^2}{N}\sum_{j=1}^{N}\tau_{ij}[\langle\Omega^0\rangle - \Omega_j^0(0)]t\exp(-Kt). \tag{4.37}$$

Here, we have taken $\Xi_i(0) = 0$ as the initial condition. This corresponds to solving the perturbed problem with the initial conditions of the unperturbed equations.

Within these approximations, the solution of Eqs. (4.34) is

$$\Omega_i(t) = \Omega[1 - \exp(-Kt)] + \Omega_i(0)\exp(-Kt) - \frac{K^2}{N}\sum_{j=1}^{N} \tau_{ij}[\langle \Omega^0 \rangle - \Omega_j^0(0)]t\exp(-Kt), \qquad (4.38)$$

where

$$\Omega = \frac{1}{N}\sum_{j=1}^{N} \Omega_j(0) \qquad (4.39)$$

is the long-time asymptotic value of $\Omega_i(t)$ for all i. In the limit of short time delays, thus, the ensemble becomes synchronized in frequency, with synchronization frequency Ω.

The phase of each oscillator is obtained by integrating Eq. (4.38) over time. At asymptotically large times, all phases move uniformly with frequency Ω, $\phi_i(t) = \Omega t + \psi_i$. Replacing in Eq. (4.33), we find

$$\psi_i = \Psi - \Omega \frac{1}{N}\sum_{j=1}^{N} \tau_{ij}. \qquad (4.40)$$

The constant Ψ can be arbitrarily chosen, due to the symmetry of the ensemble under uniform phase shifts. In general situations, the average delay time $\langle \tau_i \rangle = N^{-1}\sum_j \tau_{ij}$ is different for each oscillator i, so that phases differ between oscillators. According to solution (4.40) oscillators with smaller average delays have larger phases, and are therefore relatively ahead in the evolution.

If, on the other hand, the underlying network is geometrically uniform and all its sites are equivalent, the average time delay $\langle \tau_i \rangle$ is the same for all oscillators and the ensemble becomes fully synchronized. As an example of these uniform networks, we study below a one-dimensional regular lattice with periodic boundary conditions. Full synchronization is found for small time delays but, as delays become larger, other kinds of phase distributions can develop even while the system remains synchronized in frequency.

4.3.1 *Periodic linear arrays*

Let us consider an ensemble of N phase oscillators arranged on a regular linear array with periodic boundary conditions, namely, on a ring. If labels are orderly assigned to the oscillators around the ring, the distance between any two oscillators is $d_{ij} = N^{-1}L \min\{|i-j|, N-|i-j|\}$, where L is the linear size of the array. The time delays associated with these distances are

$$\tau_{ij} = \frac{T}{N} \min\{|i-j|, N-|i-j|\}, \tag{4.41}$$

where $T = L/v$ is the time needed by the interaction signal to travel around the whole ring at velocity v.

Extensive numerical analysis of Eqs. (4.32) with the time delays (4.41) show that the ensemble evolves always to a state of frequency synchronization [Zanette (2000)]. Though this system has also a state of full synchronization, the asymptotic distribution of phases is typically heterogeneous. Replacing $\phi_i(t) = \Omega t + \psi_i$ in Eqs. (4.32), we find that the phase shifts ψ_i must satisfy

$$\Omega = \omega - \frac{K}{N} \sum_{j=1}^{N} \sin(\Omega\tau_{ij} + \psi_i - \psi_j). \tag{4.42}$$

Note that, while the sums $S_i = \sum_j \sin(\Omega\tau_{ij} + \psi_i - \psi_j)$ are in general different for each i, their values must coincide for all the oscillators. In fact, Eq. (4.42) can be rewritten as $S_i = N(\omega - \Omega)/K$, where the right-hand side is independent of i. For a given value of the synchronization frequency Ω, this provides $N-1$ equations for the phase shifts ψ_i:

$$S_1 = S_2 = \cdots = S_N. \tag{4.43}$$

Since phases are defined up to an arbitrary constant, these equations make it possible to find all the phase shifts ψ_i by fixing, for instance, ψ_1.

To solve Eqs. (4.43), it is useful to write down the explicit expression of S_i:

$$S_i = \sum_{j=1}^{N} \sin\left(\frac{\Omega T}{N} \min\{|i-j|, N-|i-j|\} + \psi_i - \psi_j\right). \tag{4.44}$$

The delay τ_{ij} depends on i and j through the difference $i-j$ only. Consequently, S_i can be made independent on i if, in turn, the phase difference $\psi_i - \psi_j$ is a function of $i-j$. Under this condition, i is an irrelevant origin in the sum over j, and can be eliminated by redefining the summation variable, $j \to (j+i) \mod N$. The condition is met for all i and j if the phase shift ψ_i depends linearly on i, $\psi_i = Ai + B$ with A and B constants. Taking into account the periodic boundary conditions, and choosing $\psi_1 = 0$, we get

$$\psi_i = \frac{2\pi m}{N}(i-1), \tag{4.45}$$

where the integer m may be restricted to the interval $[-N+1, N-1]$. This solution represents a state where oscillator phases vary linearly around the ring. The simplest mode, $m = 0$, corresponds to full synchronization. Since the phases evolve as

$$\phi_i(t) = \Omega t + \frac{2\pi m}{N}(i-1), \tag{4.46}$$

we find that, for $m \neq 0$, the solution represents a propagating structure of velocity $V = -L\Omega/2\pi m$, whose linear profile is preserved.

The synchronization frequencies corresponding to these propagating solutions satisfy

$$\Omega = \omega - \frac{K}{N}\sum_{j=1}^{N} \sin\left[\frac{\Omega T}{N}\min\{j-1, N-j+1\} - \frac{2\pi m}{N}(j-1)\right]. \tag{4.47}$$

In the limit $N \to \infty$, the summation in this equation for Ω reduces to an integral that can be explicitly evaluated, yielding

$$\Omega = \omega + \frac{K\Omega T}{2}\frac{1-\cos(m\pi - \Omega T/2)}{m^2\pi^2 - \Omega^2 T^2/4}. \tag{4.48}$$

Numerical solutions to this equation are shown in Fig. 4.7 for $m = 0, \ldots, 3$ and $K/\omega = 1$, as functions of the time T. Since the synchronization frequency is independent of the sign of m, these solutions are also valid for the corresponding values with $m < 0$. As in the case of globally coupled ensembles with time-delayed interactions, we find coexistence of several synchronized states for a given set of parameters. It can be shown from Eq. (4.48) that solutions $\Omega \approx 1$ exist for all m if T is sufficiently small. As T becomes longer, the number of possible synchronized states increases, due to the appearance of new synchronization frequencies for each value of m. New solutions appear also when the coupling intensity K is increased.

Linear stability analysis shows that a frequency-synchronized propagating mode with phase shifts ψ_i and frequency Ω is stable if all the solutions λ to the transcendental equation $\det \mathbf{S}(\lambda) = 0$ are negative or have negative real parts. Here, the $N \times N$ matrix $\mathbf{S}(\lambda)$ has elements

$$s_{ij}(\lambda) = \left(\lambda + \frac{K}{N}\sum_k c_{ik}\right)\delta_{ij} - \frac{K}{N}c_{ij}\exp(\lambda\tau_{ij}), \tag{4.49}$$

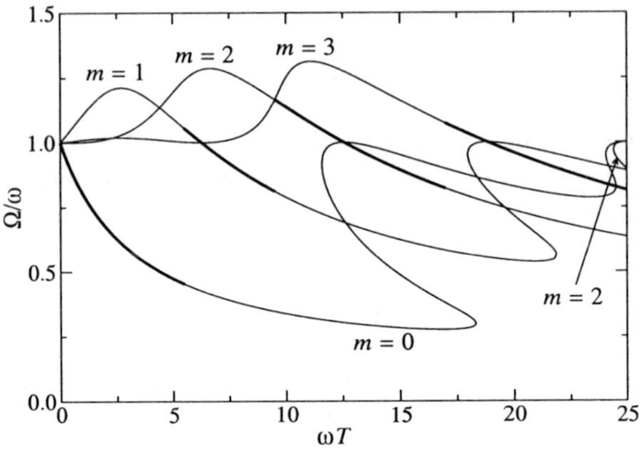

Fig. 4.7 Synchronization frequencies of propagating solutions, given by Eq. (4.48), for $m = 0, \ldots, 3$ and $K/\omega = 1$, as functions of the time T. Bold lines indicate the ranges where each mode attracts initial conditions where phases are uniformly distributed in $[0, 2\pi)$.

where $c_{ij} = \cos(\Omega\tau_{ij} + \psi_i - \psi_j)$ and δ_{ij} is the Kronecker delta symbol. This problem cannot be treated analytically. However, it is possible to show that one of the solutions is $\lambda = 0$, corresponding to the longitudinal eigenvalue associated with the symmetry under uniform phase shifts. In the limit $T \to 0$ and for $K > 0$, all modes are unstable except the fully synchronized state $m = 0$. In fact, as shown in Sec. 2.2, the $N-1$ transversal eigenvalues for the fully synchronized state are all equal to $-K$. For $m \neq 0$, $N-2$ solutions vanish while the other two are positive and equal to $K/2$.

The relative stability of the different modes can be evaluated numerically. Starting from an initial condition where phases have a uniform random distribution in $[0, 2\pi)$ we find that, for each value of T, the system approaches a well-defined mode m in all realizations. Bold curves in Fig. 4.7 show the ranges where each mode is the asymptotic state for that kind of initial conditions, in an ensemble of $N = 100$ phase oscillators with rescaled coupling constant $K/\omega = 1$. The synchronization frequencies for $N = 100$ are practically indistinguishable from those found in the limit $N \to \infty$, plotted in the figure. The fact that, for a given value of T, all the above random initial conditions approach the same propagating mode does not exclude, however, that other initial conditions may lead to modes with different m. Starting from initial conditions prepared as perturbations of the propagating modes, we find that the parameter range where each mode is stable is broader than that shown by the respective bold curve in Fig. 4.7. Hence, stable modes are simultaneously stable and the system exhibits multistability. As discussed for globally coupled ensembles in Sec. 3.4, this is a typical effect of time-delayed interactions.

4.3.2 *Local interactions with uniform delay*

Equations (4.32) describe an ensemble of phase oscillators where the strength of coupling does not depend on their relative position, but where their interactions are delayed due to the distance between them. A complementary problem is given by a network of oscillators where interaction is limited to those elements which are directly connected by a network link, and the same time delay τ is assigned to all interacting pairs. Coupling is thus local, and time delays are uniform. In the case of identical natural frequencies, the equations of motion are [Earl and Strogatz (2003)]

$$\dot{\phi}_i(t) = \omega + \frac{K}{\sum_j a_{ij}} \sum_{j=1}^{N} a_{ij} F[\phi_j(t-\tau) - \phi_i(t)], \tag{4.50}$$

where $a_{ij} = 1$ if oscillator j acts over oscillator i, and $a_{ij} = 0$ otherwise. Self-interactions are not allowed, so that $a_{ii} = 0$ for all i. The coefficient a_{ij} needs not to equal a_{ji}, indicating that interactions are not necessarily symmetric. In other words, while the evolution of oscillator i may be influenced by oscillator j ($a_{ij} = 1$), the evolution equation of j may be independent of the state of i ($a_{ji} = 0$). The underlying network is therefore a directed graph. Note that the interaction term is normalized by the number $z_i = \sum_j a_{ij}$ of neighbors that interact with oscillator i.

As in the case of globally coupled oscillators with uniform delays analyzed in Sec. 3.4, Eqs. (4.50) have a fully synchronized solution. This, however, requires that the numbers z_i of active neighbors of all oscillators are identical, $z_i = z$ for all i. Under these conditions, the fully synchronized state is given by

$$\phi_i(t) = \Omega t, \tag{4.51}$$

where the synchronization frequency satisfies the equation

$$\Omega = \omega + KF(-\Omega\tau). \tag{4.52}$$

The state of full synchronization is linearly stable if all the solutions λ of

the equation

$$\det[\mathbf{A} - \sigma(\lambda)\mathbf{I}] = 0 \tag{4.53}$$

are negative or have negative real parts. In this equation, $\mathbf{A} = \{a_{ij}\}$ is the connectivity matrix of the network under study, $\mathbf{I}$ is the identity matrix, and

$$\sigma(\lambda) = \frac{z[\lambda + KF'(-\Omega\tau)]\exp(\lambda\tau)}{KF'(-\Omega\tau)}, \tag{4.54}$$

where $F'(\phi)$ is the derivative of the interaction function. Equation (4.53) implies that σ is an eigenvalue of the matrix $\mathbf{A}$. The stability condition, thus, is related to the eigenvalue problem for $\mathbf{A}$, and therefore depends on the topology of the interaction network. However, it is possible to show that, independently of the detailed form of $\mathbf{A}$, all its eigenvalues lie within a circle of radius z centered at the origin, i.e. $|\sigma| \leq z$ [Strang (1986)]. When $\sigma(\lambda)$ satisfies this inequality, the real part of λ is negative if and only if [Earl and Strogatz (2003)]

$$KF'(-\Omega\tau) > 0. \tag{4.55}$$

This is, therefore, the stability condition for the fully synchronized state of system (4.50), on networks of arbitrary topology with a uniform number of active neighbors per node. This condition is valid for any system where each oscillator receives exactly the same number of signals, independently of other details in the pattern of interactions.

Note that both Eq. (4.52), for the synchronization frequency Ω, and the stability condition (4.55) are independent of z. They are valid for attractive ($K > 0$) and repulsive ($K < 0$) interactions. As expected, the stability condition reduces to Eq. (3.50) when $F(\phi) = \sin\phi$.

Chapter 5

Arrays of Limit-Cycle Oscillators

Dynamical systems with asymptotic periodic orbits, or limit cycles, provide a suitable model for the self-sustained elementary oscillations that give rise to macroscopic rhythms through synchronization. In the preceding chapters, we have used coupled phase oscillators as a phenomenological approach to ensembles of interacting periodic elements. In fact, phase oscillators represent a good approximation to systems with limit-cycle orbits when the strength of coupling is small. As interactions become stronger, however, the phase approximation breaks down and it becomes necessary to take into account the full dynamics of each oscillatory element, including both its phase and its amplitude. The interplay of these two variables in ensembles of coupled limit-cycle oscillators gives rise to new phenomena in the collective evolution.

In this chapter we study the synchronization properties of ensembles of weakly nonlinear limit-cycle oscillators. First, we consider globally coupled ensembles, analyzing the regimes of full and frequency synchronization, oscillation death, collective chaos, and incoherent behavior. The effect of time-delayed interactions on these dynamical regimes is also studied. Then, we discuss an interpolation between globally coupled ensembles and the locally coupled systems described by the Ginzburg–Landau equation. This model, based on intermediate-range interactions, reveals the connection between collective chaos and diffusion-induced turbulence.

5.1 Synchronization of Weakly Nonlinear Oscillators

As a model for the individual dynamics of a limit-cycle oscillator, we consider the normal form of a dynamical system at an Andronov–Hopf bifurcation, where a fixed point becomes unstable and, simultaneously, a limit cycle is created around the fixed point. The corresponding evolution equation reads

$$\dot{z} = (1 + i\omega)z - (1 + ib)|z|^2 z. \tag{5.1}$$

The variable $z(t)$ is a complex number. Thus, the limit-cycle oscillator (5.1) is a two-dimensional dynamical system, usually called *weakly nonlinear oscillator*.

The evolution of $z(t)$ is better understood if we use the expression in polar coordinates $z = r\exp(i\phi)$. Equation (5.1) is then separated into an equation for the modulus $|z| = r$,

$$\dot{r} = r(1 - r^2) \tag{5.2}$$

and an equation for the phase ϕ,

$$\dot{\phi} = \omega - br^2. \tag{5.3}$$

The equation for the modulus, restricted to the relevant domain $r \geq 0$, has fixed points at $r = 0$ and $r = 1$. The former is unstable, while the latter is stable and attracts all the orbits starting at $r > 0$. Thus, from any initial condition $z \neq 0$, $z(t)$ asymptotically approaches an orbit of constant modulus $r = 1$. On this orbit, the variation of the phase is uniform, $\dot{\phi} = \omega - b$. For sufficiently long times, therefore, Eq. (5.1) represents a two-dimensional harmonic oscillator, moving uniformly on a circular trajectory of unit radius at frequency $\omega - b$.

In this chapter we study the emergence of collective behavior in ensembles of limit-cycle oscillators whose individual dynamics is given by Eq. (5.1). We begin by considering N globally coupled oscillators,

$$\dot{z}_i = (1 + i\omega_i)z_i - (1 + ib)|z_i|^2 z_i + K(1 + i\epsilon)(\langle z \rangle - z_i). \tag{5.4}$$

Each oscillator interacts with the ensemble through the average $\langle z \rangle = N^{-1}\sum_j z_j$. Let us point out that the case of locally-coupled weakly nonlinear oscillators has been extensively studied in the literature in the continuous limit where $z_i(t)$ is replaced by a function of space and time, $Z(\mathbf{x}, t)$. In this situation, the system is described by the complex Ginzburg–Landau equation [Mikhailov and Loskutov (1996)],

$$\frac{\partial Z}{\partial t} = (1 + i\omega)Z - (1 + ib)|Z|^2 Z + K(1 + i\epsilon)\nabla_{\mathbf{x}}^2 Z. \tag{5.5}$$

First, we study Eq. (5.4) in the simpler case where $b = \epsilon = 0$. By analogy with phase oscillators, we call ω_i the natural frequency of oscillator

i. Introducing the real quantities $\sigma(t)$ and $\Phi(t)$ as

$$\sigma(t) \exp[i\Phi(t)] = \frac{1}{N} \sum_{j=1}^{N} z_j, \tag{5.6}$$

Eq. (5.4) yields

$$\dot{r}_i = (1 - r_i^2 - \sigma) r_i + K\sigma \cos(\Phi - \phi_i), \tag{5.7}$$

and

$$\dot{\phi}_i = \omega_i + \frac{K\sigma}{r_i} \sin(\Phi - \phi_i), \tag{5.8}$$

for the modulus and the phase of z_i [cf. Eq. (2.21)].

It is instructive to consider first a system of two coupled limit-cycle oscillators [Aronson *et al.* (1990)],

$$\begin{aligned} \dot{z}_1 &= (1 - |z_1|^2 + i\omega_1) z_1 + \tfrac{K}{2}(z_2 - z_1), \\ \dot{z}_2 &= (1 - |z_2|^2 + i\omega_2) z_2 + \tfrac{K}{2}(z_1 - z_2). \end{aligned} \tag{5.9}$$

When the natural frequencies are identical, $\omega_1 = \omega_2 = \omega$, these equations have a fully synchronized solution, $z_1 = z_2 = \exp(i\omega t)$. This solution coincides with the asymptotic orbit of a single oscillator and, as expected, is stable for attractive interactions, $K > 0$.

For $K < 0$, on the other hand, the anti-phase solution $z_1 = -z_2$ is stable. In contrast with fully synchronized motion, however, the orbits corresponding to this solution are circles whose radii differ from unity. The anti-phase solution is $z_1 = r\exp(i\omega t)$ and $z_2 = r\exp(i\omega t + i\pi)$, with

$$r = \sqrt{\frac{K^2}{4} + 1} - \frac{K}{2}. \tag{5.10}$$

The radius approaches unity for $K \to 0$ and grows linearly, $r \approx |K|$, as $K \to -\infty$.

For $\omega_1 \neq \omega_2$, Eqs. (5.9) have a solution corresponding to frequency synchronization, where the two oscillators rotate with the same frequency Ω and different phases, $z_1 = r\exp[i(\Omega t + \psi_1)]$ and $z_2 = r\exp[i(\Omega t + \psi_2)]$. The synchronization frequency is

$$\Omega = \frac{\omega_1 + \omega_2}{2}, \tag{5.11}$$

while the phase difference satisfies

$$\sin(\psi_2 - \psi_1) = \frac{\omega_2 - \omega_1}{K}. \tag{5.12}$$

The radius is given by

$$r = \sqrt{1 - \frac{K}{2}[1 - \cos(\psi_2 - \psi_1)]} = \sqrt{1 - \frac{K}{2} + \frac{|K|}{2}\sqrt{1 - \left(\frac{\omega_2 - \omega_1}{K}\right)^2}}. \tag{5.13}$$

The frequency-synchronized solution exists if the quantity under the square root of this equation is positive. For $K > 2$, r is well defined if, moreover, $K > 1 + (\omega_2 - \omega_1)^2/4$. For $K < 2$, on the other hand, the solution exists if $|K| > |\omega_2 - \omega_1|$. Linear stability analysis shows that these solutions are stable in their whole domain of existence. In all the frequency-synchronized states, including the anti-phase solution found in the case of identical natural frequencies for $K < 0$, there is also *amplitude synchronization*, $|z_1| = |z_2|$. For any value of the natural frequencies and of the coupling constant, Eqs. (5.9) have also a stationary solution with no parallel in the case of phase oscillators, namely, the trivial solution $z_1 = z_2 = 0$. In this state, the two oscillators are at rest at the origin of the complex plane. Such behavior is known as *oscillation death*. The corresponding solution is stable if [Aronson *et al.* (1990)]

$$2 < K < 1 + \frac{1}{4}(\omega_2 - \omega_1)^2. \tag{5.14}$$

Oscillation death corresponds to the stabilization, due to coupling, of the unstable state of the individual dynamics. This phenomenon was first described for chemical oscillators [Bar-Eli (1985); Shiino and Frankowicz (1989); Ermentrout (1990); Mirollo and Strogatz (1990a)], and requires that interactions are attractive [Strogatz (1998)].

Figure 5.1 shows the domains of stability of the different types of synchronized solutions described so far, in the parameter space $(|\omega_2 - \omega_1|, K)$. In the region not covered by these solutions, given by $|K| < |\omega_2 - \omega_1|$ for $|\omega_2 - \omega_1| < 2$ and $-|\omega_2 - \omega_1| < K < 2$ for $|\omega_2 - \omega_1| > 2$, the oscillators perform incoherent motion, with no synchronization in any of their variables. The evolution in this region is quasiperiodic, as illustrated in the inset of the figure.

Coming now to the analysis of large ensembles of limit-cycle oscillators, we first note that if all their natural frequencies are identical, $\omega_i = \omega$ for

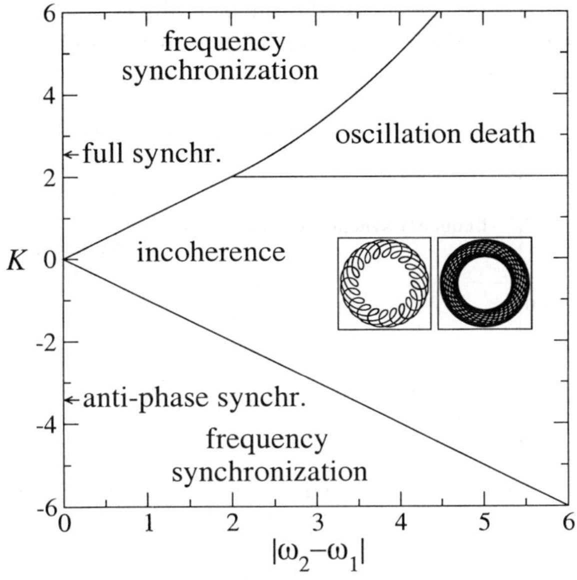

Fig. 5.1 Synchronization regimes for two coupled limit-cycle oscillators, Eqs. (5.9), in the parameter plane of coupling intensity versus difference of natural frequencies. The inset in the region of incoherence shows the orbits of two oscillators with natural frequencies $\omega_1 = 0.1$ and $\omega_2 = 0.2$, interacting with coupling intensity $K = 1$.

all i, we can fix $\omega = 0$ by redefining $z_i \to z_i \exp(i\omega t)$. As in the case of ensembles of identical phase oscillators (Sec. 2.2), this is a change to a reference frame rotating with frequency ω. In this reference frame, the state of full synchronization is given by $z_i = z_0$ for all the oscillators, where z_0 is a constant complex number. Since in a fully synchronized ensemble all the elements move along the orbit of a single uncoupled oscillator, we have $|z_0| = 1$. The stability analysis of full synchronization for an ensemble of N limit-cycle oscillators yields $2N$ eigenvalues. Besides the longitudinal eigenvalue $\lambda_0 = 0$, we find $\lambda_1 = -2$, $\lambda_2 = \cdots = \lambda_N = -2 - K$, and $\lambda_{N+1} = \cdots \lambda_{2N-1} = -K$. Thus, the stability of full synchronization is ensured if $K > 0$. For $K < 0$, numerical results show that the system approaches a state of frequency and amplitude synchronization where phases are irregularly distributed over the interval $[0, 2\pi)$, and the radius is larger than unity.

Also oscillation death, $z_i = 0$ for all i, is a possible solution to the equations of motion for the ensemble of identical oscillators. However, one of the eigenvalues given by the linear stability analysis, $\lambda_0 = 1$, is always positive. Therefore, oscillation death is unstable when all the natural frequencies are identical.

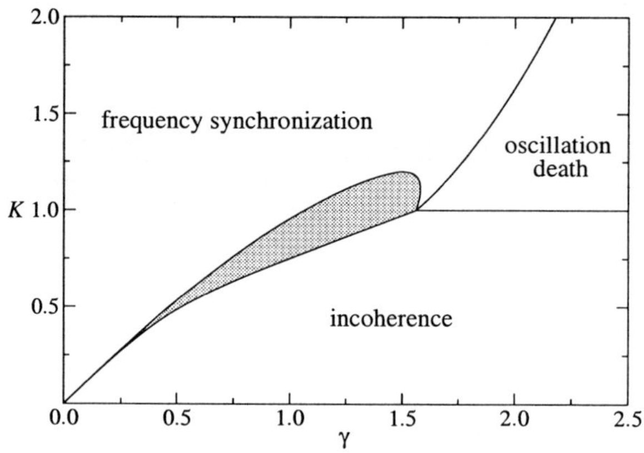

Fig. 5.2 Synchronization regimes for an ensemble of globally coupled limit-cycle oscillators, Eq. (5.4), with $b = \epsilon = 0$ and natural frequencies distributed uniformly in the interval $(-\gamma, \gamma)$. The shaded region indicates the regime of partial unsteady synchronization (adapted from [Matthews and Strogatz (1990)]).

Heterogeneous ensembles of limit-cycle oscillators have been studied for the case where the natural frequencies are uniformly distributed in the interval $(-\gamma, \gamma)$ [Matthews and Strogatz (1990); Matthews *et al.* (1991)]. Each frequency ω_i is thus chosen at random from a distribution

$$g(\omega) = \begin{cases} 1/2\gamma & \text{for } \omega < \gamma, \\ 0 & \text{otherwise.} \end{cases} \tag{5.15}$$

The analysis is performed for attractive interactions, $K > 0$.

Figure 5.2 shows the different regimes of synchronization found for this system. Comparing with Fig. 5.1, we realize that the collective behavior of the ensemble reproduces all the synchronization regimes observed for two oscillators. Moreover, there is a zone of parameter space, the shaded region of Fig. 5.2, where the system exhibits a form of partial synchronization characterized by unsteady dynamics. In this region, the amplitude $\sigma(t)$ defined in Eq. (5.6) is different from zero and its evolution depends strongly on the values of K and γ. Starting from the regime of frequency synchronization and decreasing the coupling intensity with fixed $\gamma > 1$, the frequency-synchronized state losses stability via an Andronov–Hopf bifurcation at the upper boundary of the shaded region. This bifurcation gives rise to small-amplitude periodic oscillations of $\sigma(t)$. As K is decreased further, the amplitude grows and $\sigma(t)$ oscillates between $\sigma \approx 0$ and relatively large values. If $\gamma < 1$, on the other hand, this regime of large oscillations is directly reached from the frequency synchronization region via a saddle-node bifurcation. For lower K, periodicity is lost via a new Andronov–Hopf bifurcation and quasiperiodic evolution of $\sigma(t)$ sets in. Finally, just above the lower boundary of the shaded region, we find irregular motion. Numerical results suggest that such motion is chaotic. Figure 5.3 illustrates the evolution of $\sigma(t)$ in the regimes of large oscillations, and quasiperiodic and chaotic motion.

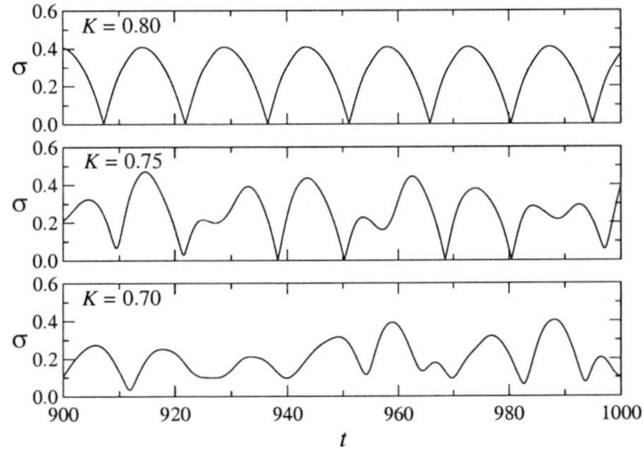

Fig. 5.3 Evolution of the amplitude $\sigma(t)$, defined in Eq. (5.6), across the region of unsteady motion ($\gamma = 0.8$), in an ensemble of $N = 10^3$ limit-cycle oscillators. We have, from top to bottom, large oscillations, quasiperiodic motion, and chaotic motion.

Let us qualitatively analyze these numerical results in the light of the conclusions drawn for globally coupled non-identical phase oscillators in Sec. 3.1. For small values of the dispersion γ and at low coupling intensities, the ensemble of limit-cycle oscillators is found in an incoherent state, $\sigma \approx 0$, where no entrainment takes place. For an infinitely large ensemble, $N \to \infty$, σ should be equal to zero. This incoherent state is also observed in phase oscillators, below the transition that leads to frequency synchronization. As K grows, a similar transition takes place for limit-cycle oscillators, above which $\sigma > 0$. This indicates that a frequency-synchronized cluster has appeared [Matthews *et al.* (1991)]. The unsteady evolution of σ, however, reveals that the degree of entrainment is still relatively low. For larger K, the evolution of σ becomes more regular, as more oscillators are entrained in periodic motion. Finally, frequency synchronization is complete for sufficiently large K. Note that all oscillators become synchronized in frequency at a finite value of K because the distribution of frequencies considered here, Eq. (5.15), has a bounded support, with $g(\omega) = 0$ for sufficiently large $|\omega|$. A finite coupling intensity is thus enough to entrain the whole ensemble into a frequency-synchronized state.

The region where the incoherent state is stable can be determined by self-consistency arguments, analogous to those presented in Sec. 3.1 for infinitely large heterogeneous ensembles of phase oscillators [Matthews *et al.* (1991)]. Assuming that the population consists of a cluster of frequency-synchronized oscillators in a background of non-entrained elements, it is found that the cluster population is different from zero for coupling intensities above the boundary determined by

$$\frac{1}{K} = \frac{\pi g(0)}{2} + (1 - K) \int_{-\infty}^{\infty} \frac{g(\omega) d\omega}{4(1 - K)^2 + \omega^2}. \tag{5.16}$$

For the distribution of natural frequencies given in Eq. (5.15), this boundary is determined by

$$\tan\left(\frac{2\gamma}{K}\right) = \frac{2}{\gamma}(K - 1). \tag{5.17}$$

The boundary ends at $\gamma = \pi/2$, where K reaches unity (cf. Fig. 5.2).

The situation is considerably different for $\gamma > \pi/2$. Increasing the coupling intensity, the system passes from incoherent collective motion to the state of oscillation death where, at long times, the state $z_i = 0$ is asymptotically reached for all i. This transition implies a drastic change in the dynamics of the ensemble, but does not affect the value of σ, which vanishes in both regimes. For larger K, a second transition leads from oscillation death to frequency synchronization, which necessarily implies an abrupt jump in σ.

The state of oscillation death exists over all the parameter space, but its stability depends on the coupling intensity and on the distribution of natural frequencies. Linear stability analysis for this state can be carried out in the limit $N \to \infty$ [Matthews and Strogatz (1990)]. Besides the longitudinal eigenvalue $\lambda = 0$, the set of discrete eigenvalues satisfies

$$\frac{1}{K} = \int_{-\infty}^{\infty} \frac{\lambda + K - 1}{(\lambda + K - 1)^2 + \omega^2} g(\omega) d\omega. \tag{5.18}$$

There is also a continuous part of the eigenvalue spectrum, given by $\lambda = 1 - K + i\omega$, where ω runs over the values of the natural frequency for which $g(\omega) \neq 0$. Stability requires that all transversal eigenvalues are negative or have negative real parts. For the distribution of natural frequencies given in Eq. (5.15), the condition derived from the discrete eigenvalues, Eq. (5.18), sets the stability boundary between oscillation death and frequency synchronization at

$$\tan\left(\frac{\gamma}{K}\right) = \frac{\gamma}{K - 1}. \tag{5.19}$$

The eigenvalues in the continuous spectrum impose the additional condition $K > 1$, which determines the boundary with the incoherent state.

5.1.1 *Oscillation death due to time delays*

While, as discussed above, oscillation death does not take place in ensembles of identical limit-cycle oscillators coupled as in Eq. (5.4), this phenomenon is possible when interactions between oscillators have time delays [Ramana Reddy *et al.* (1998); Ramana Reddy *et al.* (1999); Strogatz (1998)]. Let us first consider two oscillators subject to time-delayed coupling,

$$\begin{aligned} \dot{z}_1(t) &= [1 - |z_1(t)|^2 + i\omega_1] z_1(t) + \tfrac{K}{2}[z_2(t-\tau) - z_1(t)] \\ \dot{z}_2(t) &= [1 - |z_2(t)|^2 + i\omega_2] z_2(t) + \tfrac{K}{2}[z_1(t-\tau) - z_2(t)], \end{aligned} \tag{5.20}$$

with $K > 0$. Linear stability analysis of the oscillation death state, $z_1 = z_2 = 0$, shows that the boundaries of the stability region are

$$4a - 2aK - K^2 \sin[a\tau \pm (\omega_1 + \omega_2)\tau] = 0, \tag{5.21}$$

where

$$a = \sqrt{(\omega_2 - \omega_1)^2 - (2 - K)^2 \mp \sqrt{K^4 - 4(2 - K)^2(\omega_2 - \omega_1)^2}}. \tag{5.22}$$

For $\tau = 0$ this equation yields the same result of Eq. (5.14). In general, the boundary depends not only on the difference of natural frequencies (as in the case without delay) but also on the individual values, through their sum. We recall that, in the presence of time delays, the equations of motion are not invariant under uniform shifts in natural frequencies.

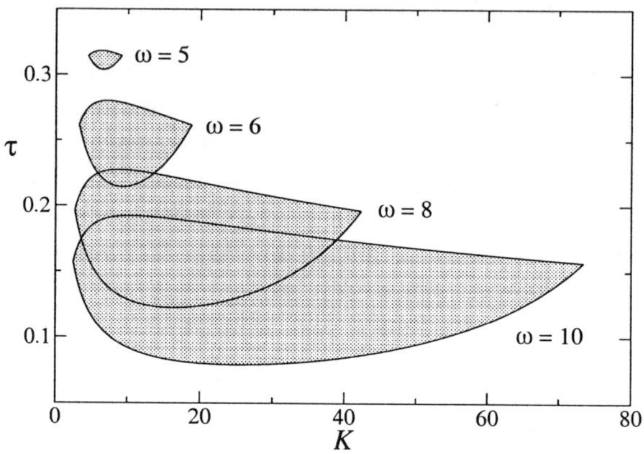

Fig. 5.4 Stability regions of oscillation death (shaded) for two coupled limit-cycle oscillators with identical natural frequencies ω, in the parameter space of coupling constant K and time delay τ.

It can be seen from Eq. (5.21) that in the case of identical oscillators, $\omega_1 = \omega_2 = \omega$, the stability region for oscillation death is bounded by the curves represented by

$$\tau = \frac{n\pi + \arccos(1 - 2/K)}{\omega - \sqrt{K-1}}, \qquad \tau = \frac{(n+1)\pi - \arccos(1 - 2/K)}{\omega + \sqrt{K-1}}, \qquad (5.23)$$

with $n = 0, 1, \ldots$. For each value of n, such region exists when the two curves intersect each other. This requires that the values of ω and K are large enough. As ω grows, the first intersections occur for $n = 0$ if $\omega > \omega_c$, with $\omega_c \approx 4.81$. Figure 5.4 shows the stability regions of oscillation death in the (K, τ)-plane, for several values of the natural frequency. All the cases represented in the figure correspond to the intersection of the curves with $n = 0$. For these frequencies and other values of n, there are no intersections. On the other hand, if the natural frequency is larger, intersections for higher n are possible, and the stability region of oscillation death for each value of ω becomes multiply connected [Ramana Reddy *et al.* (1998)].

These analytical results can be extended to an arbitrary number of coupled identical limit-cycle oscillators, which are described by the equations

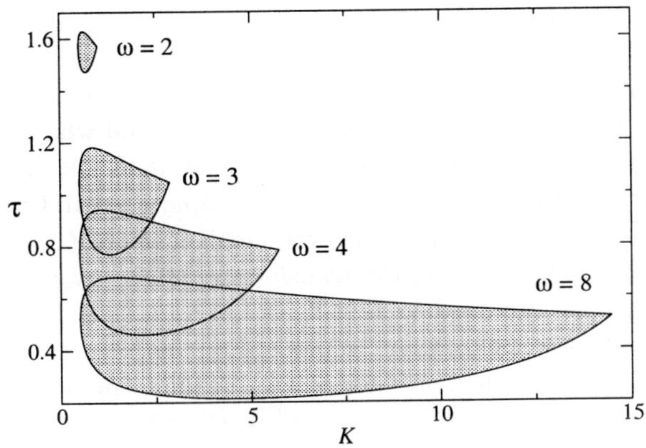

Fig. 5.5 Stability regions of oscillation death (shaded) for an ensemble of infinitely many limit-cycle oscillators with identical natural frequencies ω, in the parameter space of coupling constant K and time delay τ.

of motion [Ramana Reddy *et al.* (1998)]

$$\dot{z}_i(t) = (1 + i\omega - |z_i(t)|^2)z_i(t) + \frac{K}{N}\sum_{j\neq i}[z_j(t-\tau) - z_i(t)]. \tag{5.24}$$

The stability region for oscillation death is now determined by four curves, given by the equations,

$$\tau = \frac{2n\pi + \arccos\frac{\beta K - 1}{\beta K}}{\omega - \sqrt{2\beta K - 1}}, \qquad \tau = \frac{2(n+1)\pi - \arccos\frac{1-\beta K}{(1-\beta)K}}{\omega - \sqrt{(1-\beta)^2K^2 - (\beta K - 1)^2}}, \tag{5.25}$$

and

$$\tau = \frac{2(n+1)\pi - \arccos\frac{\beta K - 1}{\beta K}}{\omega + \sqrt{2\beta K - 1}}, \qquad \tau = \frac{2n\pi + \arccos\frac{1-\beta K}{(1-\beta)K}}{\omega + \sqrt{(1-\beta)^2K^2 - (\beta K - 1)^2}}. \tag{5.26}$$

with $\beta = 1 - N^{-1}$. For $N = 2$, the first two combine to give the first of Eqs. (5.23), and the last two combine to give the second. For $N \to \infty$, on the other hand, only the first identities in both Eqs. (5.25) and (5.26) are meaningful, since $\beta \to 1$. They determine a non-vanishing stability region if $\omega > \omega_c = \pi/2$. Figure 5.5 shows the stability regions corresponding to $n = 0$ for an infinitely large ensemble and several values of the natural frequency.

5.2 Complex Global Coupling

We now analyze Eq. (5.4) for an ensemble of identical oscillators, in the general case with complex global coupling ($\epsilon \neq 0$), and when the frequency of each oscillator depends on the radius of its orbit ($b \neq 0$) as in Eq. (5.3). We focus the attention on the effect of the parameters b and ϵ in the collective dynamics of the system [Nakagawa and Kuramoto (1993); Nakagawa and Kuramoto (1994)]. Using the transformation $z_i \to z_i \exp(i\omega t)$, all the natural frequencies are fixed to zero.

In the limit of weak coupling, $K \to 0$, Eq. (5.4) can be reduced to an equation of motion for the oscillator phases, of the form of Eq. (2.4). Using the polar representation $z_i(t) = r_i(t) \exp[i\phi_i(t)]$ in Eq. (5.4) and separating real and imaginary parts, we first obtain equations for the radius,

$$\dot{r}_i = r_i - r_i^3 + K[\langle r \cos(\phi - \phi_i)\rangle - \epsilon\langle r \sin(\phi - \phi_i)\rangle - r_i], \tag{5.27}$$

and for the phase

$$\dot{\phi}_i = -br_i^2 + \frac{K}{r_i}[\langle r \sin(\phi - \phi_i)\rangle + \epsilon\langle r \cos(\phi - \phi_i)\rangle - \epsilon r_i]. \tag{5.28}$$

Here, we have defined $\langle r \sin(\phi - \phi_i)\rangle = N^{-1}\sum_j r_j \sin(\phi_j - \phi_i)$ and $\langle r \cos(\phi - \phi_i)\rangle = N^{-1}\sum_j r_j \cos(\phi_j - \phi_i)$.

When coupling is weak, we expect that for long times the trajectory of each oscillator is close to the limit cycle of Eq. (5.1). In particular, the radius r_i should differ from unity by a correction of order K, $r_i = 1 + K\delta r_i$. Substituting this form of r_i into Eq. (5.27) and expanding up to the first order in K, we get an equation for the deviation δr_i:

$$\delta\dot{r}_i = -2\delta r_i + \langle\cos(\phi - \phi_i)\rangle - 1 - \epsilon\langle\sin(\phi - \phi_i)\rangle, \tag{5.29}$$

with $\langle\cos(\phi - \phi_i)\rangle = N^{-1}\sum_j \cos(\phi_j - \phi_i)$ and $\langle\sin(\phi - \phi_i)\rangle = N^{-1}\sum_j \sin(\phi_j - \phi_i)$. For small K, moreover, the phase difference between any two oscillators evolves slowly, over time scales of order K^{-1}. Therefore, the deviations δr_i are expected to adjust adiabatically to those phase differences. This makes it possible to neglect the time derivative of the radius deviations in Eq. (5.29) and express δr_i as a function of the phase differences:

$$\delta r_i = \frac{1}{2}[\langle\cos(\phi - \phi_i)\rangle - 1 - \epsilon\langle\sin(\phi - \phi_i)\rangle]. \tag{5.30}$$

Substituting this expression into Eq. (5.28) and expanding up to the first order in K, we obtain an autonomous equation for the oscillator phases,

$$\dot{\phi}_i = -b + K(1 + b\epsilon)\langle\sin(\phi - \phi_i)\rangle + K(\epsilon - b)[\langle\cos(\phi - \phi_i)\rangle - 1], \quad (5.31)$$

which can be rewritten as

$$\dot{\phi}_i = -b' + \frac{K'}{N}\sum_{j=1}^{N}\sin(\phi_j - \phi_i + \alpha). \quad (5.32)$$

Here, $b' = b + K(\epsilon - b)$, $K' = K\sqrt{(1 + b^2)(1 + \epsilon^2)}$, and

$$\cos\alpha = \frac{1 + b\epsilon}{\sqrt{(1 + b^2)(1 + \epsilon^2)}}, \qquad \sin\alpha = \frac{\epsilon - b}{\sqrt{(1 + b^2)(1 + \epsilon^2)}}. \quad (5.33)$$

Equation (5.32) describes a set of coupled identical phase oscillators with natural frequency $-b'$. Interactions are attractive if $0 < \alpha < \pi/2$ or $3\pi/2 < \alpha < 2\pi$, and repulsive otherwise. Full synchronization is possible, with all oscillators moving with frequency $\Omega = -b' + K'\sin\alpha = -b$. For $K > 0$, the fully synchronized state is stable if $\cos\alpha > 0$. In the limit $N \to \infty$, moreover, we have a stationary incoherent state with a homogeneous distribution of phases, $n(\phi) = (2\pi)^{-1}$, which is marginally stable for $\cos\alpha < 0$ and unstable otherwise. Therefore, the curve

$$1 + b\epsilon = 0 \quad (5.34)$$

defines the boundary between the stability regions of full synchronization and the incoherent state.

If the coupling intensity is large, the limit of weak coupling does not hold and we must go back to Eq. (5.4). Full synchronization and the stationary incoherent state are also solutions to this equation for an ensemble of identical oscillators. In the state of full synchronization we have $z_i(t) = \exp(-ibt + \phi_0)$ for all i, while the incoherent state corresponds to a homogeneous distribution of phases on a circle of radius $r = \sqrt{1 + K}$. In contrast with the limit of weak coupling, however, the boundaries of their stability regions do not coincide. Full synchronization is stable if

$$1 + b\epsilon > \frac{1 + \epsilon^2}{2}K, \quad (5.35)$$

while the homogeneous incoherent state is stable if

$$\frac{4}{K}(K - 1)(2K - 1)(1 + b\epsilon) < (K - 1)b^2 - (2K - 1)\epsilon^2 - 1. \quad (5.36)$$

Figure 5.6 shows the stability regions for the two solutions, in the planes (ϵ, K) and (ϵ, b). Since the boundaries intersect each other, there is a region of parameter space where both states are stable, as well as a region where they are simultaneously unstable. The collective behavior of the oscillator ensemble in this latter zone turns out to be quite complex, as described below.

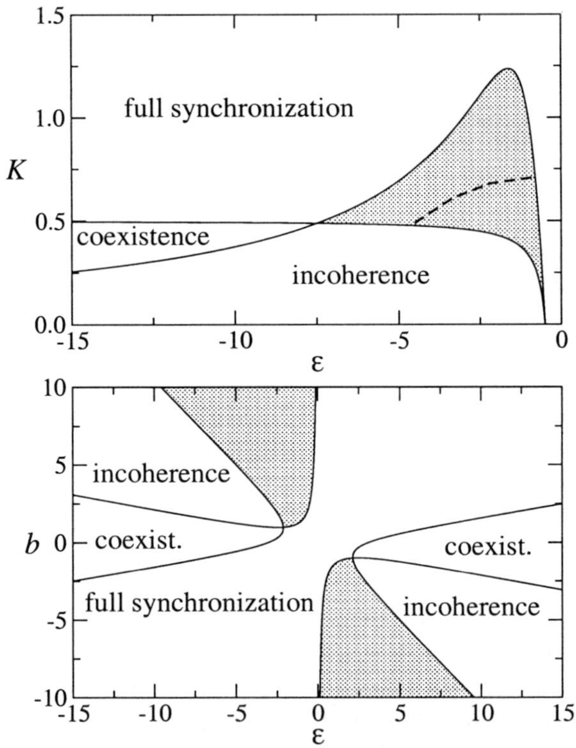

Fig. 5.6 Stability regions of full synchronization and incoherence in the parameter spaces (ϵ, K) for $b = 2$, and (ϵ, b) for $K = 0.4$. The regions of coexistence of these two solutions are also indicated. Shading shows the zone of instability, where complex collective behavior takes place. The dashed line in the upper plot shows the boundary below which collective chaos is observed.

Let us fix $b = 2$ and $\epsilon = -1$, and decrease K from the stability region of full synchronization to that of incoherence, through the zone of complex collective evolution [Nakagawa and Kuramoto (1993)]. In the upper plot of Fig. 5.6, this corresponds to moving downwards along a vertical line ($\epsilon = -1$) and across the shaded region. Crossing the upper boundary of this region from above, at $K = 1$, full synchronization loses stability and is replaced by a two-cluster state. As a matter of fact, for certain initial conditions, two clusters are also found just above the boundary ($1 < K < 1.05$) where full synchronization is stable. This implies that the system is multistable in that zone. The sizes of the clusters depend also on the initial condition. Close to the boundary, the two clusters can be very different. This difference, however, becomes smaller as the coupling intensity decreases further. While for large values of K the two clusters are stationary, a new transition occurs for smaller values. Afterwards, the two clusters perform oscillations of small amplitude with finite frequency, revealing an Andronov–Hopf bifurcation. For even lower coupling intensity, the two-cluster state becomes unstable and is replaced by three clusters. Near the transition, the partition consists of two big clusters and a smaller one, which subsequently grows as K decreases. The motion of the three clusters is chaotic for most initial conditions.

Then, at $K \approx 0.7$, a sudden change takes place. Three-cluster states become unstable and the ensemble is entrained in a form of partially coherent motion, which has been called *collective chaos* [Nakagawa and Kuramoto (1993); Nakagawa and Kuramoto (1994)]. The dashed line in the upper plot of Fig. 5.6 shows the boundary below which this form of complex collective behavior is observed [Chabanol *et al.* (1997)]. In this state, oscillators are arranged in the complex plane along a curve or string. Depending on the value of K, this string may evolve in a complex manner. Figure 5.7 shows three successive snapshots of the distribution over phase space of an ensemble of $N = 10^3$ oscillators. As time elapses, the string stretches and folds, in a way strongly reminiscent of chaotic phase-space dynamics. Here, however, the role of different points in phase space is played by different oscillators. The analogy suggests that the distance between two initially close oscillators grows, on the average, exponentially with time. Stretching and folding also implies shuffling and mixing of oscillators. This is due to the dependence of the frequency on the amplitude, given by the nonlinear term in the individual dynamics of the oscillators. Oscillators with larger amplitudes move faster, and can outrun slower oscillators of small amplitude.

The fact that the motion of each individual oscillator is chaotic is verified by studying the Lyapunov exponents of the system, which can be calculated numerically [Nakagawa and Kuramoto (1995)]. It turns out that approx-

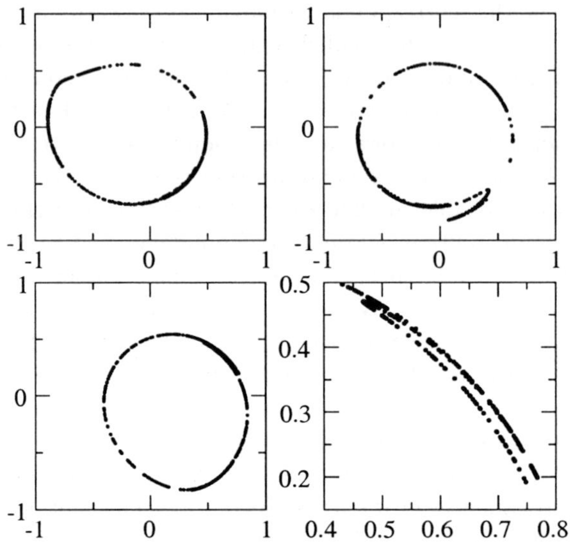

Fig. 5.7 Three snapshots in the complex plane of an ensemble of 10^3 identical limit-cycle oscillators, with $b = 2$, $\epsilon = -1$ and coupling constant $K = 0.6$. The snapshots are separated by 10 time units. The fourth plot shows a close-up of the last snapshot, illustrating the effect of stretching and folding in the distribution of oscillators.

imately one half of the $2N$ Lyapunov exponents are positive. Therefore, the system is in a state of high-dimensional chaos and, as a consequence, the motion of single elements is aperiodic. This contrasts with the low-dimensional chaos found when the ensemble is divided into three clusters, where oscillators of a given cluster are entrained in synchronous chaotic motion and the effective number of degrees of freedom is thus drastically reduced.

It is interesting to point out that distributions like those of Fig. 5.7 can be obtained as Poincaré sections of the trajectory of a single oscillator subject to the effect of periodic external forcing [Nakagawa and Kuramoto (1993)]. The equation

$$\dot{z} = z - (1 + ib)|z|^2 z + K(1 + i\epsilon)[\zeta(t) - z] \tag{5.37}$$

describes the motion of an oscillator subject to an external force $\zeta(t)$. This external force plays the same dynamical role as the average $\langle z \rangle$ in Eq. (5.4). When both the modulus and the phase of $\zeta(t)$ vary periodically with time, a Poincaré section of $z(t)$ at fixed values of the oscillation phase of $|\zeta|$ shows the same kind of distribution on the complex plane as the whole ensemble of limit-cycle oscillators. This analogy suggests that complex individual dynamics can take place even when global averages exhibit periodic motion. Slightly different versions of Eq. (5.37) show the same qualitative behavior [Hakim and Rappel (1992); Chabanol *et al.* (1997)].

5.3 Non-local Coupling

Consider now an ensemble of identical limit-cycle oscillators distributed in space, whose evolution is governed by the equation

$$\dot{z}_i = (1 + i\omega)z_i - (1 + ib)|z_i|^2 z_i + K(1 + i\epsilon)(\langle z \rangle_i - z_i). \tag{5.38}$$

The average $\langle z \rangle_i$ is in general different for each oscillator. It is defined as

$$\langle z \rangle_i = \sum_{j=1}^{N} J(\gamma|\mathbf{r}_j - \mathbf{r}_i|) z_j, \tag{5.39}$$

where $\mathbf{r}_i$ is the spatial position of oscillator i. The function $J(\gamma|\mathbf{r}_j - \mathbf{r}_i|)$ weights the effect of the state of oscillator j on the evolution of oscillator i. It is symmetric with respect to the two oscillators, and is normalized according to

$$\sum_{j=1}^{N} J(\gamma|\mathbf{r}_j - \mathbf{r}_i|) = 1 \tag{5.40}$$

for all i. We assume that its magnitude decays on a spatial scale of order γ^{-1}, so that $J(u) \ll 1$ for $u \gg 1$.

This model has been proposed as an interpolation between the space-continuous, locally coupled Ginzburg–Landau equation (5.5) and globally coupled oscillator ensembles, Eq. (5.4) [Kuramoto (1995); Kuramoto and Nakao (1997)]. When the variation of z_i as a function of i takes place over sufficiently long scales as compared with γ^{-1} and with the typical distance between neighbor oscillators, Eq. (5.38) is a good approximation of the continuous Ginzburg–Landau equation. On the other hand, for $\gamma \to 0$ it reduces to the equation for a globally coupled ensemble of identical limit-cycle oscillators.

The study of Eq. (5.38) has been focused on ensembles arranged over a regular linear lattice with periodic boundary conditions. This periodic lattice is represented by an infinite array where $z_i = z_{i+N}$ for all $i \in (-\infty, \infty)$. The weight functions are chosen to have the form

$$J(\gamma|x_j - x_i|) = J_0 \exp(-\gamma a|j-i|), \tag{5.41}$$

where the position of oscillator i is $x_i = ai$, and a is the lattice spacing. The normalization constant is $J_0 = [1-\exp(-\gamma a)]/[1+\exp(-\gamma a)]$.

This choice of $J(\gamma|x_j - x_i|)$ is especially suitable for numerical analysis, since the local averages $\langle z\rangle_i$ can be obtained recursively and do not require the calculation of the sum in Eq. (5.39) for each i [Kuramoto (1995)]. To show this, we note that $\langle z\rangle_i$ can be written as $\langle z\rangle_i = \langle z\rangle_i^- + \langle z\rangle_i^+$, with

$$\langle z\rangle_i^- = J_0 \sum_{j=-\infty}^{i} \exp[-\gamma a(i-j)], \qquad \langle z\rangle_i^+ = J_0 \sum_{j=i+1}^{\infty} \exp[-\gamma a(j-i)]. \tag{5.42}$$

These quantities satisfy the recursion relation

$$\langle z\rangle_{i+1}^{\pm} = \mp J_0 z_{i+1} + \exp(\pm\gamma a)\langle z\rangle_i^{\pm}, \tag{5.43}$$

which imply

$$\langle z\rangle_{i+1} = \exp(-\gamma a)\langle z\rangle_i^- + \exp(\gamma a)\langle z\rangle_i^+. \tag{5.44}$$

This makes it possible to calculate $\langle z\rangle_i$ for $i = 2, 3, \ldots, N$ given the value of $\langle z\rangle_1$ at each time. Taking into account the condition of periodic boundaries, we find

$$\langle z\rangle_1^{\pm} = \frac{J_0}{1-\exp(\pm N\gamma a)}\left(z_1 + \sum_{j=2}^{N} \exp[\pm(N+1-j)\gamma a]z_j\right). \tag{5.45}$$

The initial condition for the recursion process is then given by $\langle z\rangle_1 = \langle z\rangle_1^- + \langle z\rangle_1^+$.

The most interesting solutions to Eqs. (5.38) are found in the region of parameter space where the state of full synchronization is unstable. In the case of the continuous Ginzburg–Landau equation, this is the regime of diffusion-induced chemical turbulence, characterized by the development of intricate spatiotemporal patterns [Kuramoto (1984)]. For globally coupled oscillators it corresponds to collective chaos.

Figure 5.8 shows snapshots of the distribution of $|z_i|$ in an array of $N = 10^3$ oscillators, coupled as in Eqs. (5.38). The coupling weight is given by Eq. (5.41), with $N\gamma a = 8$. Thus, the coupling range equals 1/8 of the system length. The values of b and ϵ are such that full synchronization is

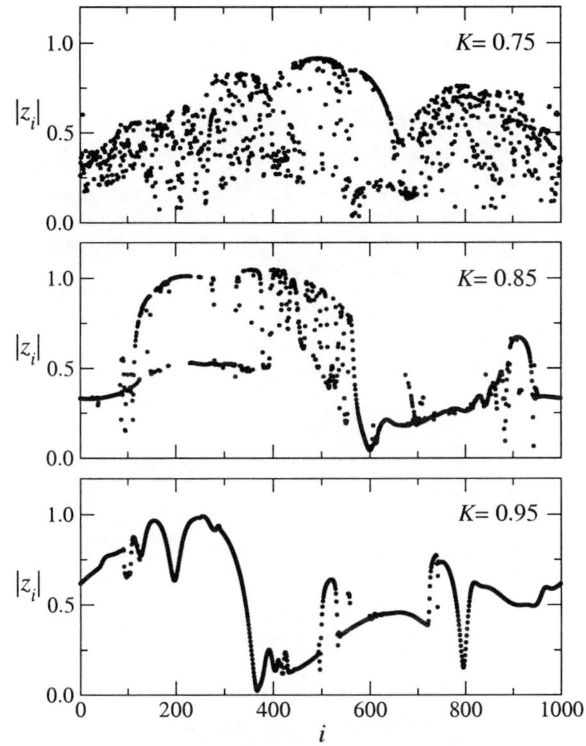

Fig. 5.8 Snapshots of the distribution of amplitudes $|z_i|$ over an array of $N = 10^3$ limit-cycle oscillators subject to non-local coupling, for three values of the coupling intensity K. In all three cases, $b = 2$, $\epsilon = -2$, and $N\gamma a = 8$.

unstable both in the limit of global coupling and for the Ginzburg–Landau equation. As the coupling intensity K grows, locally ordered domains, where the value of $|z_i|$ varies smoothly with i, become more extended. Within these domains the temporal evolution of z_i is also coherent, so that a given domain persists over long times. The ordered domains are separated by regions where $|z_i|$ varies with i in a seemingly irregular manner. Here, the time dependence of z_i for contiguous oscillators is essentially uncorrelated [Kuramoto (1995); Kuramoto and Nakao (1997)].

Figure 5.9 illustrates the time evolution of $|z_i|$ for the same sets of parameters as in Fig. 5.8 in a system of 400 oscillators. The intermittent behavior observed for low coupling intensities gradually disappears as K is increased. For large K, most of the system is occupied by ordered domains, which exhibit slow but complex changes of size as time elapses.

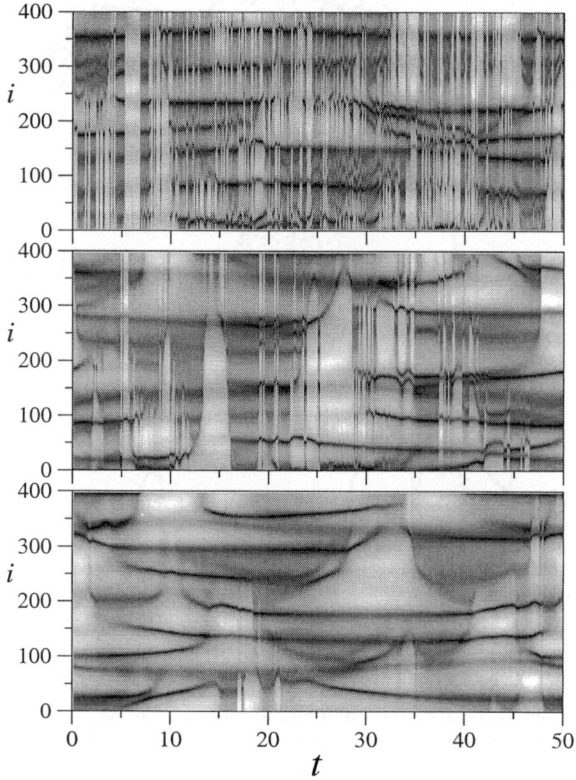

Fig. 5.9 Time evolution of the amplitudes $|z_i|$ in an array of $N = 400$ limit-cycle oscillators subject to non-local coupling, for three values of the coupling intensity (from top to bottom, $K = 0.75$, 0.85 and 0.95). The parameters are the same as in Fig. 5.8. Darker tones correspond to lower values of $|z_i|$.

The spatial structure of z_i can be statistically characterized by means of a correlation function

$$G(x) = \overline{z^*_{i+x} z_i}, \tag{5.46}$$

where the bar indicates average over space, time, and different realizations of the evolution, and z^* is the complex conjugate of z. It is found that over a substantial range of coupling intensities and in the limit of small x

($x \neq 0$), the correlation $G(x)$ behaves as

$$G(x) \approx G_0 - G_1 x^\alpha, \tag{5.47}$$

where α depends on K [Kuramoto (1995); Kuramoto and Nakao (1997)]. This is illustrated in Fig. 5.10, where $G_0 - G(x)$ is plotted in logarithmic scales against x. The slope of the straight lines defined by the data for each value of K gives the exponent α, whose dependence on the coupling intensity is shown in the inset. The approximate form of $G(x)$ given by Eq. (5.47) can be independently verified by calculating the power spectrum $I(q) = |\tilde{z}_q|^2$, where $\tilde{z}_q$ is the discrete Fourier transform of z_i. In fact, for large values of q, the power spectrum behaves as $I(q) \sim q^{-1-\alpha}$.

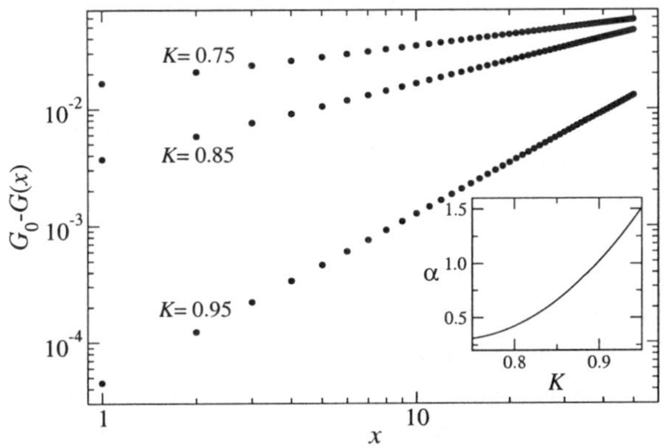

Fig. 5.10 The correlation difference $G_0 - G(x)$ as a function of the distance x for three values of the coupling intensity. The inset represents the dependence on K of the exponent α in Eq. (5.47) (adapted from [Kuramoto (1995)]).

For large values of K, where $\alpha > 1$, $G(x)$ has a flat maximum at $x = 0$. This is an indication of coherent evolution of neighbor elements. If on the other hand, $\alpha < 1$, $G(x)$ shows a cusp as $x \to 0$. For sufficiently low coupling intensities, moreover, it is found that $G(x)$ develops a sharp peak just at $x = 0$. Under these conditions, $G(x)$ is discontinuous at the origin, since $G(0)$ is different from G_0. The value of G_0 must be found by extrapolating the correlation function from $x > 0$. The presence of such peak for small K is related to the fact that the motions of nearest-neighbor oscillators are weakly correlated.

The power-law dependence of the correlation $G(x)$ with the distance

x suggests that the profile of z_i along the system may have self-similar properties. Such properties are revealed, for instance, by measuring the length of the curve defined by the graph of $|z_i|$. Instead of working with the actual length, it is convenient to define the closely related quantity [Kuramoto (1995); Kuramoto and Nakao (1997)]

$$S_m = \sum_{n=1}^{N/m} |z_{nm} - z_{(n-1)m}| \tag{5.48}$$

($m = 1, 2, \dots$). This quantity is a measure of the accumulated change in z_i determined with a spatial resolution equal to the length of a segment containing m contiguous oscillators. If the fractal dimension d_f of the curve is larger than unity, we expect that S_m behaves as

$$S_m \sim m^{1-d_f}, \tag{5.49}$$

where, moreover, $d_f = 2-\alpha$ [Mandelbrot (1982)]. The numerical evaluation of S_m is in agreement with this prediction and, therefore, confirms the fractal nature of the profile of z_i along the oscillator array.

A possible origin for the power-law dependence of the spatial correlation of non-locally coupled oscillators has been proposed by invoking a simple but quite general model [Kuramoto and Nakao (1996)]. The explanation applies also to the case of chaotic oscillators, for which the same kind of correlations are found [Kuramoto and Nakao (1997)]. In this model, the effect of coupling on each oscillator is represented as an external force varying with time. This force may induce occasional changes of sign in the largest Lyapunov exponent of a given oscillator, giving rise to transitions between regular and chaotic evolution. While during periods of regular evolution it is expected that neighbor oscillators approach similar states, with $|z_i - z_{i+x}| \sim x$, chaotic motion will lead to the exponential separation of those states. Noting that

$$G(x) = G(0) - \frac{1}{2}\overline{|z_i - z_{i+x}|^2}, \tag{5.50}$$

the main contribution to this correlation function will originate in the large deviations of z_i between neighbor oscillators during chaotic periods. It can be assumed that the changes of sign of the Lyapunov exponents can be approximately described by a Markovian random process, where the probability of having a chaotic period decreases exponentially with its duration. In this case, it is possible to show that the contribution of chaotic motion to the correlation function will grow as a power of x, with a nontrivial exponent.

The effects of non-local coupling on linear arrays of limit-cycle oscillators have also been analyzed in the regime where full synchronization is stable [Battogtokh and Kuramoto (2000)]. Numerical results suggest that, under sufficiently strong perturbations of the fully synchronized state, the system can develop a regime of spatiotemporal intermittency. This regime is characterized by the recursive appearance of pairs of "holes," where the amplitude $|z_i|$ is strongly depleted. Across each hole the phase of z_i changes by 2π. The two holes are created at the same point of the array. As time elapses, they move apart and propagate over a certain distance, but gradually decelerate and finally vanish. In large systems, hole pairs can appear at different points and their trajectories may cross each other. This gives rise to disordered spatiotemporal patterns. Remarkably, their statistical properties over a considerable range of coupling intensities are again characterized by a correlation function of the form of Eq. (5.47).

PART 2

Synchronization and Clustering in Chaotic Systems

Chapter 6

Chaos and Synchronization

Many deterministic nonlinear systems display, apart from fixed-point solutions and limit cycles, more complex invariant sets which act as attractors for their dynamics. Among them we find chaotic attractors. Chaotic dynamics is unpredictable in the long run. Tiny differences in the initial conditions are exponentially enhanced as time elapses. This qualitative signature of chaotic dynamics is known as sensitivity to the initial conditions. However, chaotic systems can be synchronized despite their seemingly irregular dynamics. When ensembles of chaotic oscillators are coupled, the attractive effect of the coupling makes the individual trajectories approach. If the coupling strength is large enough, it can counterbalance the trend of the trajectories to separate due to chaotic dynamics. As a result, it is possible to reach full synchronization also in chaotic systems.

The first studies on chaos synchronization were purely theoretical. However, it was later demonstrated that such synchronous behavior could be achieved in experimental systems as well. Since then, intensive systematic investigations of different systems have led to a coherent picture of chaos synchronization. The larger number of variables available has also permitted the analysis of many different coupling schemes, and new forms of synchronization have been identified. In this chapter we review general concepts related to synchronization of chaotic systems by using as examples systems formed by only two maps or two oscillators.

6.1 Chaos in Simple Systems

Probably the simplest chaotic system is that described by the logistic map,

$$x(t+1) = 1 - a[x(t)]^2. \tag{6.1}$$

The variable $x(t)$ takes values between -1 and 1, and the parameter $a \in [0, 2]$ determines the type of behavior of the system. For small values of a, only fixed point solutions to (6.1) are found. At $a = 1$ the fixed point becomes unstable and is replaced by period-2 dynamics, where the state $x(t)$ alternates between two values. This period-2 solution becomes itself unstable under further increase in a, and is replaced by a period-4 orbit. As the parameter a increases, the cascade of period doubling bifurcations continues and eventually leads to the onset of chaotic dynamics at the accumulation point $a_\infty = 1.40115\ldots$. Figure 6.1 shows the bifurcation diagram for the map (6.1), where the sequence of attractors of increasingly high periods is seen, together with chaotic regions and periodic windows.

The origin of the logistic map goes back to theoretical ecology. As early as in the 1970's, it was realized that an equation equivalent to (6.1) could describe the successive values of the density of individuals in a population with non-overlapping generations [May (1974)]. Until then, in the theoretical analysis of the evolution of populations it was always assumed that their fate was to end in a fixed point or at most to oscillate regularly. The introduction of the logistic map opened the possibility, not yet considered, that the population of a species in an ecosystem could behave chaotically. The precise equation proposed had the form

$$\rho(t+1) = r\rho(t)(1 - \rho(t)), \tag{6.2}$$

with $\rho(t) \in [0, 1)$ representing the density of a given population, and $r \in [0, 4]$ specifying its intrinsic growth rate. In this way, the relative number of individuals at generation t was mapped to the density one generation later. This equation is related to (6.1) through rescaling of the variables. The study of the logistic map showed that small changes in the parameters of the system (for example in the growth rate r) could trigger major qualitative changes in the dynamics. This is an important property of chaotic systems displaying period-doubling (the behavior represented in Fig. 6.1) when the parameter a (or equivalently r) increases.

In order to have chaotic dynamics in a time-continuous system, it is required that its state is determined by three or more variables and that at least one nonlinear term appears in one of the evolution equations. This

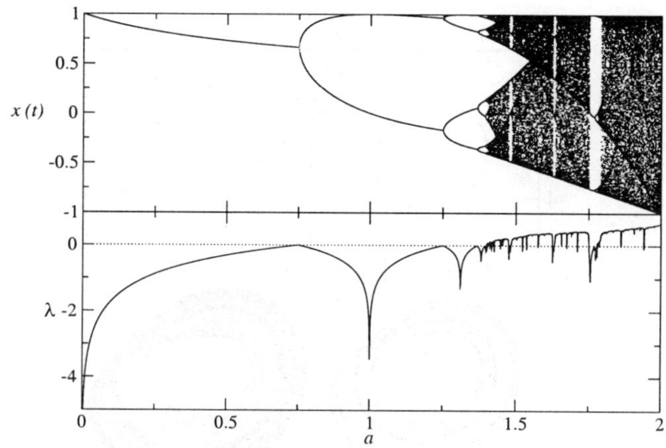

Fig. 6.1 Cascade of period doubling bifurcations for the logistic map. Below, the Lyapunov exponent corresponding to each of the values of the parameter a is shown. Note that the intervals where $\lambda \leq 0$ correspond to periodic behavior, with $\lambda = 0$ at the bifurcation points.

is due to the fact that trajectories do not cross in a continuous-time system, and only in dimension 3 or higher can the invariant attractor show a topology qualitatively more complex than a limit cycle.

A paradigmatic example that we will use throughout the next chapters is the Rössler system [Rössler (1976)], whose dynamical equations are

$$\begin{aligned} \dot{x} &= -\nu y - z \\ \dot{y} &= \nu x + ay \\ \dot{z} &= b + z(x - c). \end{aligned} \tag{6.3}$$

The parameters ν, a, b, and c determine the dynamics of the system and the properties of the corresponding attractor. In the Rössler system, the chaotic behavior is typical and associated with stretching and folding of the trajectories. Later in this chapter we will introduce the Lorenz system where the chaotic attractor is different.

Fig. 6.2 illustrates different dynamics in the Rössler system, depending on its parameters. Though for low values of c limit cycles are observed, increasing its value causes a series of period doubling bifurcations, similarly to what we have seen for the logistic map, which eventually lead to chaotic behavior [Rössler (1976); Mikhailov and Loskutov (1996)].

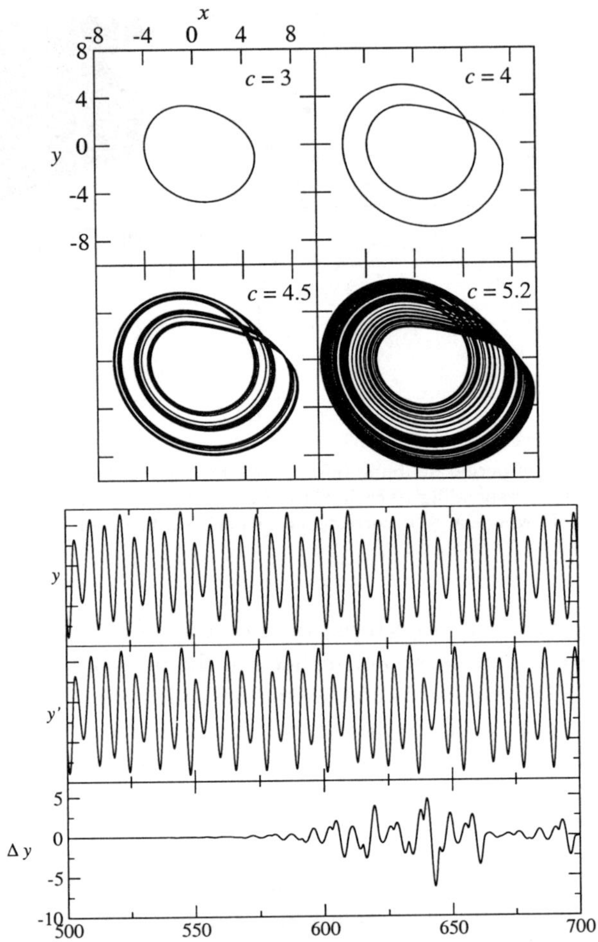

Fig. 6.2 Four different projections in the (x, y) plane of attractors for the Rössler system in its route to chaos. In all cases shown, $\nu = 1$, $a = 0.2$ and $b = 0.3$, all plots share the scale. Below, we show the variable $y(t)$ for two chaotic trajectories with $c = 5.2$. Initially, they had close initial values on the chaotic attractor (equal values for the coordinates x and y and $\Delta z = 0.005$). The lowermost series represents the difference Δy between their coordinates y. The scale is the same for all three series.

6.1.1 *Lyapunov exponents*

Sensitivity to initial conditions is quantified through the measurement of the rate of divergence of two trajectories on a chaotic attractor. Consider an autonomous dynamical system of the form $\dot{\mathbf{r}}(t) = \mathbf{f}(\mathbf{r}(t))$, where $\mathbf{r}(t)$ is a vector of dimension n whose elements take real values. Starting with a random initial condition $\mathbf{r}(t=0)=\mathbf{r}_0$, and after a transient has elapsed, the dynamics eventually reaches the chaotic attractor. Let us call $s(t)$ the trajectory of the system on that attractor, and take two arbitrarily close points on the attractor separated by a distance $\boldsymbol{\eta}$. As long as the two trajectories starting at each of these two points remain close enough, the evolution of their separation $\boldsymbol{\eta}$ obeys a linearized equation,

$$\dot{\boldsymbol{\eta}}(t)=\mathbf{Df}(s)\boldsymbol{\eta}(t). \tag{6.4}$$

The elements of the Jacobian matrix $\mathbf{Df}(s)$ are defined as

$$\mathbf{Df}(s)=\left\{\frac{\partial f_i}{\partial r_j}\right\} \tag{6.5}$$

and have to be evaluated along the solution $s(t)$ of the dynamical system. In the most general case, the Jacobian matrix is time-dependent and its values should be numerically computed. In the particular case when $\mathbf{Df}(s)$ is constant (for instance when the solution s is a simple fixed point), Eq. (6.4) has the formal solution

$$\boldsymbol{\eta}(t)=\sum_j C_j\mathbf{e}_j\exp(\lambda_j^*t), \tag{6.6}$$

where C_j are coefficients that depend on the initial conditions and λ_j^* and $\mathbf{e}_j$ represent the eigenvalues and the eigenvectors of the constant matrix $\mathbf{Df}$, respectively. In this case, the spectrum λ^j of Lyapunov exponents for the dynamical system is simply the set of real parts of the eigenvalues. However, for a time-dependent matrix $\mathbf{Df}(s)$ the two sets do not have such a simple relationship, and in general the spectrum λ^j is defined as

$$\lambda^j(\boldsymbol{\eta})=\lim_{t\to\infty}\frac{1}{t}\ln|(\mathbf{Df}^t)\boldsymbol{\eta}|, \tag{6.7}$$

where $\mathbf{Df}^t$ is the Jacobian matrix of the $t-$th iteration of the dynamical system on the infinitesimal perturbation $\boldsymbol{\eta}$. There are $j=1,\dots,n$ exponents, as many as the dimension of the dynamical system, and they correspond to different directions along which the system can be perturbed. Each of these directions are associated to a particular choice of the vector $\boldsymbol{\eta}$.

The nature of the dynamics along the actual trajectory on the chaotic attractor is dominated by the largest Lyapunov exponent λ^1. In the following, we call it simply λ. Fixed points are characterized by $\lambda < 0$, limit cycles have $\lambda = 0$, and chaotic attractors return $\lambda > 0$, thus quantifying the exponential separation between two neighboring trajectories. After a characteristic time of order λ^{-1} the precise position of the system in phase space becomes essentially unpredictable. In three dimensional oscillators such as the Rössler system, there are two more Lyapunov exponents characterizing the dynamics of the system. The second exponent takes always value zero ($\lambda^2 = 0$) while the third one is negative and larger in absolute value than the first one in dissipative systems ($|\lambda^3| > \lambda$). The presence of a vanishing Lyapunov exponent is a direct consequence of the invariance of the evolution equations with respect to time shifts in autonomous dynamical systems. The third exponent quantifies the rate at which an arbitrary trajectory out of the attractor approaches it.

Eq. (6.7) takes a particularly simple form for one-dimensional maps. If $x(0)$ is the initial value on the trajectory, differentiation of the t–th iteration of the matrix f with respect to $x(0)$ can be performed as

$$\begin{aligned} Df^t(x(0)) &\equiv \tfrac{d}{dx} f^t(x(0)) \\ &= \tfrac{d}{dx} f(x(t-1)) \tfrac{d}{dx} f(x(t-2)) \dots \tfrac{d}{dx} f(x(0)). \end{aligned} \tag{6.8}$$

Let us take as an example the logistic map (6.1), for which $Df(x) = -2ax$ is an aperiodic, time-dependent quantity for any $a > a_\infty$ (except when a periodic window is found). The Lyapunov exponent can then be calculated as

$$\lambda = \lim_{t\to\infty} \frac{1}{t} \sum_{j=1}^{t} \ln|-2ax(j)|, \tag{6.9}$$

where $x(j)$ is the j–th iterate of the map, that is, the actual trajectory. The values of λ obtained for different values of a are shown in Fig. 6.1.

In general, when two or more chaotic oscillators are coupled and synchronization is achieved, the number of dynamical degrees of freedom for the whole system effectively decreases. A system formed by N oscillators, each described through n variables, has dimension nN. When full synchronization is achieved, the dynamics takes place in the subspace $s(t) = \mathbf{r}_1(t) = \mathbf{r}_2(t) = \cdots = \mathbf{r}_N(t)$, such that the global dynamics becomes restricted to a sub-manifold of dimension n. There is a set λ^j of Lyapunov exponents measuring the rate of divergence of two realizations on the synchronous, chaotic attractor, and a second set of exponents $\lambda_{\perp}^j$ that characterize the stability of the synchronous state. The former case corresponds to perturbations along the invariant manifold $s(t)$. These are rigid displacements of the whole synchronous cluster which do not destroy the stability of the synchronous state. The latter case corresponds to perturbations in directions *transversal* to the actual synchronous trajectory, which explicitly separate the trajectories of the oscillators from the synchronous cluster. In some cases, the sets λ^j and $\lambda_{\perp}^j$ are related and to determine the stability of synchronization only the dynamical properties of the synchronous state should be known. Two relevant examples are presented in this chapter, and later in the forthcoming chapters we show that there is a whole class of coupling schemes which permit to establish this relation.

Further discussion on formal and constructive definitions for Lyapunov exponents, as well as analysis of other issues related to chaotic dynamical systems, can be, e.g., found in [Guckenheimer and Holmes (1983); Eckmann and Ruelle (1985); Wiggins (1990); Mikhailov and Loskutov (1996)].

6.1.2 *Phase and amplitude in chaotic systems*

A universal and unambiguous definition of phase for chaotic systems does not exist. This might be a difficulty when trying to establish if two chaotic systems are synchronized, since, as we have seen in all the cases studied in the first part of this book, the comparison of phases is essential to detect and quantify the synchronous state. Nevertheless, in many cases one can clearly see that there is an oscillatory variable related to the dynamics of the system.

For example, in the case of the Rössler attractor (see Fig. 6.2), it seems plausible that the rotations in the (x, y) plane should allow an operational definition of phase under a suitable choice of the Poincaré map. In the following, we define the phase in this system as

$$\phi(t) = \arctan\left(\frac{y(t) - y_0}{x(t) - x_0}\right) + 2\pi m, \tag{6.10}$$

where x and y are the variables corresponding to a plane on which oscillations of the projected dynamics are observed. The point (x_0, y_0) is an interior point around which the system rotates. In the case of the Rössler oscillator, there is an unstable point with $(x_0, y_0) = (0, 0)$ which allows a simple definition of the phase. The last term in Eq. (6.10) adds a factor 2π every time that the system crosses the section $y = y_0$ for $x > x_0$. If this were not taken into account, the phase ϕ would take values in the interval $[0, 2\pi)$. With this definition, it increases monotonically. The amplitude associated with the phase ϕ is

$$A = \sqrt{(x - x_0)^2 + (y - y_0)^2}. \tag{6.11}$$

The new coordinates ϕ and A are occasionally more convenient than x and y to describe the system's dynamics.

A general discussion on the definition of phase in chaotic systems can be found in [Pikovsky *et al.* (1997)]. For a chaotic system, the evolution of the phase depends on the mean frequency of the oscillations ω as well as on their amplitude A through a function that depends on the system considered,

$$\dot{\phi} = H(\omega, A). \tag{6.12}$$

The Rössler attractor is special in this respect, since the time employed to turn once in the (x, y) plane is almost independent of the amplitude of the oscillations [Crutchfield *et al.* (1980)], such that to a good approximation $\dot{\phi} \simeq \omega$. As a consequence, any number of identical Rössler systems behaves similarly to an ensemble of uncoupled phase oscillators. This is a special property of this system responsible for some forms of coherent behavior to be described in the following.

6.2 Synchronization of Two Coupled Maps

The study of coupled systems formed by only two oscillators will allow us to illustrate a number of properties of synchronous behavior in chaotic systems. Some of the features to be introduced in this section will later find their counterpart when we study clustering or other regimes found if full synchronization cannot be achieved.

Let us consider the system

$$\begin{aligned} x_1(t+1) &= (1-K)f(x_1(t)) + KF(x_1, x_2) \\ x_2(t+1) &= (1-K)f(x_2(t)) + KF(x_1, x_2) \end{aligned} \tag{6.13}$$

where $f(x) = 1 - ax^2$, the coupling strength K takes values in $[0, 1]$, and the coupling term has the form

$$F(x_1, x_2) = \frac{1}{2}\left[f(x_1) + f(x_2)\right]. \tag{6.14}$$

With this coupling scheme, large enough coupling strength K should eventually bring about the synchronization of the considered two-oscillator system for any value of a. In particular, when $K = 1$ the two trajectories become identical after the first iteration. When the synchronous state is reached, the dynamics of both oscillators corresponds to that of the single logistic map.

The synchronization threshold for the two maps (6.13) can be computed by linearizing around the synchronous state, where $s \equiv x_1 = x_2$. Let us define the two variables $x_\perp = x_1 - x_2$ and $x_\parallel = x_1 + x_2$. In the synchronous state, $x_\perp \equiv 0$ and $x_\parallel \equiv 2s$. Close to the synchronous state, the variable $x_\perp$ is small and its evolution, determining the stability of the synchronous state, is given by the mapping

$$x_\perp(t+1) = (1-K)[Df(s)]x_\perp(t), \tag{6.15}$$

where the derivative $Df(s)$ is evaluated along the fully synchronous trajectory. Thus, the variable $x_\perp$ corresponds to perturbations transversal to the orbit $s(t)$, and the analogous quantity $x_\parallel$ corresponds instead to perturbations along the synchronous orbit. While the rate of growth of a small difference along two trajectories on the chaotic attractor is measured through the Lyapunov exponent λ, the evolution of perturbations in the perpendicular direction, which determines the stability of the synchronous attractor, is characterized by the *transversal Lyapunov exponent* $\lambda_\perp$.

The fully synchronous state of the two maps is stable if

$$\lambda_\perp = \lim_{t\to\infty} \frac{1}{t} \sum_{j=1}^{t} \ln|2(1-K)ax(j)| < 0. \tag{6.16}$$

In the considered case, the transversal exponent is related to the Lyapunov exponent λ in a simple way,

$$\lambda_\perp = \lambda + \ln(1-K). \tag{6.17}$$

Because of this simple dependence on λ, the boundary of stability of the fully synchronous state can be immediately deduced from Fig. 6.1. The state of full synchronization for two oscillators coupled according to (6.13) is stable for all values of the coupling parameter larger that $K_{\min} = 1 - e^{-\lambda}$. For example, the largest Lyapunov exponent $\lambda = \ln 2$ is reached for $a = 2$, and in this case the fully synchronous state is stable for $K > 1/2$. As we will see later, this result actually holds for the fully synchronous state of any number of globally coupled maps.

6.2.1 *Saw-tooth maps*

The saw-tooth map represents a classical example of a system exhibiting intermittent dynamics and has dynamical properties that are significantly different from those of the logistic map. It is defined by

$$x(t+1) = \{\gamma x(t)\}. \tag{6.18}$$

Here, the curly brackets $\{\cdot\}$ denote the operation of taking the fractional part of the argument. For $\gamma > 1$ this is a piecewise linear function. In the interval $1 < \gamma < 2$, the map could be written as $x(t+1) = \gamma x(t)$, for $x(t+1) < x^*$, and $x(t+1) = \gamma x(t) - 1$ for $x(t+1) > x^*$, with a discontinuity at $x^* = \gamma^{-1}$. For any $\gamma < 1$ the only solution to the dynamics is a fixed point at $x = 0$. For $\gamma > 1$ the dynamics is chaotic, and at $\gamma = 1$ all the values of x are fixed-point solutions. The route to chaos of the saw-tooth map belongs to a universality class different from that of the logistic map. The period doubling cascade is not found here, and fully developed chaos appears as γ crosses the value $\gamma = 1$.

We introduce in the following a slightly modified version of the classical saw-tooth map where coupling appears in the place of the γ term. The system of two coupled maps to be studied is

$$\begin{aligned} x_1(t+1) &= \{e^{KF(x_1,x_2)}x_1(t)(1-\beta x_1(t)\} \\ x_2(t+1) &= \{e^{KF(x_1,x_2)}x_2(t)(1-\beta x_2(t)\} \\ F(x_1,x_2) &= \tfrac{1}{2}\left[x_1(t)+x_2(t)\right], \end{aligned} \tag{6.19}$$

where $\beta < 1/2$, so that there is no maximum inside the unity interval (and the universality class of the usual saw-tooth map is retained). It has been shown that this system can undergo full synchronization [Manrubia and Mikhailov (2000a)]. Note that, however, the dynamics of the synchronous state $s \equiv x_1 = x_2$ is not equivalent to that of the single oscillator, since the coefficient γ is replaced by a time-dependent variable,

$$s(t+1) = \{e^{Ks(t)} s(t)(1 - \beta s(t))\}. \tag{6.20}$$

This equation has a fixed point solution for small the coupling intensity $K < -\ln(1-\beta)$, while for larger values the dynamics is chaotic. If the coupling strength K exceeds this threshold (which depends on β), the fully synchronous state looses its stability. The stability of the fixed point $s = 0$ is marginal, since the derivative at the origin is always equal to unity. This results in long intervals of time spent close to the origin before a "spike" occurs (see Figs. 6.3 and 6.4).

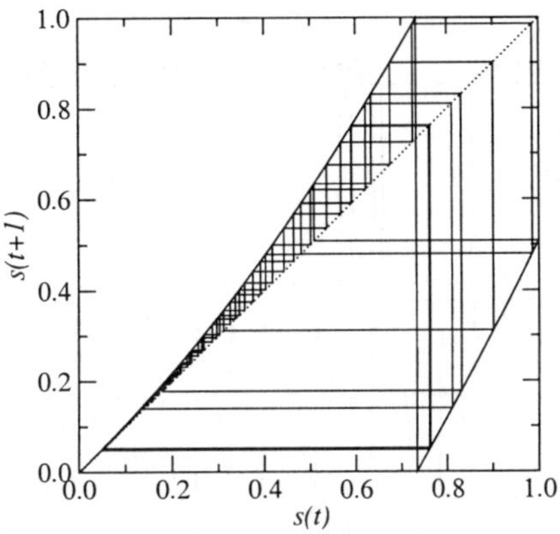

Fig. 6.3 Chaotic synchronous trajectory for $K = 0.7$ and $\beta = 0.25$. The phase space is bound by the function $\{e^{Ks} s(1 - \beta s)\}$.

To quantify correlations in this system, the average difference between the positions of the two oscillators, $\Delta x = |x_1 - x_2|$, can be used. This permits to distinguish three different phases, which after inspection of the dynamics turn out to be: (i) full synchronization corresponding to $\Delta x \simeq 0$ at any time step, (ii) antiphase dynamics with $\Delta x \simeq 0.5$, and small fluctuations around this value as time elapses, and (iii) an asynchronous phase with average distance between the states $\langle \Delta x \rangle \simeq 0.25$ and large fluctuations in its magnitude. These three regimes are illustrated in Fig. 6.4, where the average distance Δx as a function of the coupling intensity K is also displayed.

In this model, increasing K leads to an acceleration of the motion. The oscillators spike at a higher rate the larger K is, and the stable fixed point at low K transforms into chaotic (synchronous) dynamics and eventually into asynchronous dynamics for large enough K.

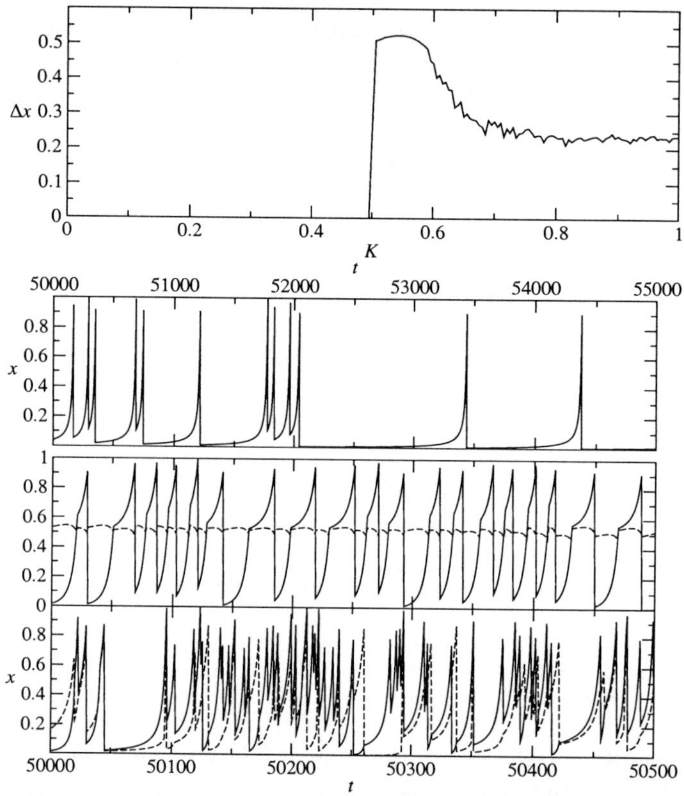

Fig. 6.4 Different dynamical behaviors of the two coupled saw-tooth maps (6.19). Above, the time-averaged difference $\langle \Delta x(t) \rangle = \langle x_1(t) - x_2(t) \rangle$ is shown. Below, three representative time series in the synchronous state (top, $K = 0.4$), in the antiphase state (middle, $K = 0.55$), and in the asynchronous phase (bottom, $K = 0.9$). The solid line corresponds to the state of one of the oscillators; the dashed line stands for the difference $\Delta x(t)$. Note the change in the temporal scale: due to the higher frequency of spikes, the two plots at the bottom represent a ten-fold enlargement with respect to the top one.

6.3 Synchronization of Two Coupled Oscillators

As we have already said, in order to have chaotic dynamics in a continuous-time system, at least three variables describing the internal state of an oscillator are needed. In this section we discuss briefly the joint dynamics of two coupled oscillators. Such studies have to be performed numerically, and in the considered case amount to the study of six coupled nonlinear differential equations.

Let us consider two identical Rössler oscillators $\mathbf{r}_1(t)$ and $\mathbf{r}_2(t)$ which are symmetrically coupled with respect to their variables x,

$$\begin{aligned} \dot{x}_{1,2} &= -y_{1,2} - z_{1,2} + K(x_{2,1} - x_{1,2}) \\ \dot{y}_{1,2} &= x_{1,2} + ay_{1,2} \\ \dot{z}_{1,2} &= b + z_{1,2}(x_{1,2} - c). \end{aligned} \tag{6.21}$$

These oscillators are able to fully synchronize above a coupling threshold which depends on the parameters a, b, and c. Figure 6.5 shows the difference $\Delta x = |x_1 - x_2|$ between the variables x of the oscillators for three values of K. Whenever the coupling is acting (i.e., for any $K > 0$), some dependence between the trajectories of the oscillators is present. For low coupling strength, this dependence might be subtle and hard to recognize. As the intensity of coupling increases, it is possible to identify a number of different states characterized by an increasing degree of correlation between the two oscillators.

In fact, the different types of synchronous behavior that have been described in the literature refer to different kinds of functional relationships (correlations) among the involved elements. In the case of two Rössler oscillators, we observe that full synchronization is not achieved for $K = 0.1$ (see Fig. 6.5), though it seems clear that the coupling induces some form of dependent evolution. This correlated dynamics could be quantified as the decrease in the average Δx with respect to the uncoupled case. Such a decrease can be also detected for a coupling strength as low as $K = 0.01$.

For two coupled oscillators, the invariant manifold corresponding to the synchronous state is stable if the differences in the individual variables $x_\perp(t) \equiv x_1(t) - x_2(t)$ vanish as $t \to \infty$, as well as the differences of the analogously defined $y_\perp(t)$ and $z_\perp(t)$ [Pecora *et al.* (1997)]. In order to analyze the stability of the synchronous state, it is useful to work with the new variables $\mathbf{r}_\perp(t)$ (on the transverse manifold) and $\mathbf{r}_\parallel(t) \equiv \mathbf{r}_1(t) + \mathbf{r}_2(t)$ (on the synchronization manifold). The transverse manifold has its own equations

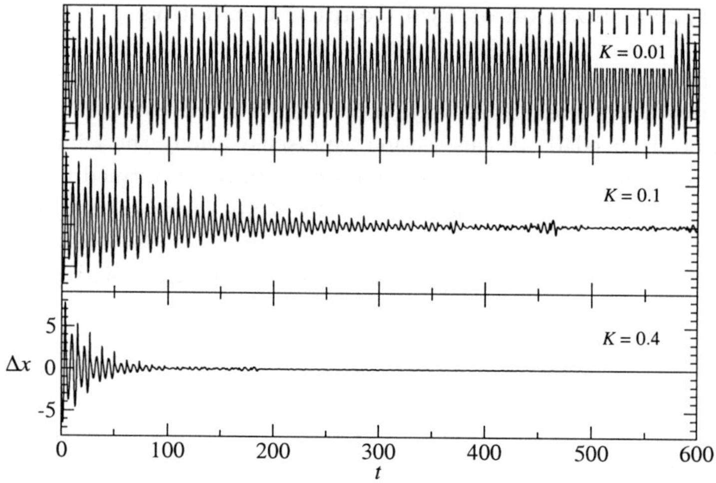

Fig. 6.5 Average difference Δx between the coordinates x of two coupled Rössler oscillators for three values of the coupling intensity K. Parameters are $\nu = 1$, $a = 0.2$, $b = 0.3$, $c = 4.5$, which yield two-banded chaotic dynamics (see also Fig. 6.2). The coupling was turned on at $t = 0$ after the two systems had relaxed to the invariant chaotic attractor.

of motion in these coordinates, and the analysis of its stability amounts to showing that the dynamical subsystem $\mathbf{r}_\perp(t)$ is stable at the point $\{0, 0, 0\}$. This immediately implies that the reduced manifold on which the synchronized dynamics takes place is an attractor for the dynamics of the coupled system.

Let us thus change to the variables $\{\mathbf{r}_\perp, \mathbf{r}_\parallel\}$ and linearize around the fixed point $\{x_\perp, y_\perp, z_\perp\} = \{0, 0, 0\}$. This operation yields the evolution equation

$$\begin{pmatrix} \dot{x}_\perp \\ \dot{y}_\perp \\ \dot{z}_\perp \end{pmatrix} = \begin{pmatrix} -2K & -1 & -1 \\ 1 & a & 0 \\ z_\parallel/2 & 0 & x_\parallel/2 - c \end{pmatrix} \begin{pmatrix} x_\perp \\ y_\perp \\ z_\perp \end{pmatrix}. \tag{6.22}$$

The matrix on the right-hand-side of this equation corresponds to the Jacobian matrix $\mathbf{Df}(s)$. Note its explicit dependence with the synchronous trajectory $s(t)$ through the variables $x_\parallel$ and $z_\parallel$. If the largest transverse exponent $\lambda_\perp$ corresponding to $\mathbf{Df}(s)$ is negative, then any perturbation perpendicular to the synchronized manifold will be damped down and the synchronous state will be stable. Since, in general, $\lambda_\perp$ depends on the variables along the synchronous manifold, it is also called *conditional Lyapunov exponent.* This example highlights the non-trivial interplay between the coupling strength, the nature of the synchronous dynamics, and its stability. Here, an exact analytical study of the stability is not feasible, and at this point one has to resort to the numerical computation of the Lyapunov exponents for the system (6.22).

6.3.1 *Phase synchronization*

Phase synchronization is a weaker kind of coherent motion possible in chaotic systems. The oscillators of a coupled system are synchronized in phase when their phases are locked, though their amplitudes evolve chaotically and remain weakly correlated. We describe this effect and some properties of the transition following the original work where it was first reported [Rosenblum *et al.* (1996)].

Consider two coupled Rössler systems whose rotation frequencies on the (x, y) plane take different values, $\nu_1 \neq \nu_2$,

$$\begin{aligned} \dot{x}_{1,2} &= -\nu_{1,2} y_{1,2} - z_{1,2} + K(x_{2,1} - x_{1,2}) \\ \dot{y}_{1,2} &= \nu_{1,2} x_{1,2} + 0.15 y_{1,2} \\ \dot{z}_{1,2} &= 0.2 + z_{1,2}(x_{1,2} - 10). \end{aligned} \tag{6.23}$$

The phase of each of the oscillators is calculated according to Eq. (6.10) with $(x_0, y_0) = (0, 0)$, and the phase difference is defined as

$$\Delta\phi(t) \equiv \phi_1(t) - \phi_2(t). \tag{6.24}$$

Figure 6.6 illustrates the onset of the phase synchronization transition for different values of the coupling, as well as the relation between the variables x of both oscillators.

Before phase synchronization is reached, the phases of the two oscillators stay locked only within finite time intervals whose length increases monotonously with the coupling strength K. The difference between the phases experiences regularly spaced jumps of size 2π. Above a critical value K_c the phases remain permanently locked, while the amplitudes of the individual oscillators evolve in a quite independent way. This can be seen in Fig. 6.6, where trajectories are represented in the plane (x_1, x_2). As

K increases, there is as well an increase in the correlations between the amplitudes.

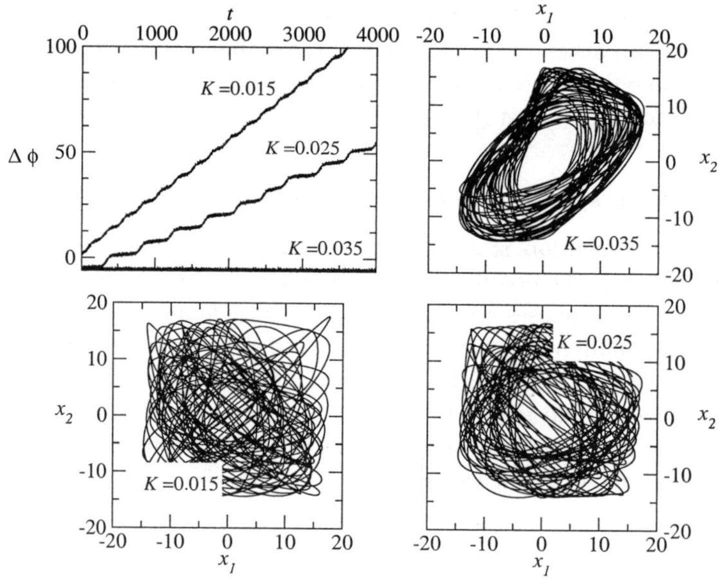

Fig. 6.6 Phase synchronization of two coupled Rössler oscillators. Upper left: evolution of the phase difference between the oscillators for different values of the coupling constant K. Other plots represent the phase space of the variables $x_{1,2}$ for the same parameters. The individual frequencies are $\nu_1 = 1.015$ and $\nu_2 = 0.985$.

The evolution equation (6.12) can be generalized to coupled systems. Under certain conditions, a simplified evolution equation for the difference in the phases of the two oscillators can be derived, which helps to qualitatively understand the nature of phase synchronization. First, one should write the dynamical equations for each of the oscillators in terms of the variables $\phi_{1,2}$ and $A_{1,2}$. Then, the fact that in the Rössler system the rotation frequency is only weakly dependent on the amplitude can be taken into account to average over rotations of the phases $\phi_{1,2}$. Next, the slowly varying phases $\theta_{1,2}$, defined as $\theta_{1,2} = \phi_{1,2} - \omega t$, where $\omega = (\omega_1 + \omega_2)/2$, are introduced. The final equation reads

$$\dot{\theta}_1 - \dot{\theta}_2 = 2(\omega_1 - \omega_2) - \frac{K}{2}\left(\frac{A_2}{A_1} + \frac{A_1}{A_2}\right)\sin(\theta_1 - \theta_2). \qquad (6.25)$$

If the amplitudes are considered as approximately constant, this equation has a stable fixed point at

$$\theta_1 - \theta_2 = \arcsin\left(\frac{4(\omega_1 - \omega_2)A_1A_2}{K(A_1^2 + A_2^2)}\right) \tag{6.26}$$

which corresponds to the phase locking of the Rössler system. Note that the last term in (6.25) can be effectively viewed as a noisy external source that perturbs the coherent evolution of the system [Pikovsky *et al.* (2001)]. In a way, chaotic systems generate internally there own noise, and in some situations the response of the dynamics to these different types of disorder (chaos and noise) is qualitatively similar. For the Rössler system, however, $\dot{\phi} \simeq \omega$, so the effective noise is negligible. Hence, Eq. (6.25) can be compared to Eq. (2.8) derived in the context of phase oscillators.

The spectrum of Lyapunov exponents has a quantitative change when phase synchronization appears [Rosenblum *et al.* (1996); Rosenblum *et al.* (1997)]. A single Rössler oscillator in its chaotic phase has one positive, one zero, and one negative Lyapunov exponent. When two oscillators are coupled, the coupling term introduces attraction between their dynamics, and the values of the Lyapunov exponents are lowered as the coupling strength K increases. The first degree of freedom that the system looses is that of independent shifts in the phase variable for each of the oscillators. When the two phases become locked, the Lyapunov exponent corresponding to relative phase shifts becomes negative, quantifying in this way the onset of the phase synchronization transition. The other exponents keep their previous signs.

There is also a complementary explanation of the mechanism of the phase synchronization transition [Lee et al. (1998)]. It has been observed that, before reaching full phase locking, the phase difference between the two oscillators showed regularly spaced jumps of size 2π. As the transition is approached, these jumps lose their regularity and happen at random intervals. This behavior can be seen in Fig. 6.6. Using a reduced effective model with the slow phases of Eq. (6.25) as variables, it can be shown that, before the transition takes place, the variable $\Delta\theta = \theta_1 - \theta_2$ behaves like a coordinate of an over-damped particle moving in a "noisy washboard potential." In this explanation, the noise indeed comes from the amplitudes, and the change from regular to irregular jumps is due to a saddle-node bifurcation preceding the full phase locking.

The range of frequency mismatch $\Delta\nu = \nu_1 - \nu_2$ for which phase synchronization can be achieved in model (6.23) is broad. Differences up to 20% can be overcome with a coupling intensity $K \simeq 0.17$. When the strength of coupling is further increased above the phase synchronization threshold, correlations between the amplitudes develop. Eventually, a new type of transition sets in: the two oscillators follow closely the same orbit but there is a time delay between their states. This type of synchronization is known as lag synchronization.

6.3.2 *Lag synchronization*

Lag synchronization constitutes a different kind of coherent evolution of coupled chaotic systems. As K is increased, the effect of the coupling grows and at a certain point one of the oscillators begins to "echo" the dynamics of the other: their trajectories are almost the same, but the precise position of the first oscillator is only reached by the second oscillator after a fixed time delay, such that $r_1(t) \simeq r_2(t-\tau)$. Both the time delay τ and the difference in amplitude between the two oscillators can be measured through the comparison of the consecutive crossings of a given Poincaré section. As an example, consider the section $y_{1,2} = 0$ of the chaotic attractor, where the amplitude is directly given by the coordinate x, $A_{1,2} = x_{1,2}$. The two oscillators cross this section at values $x_1(t_1)$ and $x_2(t_2)$. Let us consider the differences $\Delta x_P = |x_1(t_1) - x_2(t_2)|$ and $\tau_P = |t_1 - t_2|$. The time difference τ_P is proportional to the phase difference $\Delta\phi$ between the two oscillators, $\tau_P \propto \Delta\phi \equiv |\phi_1(t_1) - \phi_2(t_2)|$. When the system is phase synchronized, such that $\Delta\phi$ becomes constant, τ_P takes a constant value as well, irrespectively of the crossing times $t_{1,2}$. When lag synchronization is achieved, the value $\Delta x_P \simeq 0$ while $\tau_P > 0$. If the system eventually attains full synchronization, then also $\tau_P \simeq 0$. Fig. 6.7 shows schematically the three types of synchronization according to this criterion.

In the lag synchronized regime, the time τ_P directly corresponds to the minimum of the cross-correlation function, and this can be used to obtain the time delay between the dynamics of the two oscillators [Rosenblum *et al.* (1997)]. In Fig. 6.8 we show how Δx_P and τ_P vary with increasing coupling strength K. The transition to lag synchronization is observed as a jump in the average value of Δx_P, which afterwards attains values close to zero (note the logarithmic scale). The time difference between two successive crossings of the Poincaré section decreases as $\tau_P \propto K^{-1}$, and, apparently, does not stabilize around any finite value. This means that, even for large values of the coupling, full synchronization is only asymptotically approached for

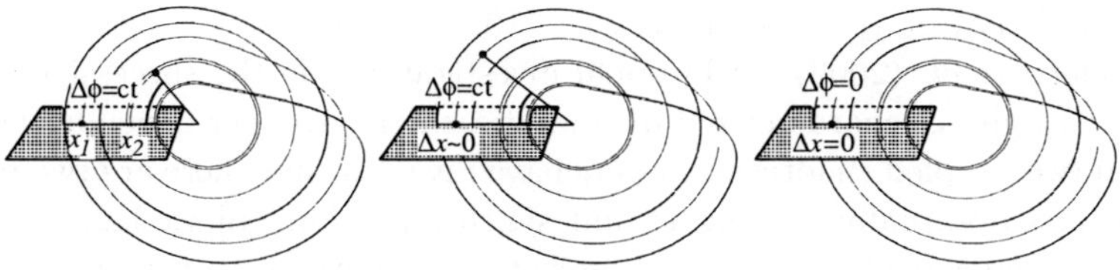

Fig. 6.7 Schematic representation of phase, lag, and full synchronization using consecutive crossings of the Poincaré section for the two oscillators. In phase synchronization, the difference $\Delta\phi \propto \tau_P$ takes a fixed value but the amplitudes at the crossing points are basically uncorrelated; in lag synchronization $\Delta x \simeq 0$. If full synchronization is achieved, both oscillators cross the section simultaneously and $\Delta\phi = \Delta x = 0$.

two Rössler oscillators with $\nu_1 \neq \nu_2$, but never exactly reached.

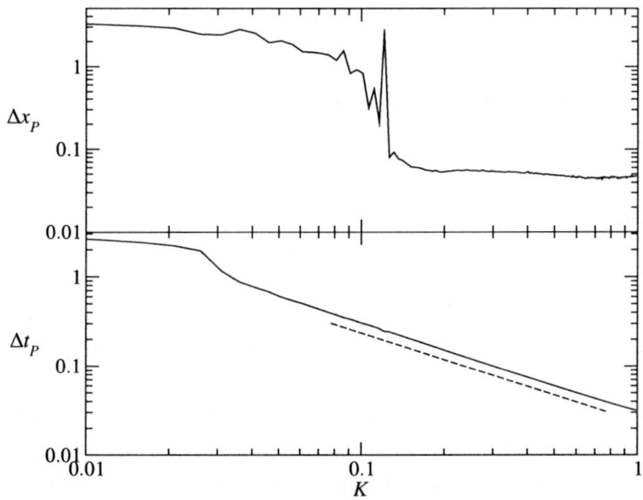

Fig. 6.8 Lag synchronization of two coupled Rössler oscillators is achieved when $\Delta x_P \simeq 0$. Below, the decrease of Δt_P with K is shown. The dashed line has slope -1. The parameter values are the same as in the previous section.

Similar to phase synchronization, the onset of lag synchronization is related to modifications of the Lyapunov spectrum. Recall that the Lyapunov exponent corresponding to relative phase shifts becomes negative when the phase synchronization transition begins. Close to the lag synchronization transition, one of the positive exponents changes sign. However, the appearance of lag synchronization is not exactly signaled by the quantitative change in the exponent due to the presence of intermittent behavior [Rosenblum *et al.* (1997); Sosnovtseva *et al.* (1999); Boccaletti *et al.* (2002)]. Additional investigations on the successive transitions to phases characterized by an increasingly stronger dependence between two coupled chaotic oscillators have revealed that there are two main sequences eventually leading to full synchronization: phase-lag-full synchronization and phase-full synchronization mediated by a transient phase characterized by large intermittent bursts [Rim *et al.* (2002)]. In the first case the full synchronization phase is attained only asymptotically, while in the second route it is observed for a broad domain of parameters.

The existence of regimes where the two oscillators bear an increasingly large similarity in their dynamics as the coupling becomes stronger has its parallel in large ensembles of coupled oscillators. The increase in similarity in this latter case is manifested as clustering behavior, already discussed for periodic oscillators in the first part of this book.

6.3.3 *Synchronization in the Lorenz system*

The first numerical investigations of chaotic behavior were carried out in the early 1960's. The meteorologist Edward Lorenz was studying a simplified model of atmospheric convection when he numerically observed a qualitative signature of chaotic behavior: the system presented extreme sensitivity to initial conditions [Lorenz (1963)]. The Lorenz model is a set of three differential equations sharing some properties with the Rössler system. However, in this model a phase cannot be easily identified and the frequency of the oscillations depends strongly on their amplitude.

Theoretical and numerical analysis of two identical, coupled Lorenz oscillators, showed that full synchronization can be achieved in this system [Fujisaka and Yamada (1983)]. More recently, two different Lorenz oscillators with a coupling analogous to that discussed for the Rössler system have been studied [Lee et al. (1998); Rim *et al.* (2002)]:

$$\begin{aligned} \dot{x}_{1,2} &= 10(y_{1,2} - x_{1,2}) + K(x_{2,1} - x_{1,2}) \\ \dot{y}_{1,2} &= b_{1,2}x_{1,2} - y_{1,2} - x_{1,2}z_{1,2} \\ \dot{z}_{1,2} &= -\tfrac{8}{3}z_{1,2} + x_{1,2}y_{1,2}. \end{aligned} \tag{6.27}$$

Typical trajectories of a single Lorenz oscillator are shown in Fig 6.9. When two non-identical oscillators are coupled, coherent behaviors more weakly correlated than full synchronization may be expected. A difficulty encountered in the investigation of this system is how to suitably define the phase.

One possibility consists in taking advantage of the reflection symmetry of the equations (note that they are invariant under the change $x \to -x$, $y \to -y$) and defining a new variable $u = \sqrt{x^2 + y^2}$. Figure 6.9 shows how the two characteristic lobes of the attractor in the (x, z) projection are mapped into a single one in the (u, z) plane.

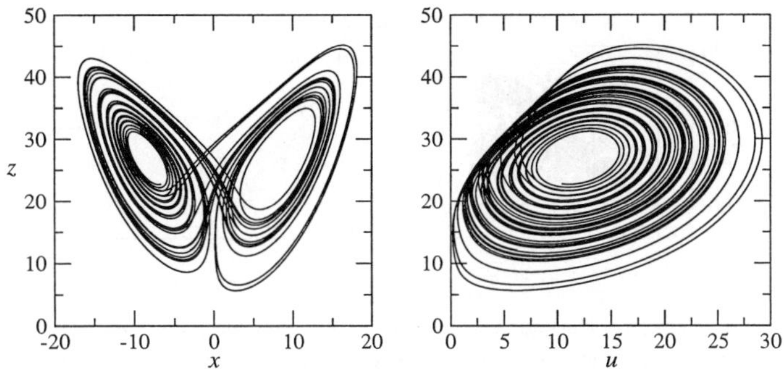

Fig. 6.9 Chaotic attractor of the Lorenz system. The projection on the (x, z) plane shows a bi-lobular figure where no clear phase is evident. Using the coordinate $u = \sqrt{x^2 + y^2}$ a topologically equivalent attractor with a clear rotation center can be obtained. In this example we have used Eqs. (6.27) with $K = 0$ and $b = 28$.

A major difference with the Rössler system consists in the irregular pace of rotations around the lobes in the Lorenz system. To be synchronized in phase, two Lorenz oscillators need to adjust much more strongly their individual dynamics as compared to those systems where the principal rotation frequency is sharply defined. Indeed, numerical investigations reveal that when the phases of two Lorenz oscillators are locked, also the amplitudes are. This coherent dynamical state is however interrupted by intermittent bursts where the two oscillators desynchronize. Phase synchronization is not observed in the Lorenz system.

An example of how the transition to synchronization in the Lorenz system proceeds is shown in Fig. 6.10. Let us consider a Poincaré section at $u = 10$. The differences in the amplitudes when crossing this section can be represented by the variable z. If, as for two coupled Rössler oscillators, the crossing times are significantly different, while the amplitudes are almost equal, lag synchronization would have been present. This is however not the case for the Lorenz system. As seen in Fig 6.10, there is a sudden decrease in both variables at a certain coupling intensity. It is also remarkable that the values of these variables after the transition remain significantly larger

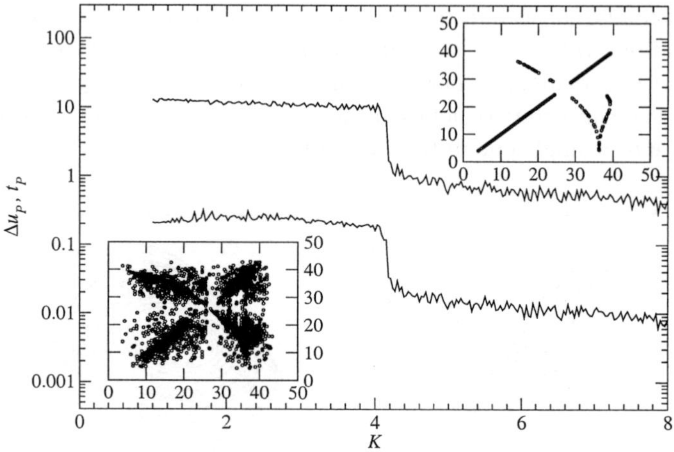

Fig. 6.10 Transition to the synchronous state for two coupled, non-identical Lorenz systems. We have used the Poincaré section method described with the section defined by the plane $u = 10$. A large decrease in the average difference in time crossing (upper curve) and amplitude at crossing (lower curve) for the two oscillators occurs simultaneously at $K \simeq 4.13$. The two insets show the two values of the coordinates x_1 and x_2 before the lag synchronization transition (left) and after it (right). The time intervals of full synchronization yield values along the diagonal $x_1 = x_2$, while off-diagonal values correspond to intermittent bursts.

than zero. The insets in Fig. 6.10 display the two values of the amplitudes when crossing the Poincaré section. Being highly uncorrelated before the transition, they almost fall on the diagonal line once the oscillators become entrained. Occasional intermittent bursts explain the appearance of non-diagonal points in the insets.

Different investigations have shown that the synchronous state where both phases and amplitudes are locked exhibits on-off intermittency: desynchronization bursts interrupt periods of laminar behavior where the two oscillators display almost full synchronization. The probability $P(l)$ that the length of time not interrupted by one of such bursts is l behaves as $P(l) \propto l^{-3/2}$, which is characteristic for on-off intermittency [Rim *et al.* (2002)].

Chapter 7

Synchronization in Populations of Chaotic Elements

Chaotic dynamics introduces new degrees of freedom in coupled ensembles. At least three variables are required for a continuous time system to behave chaotically, and the orbit followed by each element cannot be predicted beyond a time inversely proportional to the corresponding Lyapunov exponent. This makes difficult the analytical treatment of ensembles of chaotic elements, since there are no exact expressions describing their position at a given time. Often, numerical simulations are the only way to approach the study of these populations.

The introduction of coupling in chaotic systems can drastically change the qualitative properties of the dynamics. It is not possible to predict beforehand the consequences of coupling. It can stabilize periodic behavior, occasionally produces hidden correlations among the elements —though the dynamics is apparently turbulent— or it may induce the synchronization of a subset of dynamic variables.

Still, despite the many degrees of freedom involved and the complex phenomenology of ensembles of chaotic elements, it is possible to obtain a number of analytical results when the oscillators are all identical. Under very general conditions, full synchronization arises. The first part of this chapter presents stability analysis for systems of identical elements and discusses the powerful method of the master stability function.

Numerical analysis of heterogeneous, chaotic populations, reveals that synchronization is still possible and holds frequently. Neither chaos nor disorder can destroy the trend of the elements in coupled ensembles to follow each other, provided the coupling is strong enough. Synchronization of chaos is thus a robust property expected to hold both in natural and model systems.

7.1 Ensembles of Identical Oscillators

The examples presented in Chapter 6 constitute a small sample out of a wealth of different dynamical behaviors that can be observed in coupled chaotic systems. The first general studies on chaos synchronization were mainly analytical and addressed ensembles of identical oscillators [Yamada and Fujisaka (1983); Yamada and Fujisaka (1984)].

We begin this chapter with the study of arbitrarily large arrays of identical chaotic oscillators [Fujisaka and Yamada (1983)]. A number of analytical results can be obtained for these systems if the coupling function fulfills certain symmetry properties, in which case the stability of the synchronization manifold can be related to the Lyapunov exponent of the single oscillator. In this section we present some results obtained for the case of *shift-invariant* coupling among the elements and describe the method of master stability functions.

Standard Lyapunov exponents have to be estimated numerically in all but a few particular cases. The same holds for the Lyapunov exponents of perturbations transverse to the full synchronization manifold: there is no general procedure to predict their values. However, some relations between the Lyapunov exponent of the single oscillator and the stability of the fully synchronous state can be drawn if the coupling among a set of identical dynamical elements is of the shift-invariant type. Under this general coupling scheme, the number of independent quantities characterizing the dynamical state is effectively reduced.

The initial theory on the stability of synchronized, chaotic motion [Fujisaka and Yamada (1983)], was generalized later to arbitrary dynamical systems [Heagy *et al.* (1994a)]. Consider the set of autonomous equations of motion defined by

$$
\begin{aligned}
\dot{\mathbf{r}}_0(t) &= \mathbf{f}(\mathbf{r}_0) + K\mathbf{F}_0(\mathbf{r}_0, \mathbf{r}_1, \ldots, \mathbf{r}_{N-1}) \\
\dot{\mathbf{r}}_1(t) &= \mathbf{f}(\mathbf{r}_1) + K\mathbf{F}_1(\mathbf{r}_0, \mathbf{r}_1, \ldots, \mathbf{r}_{N-1}) \\
&\vdots \\
\dot{\mathbf{r}}_{N-1}(t) &= \mathbf{f}(\mathbf{r}_{N-1}) + K\mathbf{F}_{N-1}(\mathbf{r}_0, \mathbf{r}_1, \ldots, \mathbf{r}_{N-1}).
\end{aligned}
\tag{7.1}
$$

Each of the variables $\mathbf{r}_i$ is a vector of dimension n whose components take real values. Hence, the system (7.1) has dimension nN. In order to ensure that the single oscillator dynamics is a solution of the complete system when full synchronization is achieved, the coupling functions are chosen in such a way that $\mathbf{F}_i(\mathbf{s}, \mathbf{s}, \ldots, \mathbf{s}) \equiv 0$, where $\mathbf{s}(t) \equiv \mathbf{r}_1(t) = \mathbf{r}_2(t) = \cdots = \mathbf{r}_N(t)$ is the fully synchronous trajectory. We say that a configuration is shift-invariant if the interaction functions $\mathbf{F}_i$ do not vary from one oscillator to another,

$$\mathbf{F}_i(\mathbf{r}_j, \mathbf{r}_{j+1}, \ldots, \mathbf{r}_{j+N-1}) = \mathbf{F}_{i+1}(\mathbf{r}_{j-1}, \mathbf{r}_j, \ldots, \mathbf{r}_{j+N-2}), \tag{7.2}$$

where indices have to be taken mod N. Note that two important cases where this property is fulfilled are

(1) nearest neighbor diffusive coupling in a linear array of oscillators with periodic boundary conditions: $\mathbf{F}_i = \mathbf{r}_{i-1} - 2\mathbf{r}_i + \mathbf{r}_{i+1}$, and
(2) global coupling through differences in the coordinates, $\mathbf{F}_i = N^{-1} \sum_{k=1}^{N} \tilde{\mathbf{F}}(\mathbf{r}_k - \mathbf{r}_i)$, with $\tilde{\mathbf{F}}$ a vector function satisfying $\tilde{\mathbf{F}}(\mathbf{0}) = \mathbf{0}$.

The basic step in determining the stability of the state $\mathbf{s}(t)$ consists in transforming the dynamical system (7.1) to a coordinate system where perturbations along the synchronous manifold and perturbations perpendicular to it are expressed through different variables. In this respect, the approach that follows is a generalization of the simple case outlined in Sec. 6.3.

Consider first a linearization of (7.1) about the synchronized state $\mathbf{s}(t)$,

$$\frac{d\boldsymbol{\zeta}_i}{dt} = \mathbf{Df}(\mathbf{s})\boldsymbol{\zeta}_i + K \sum_{j=0}^{N-1} \mathbf{D}_j \mathbf{F}_i(\mathbf{s}, \mathbf{s}, \ldots, \mathbf{s})\boldsymbol{\zeta}_j, \tag{7.3}$$

for $i = 0, \ldots, N-1$, and where the variables $\boldsymbol{\zeta}_i = \mathbf{r}_i - \mathbf{s}$ represent small deviations for each of the variables with respect to the synchronous trajectory; $\mathbf{Df}(\mathbf{s})$ is the Jacobian of the vector field $\mathbf{f}$ evaluated along the synchronous trajectory and, similarly, $\mathbf{D}_j\mathbf{F}_i(\mathbf{s}, \mathbf{s}, \ldots, \mathbf{s})$ is the differential operator acting on the coordinates of the j−th oscillator in the function $\mathbf{F}_i$ and evaluated along $\mathbf{s}(t)$. The shift-invariant property (7.2) permits to use only the derivatives of the coupling function of one of the oscillators, yielding

$$\frac{d\boldsymbol{\zeta}_i}{dt} = \mathbf{Df}(\mathbf{s})\boldsymbol{\zeta}_i + K \sum_{j=0}^{N-1} \mathbf{D}_{j-i} \mathbf{F}_0(\mathbf{s}, \mathbf{s}, \ldots, \mathbf{s})\boldsymbol{\zeta}_j. \tag{7.4}$$

Now we define the circular sequence

$$\{\mathbf{H}_i\}_{i=0}^{N-1} = \{\mathbf{Df}(\mathbf{s}) + K\mathbf{D}_0\mathbf{F}_0|_\mathbf{s}, K\mathbf{D}_{N-1}\mathbf{F}_0|_\mathbf{s}, K\mathbf{D}_{N-2}\mathbf{F}_0|_\mathbf{s}, \ldots, K\mathbf{D}_1\mathbf{F}_0|_\mathbf{s}\}, \tag{7.5}$$

where $\cdot|_\mathbf{s}$ indicates evaluation of the function along $\mathbf{s}(t)$, in order to write the equations for small perturbations (7.3) in the more compact form

$$\frac{d\zeta_i}{dt} = \sum_{j=0}^{N-1} \mathbf{H}^{i-j}\zeta_j. \tag{7.6}$$

Introducing discrete Fourier transforms of (7.5) and the sequence $\{\zeta_0, \zeta_1, \ldots, \zeta_{N-1}\}$,

$$\mathbf{Y}_k = \frac{1}{\sqrt{N}} \sum_{j=1}^{N} \mathbf{H}_j \mathrm{e}^{2\pi ijk/N}$$

$$\eta_k = \frac{1}{\sqrt{N}} \sum_{j=1}^{N} \zeta_j \mathrm{e}^{2\pi ijk/N}, \tag{7.7}$$

the convolution (7.6) becomes block-diagonal, and the transformed variational equations are given by

$$\frac{d\eta_k}{dt} = \sqrt{N}\mathbf{Y}_k\eta_k, k = 0, 1, \ldots, N-1. \tag{7.8}$$

When full synchronization is achieved, the dynamics of the whole system is reduced to a manifold of dimension n, corresponding to the dimensionality of the single oscillator. Since the vector $\mathbf{r}_\parallel = \sum_{i=0}^{N-1} \mathbf{r}_i$ is within the synchronization manifold (compare it with the analogous variable defined in Sec. 6.3), the variable that now determines perturbations along the synchronous trajectory is η_0 (this can be seen by setting $k = 0$ in Eq. (7.7)), while the rest of the variables η_i, $i = 1, 2, \ldots, N-1$ stand for perturbations transversal to $\mathbf{s}(t)$ and thus determine its stability.

We are particularly interested in studying those systems where the variation of η_0 obeys the dynamics of the single, uncoupled oscillator. This happens when the condition

$$\sum_{j=0}^{N-1} \mathbf{F}_j(\mathbf{s}, \mathbf{s}, \ldots, \mathbf{s}) = \text{const}, \tag{7.9}$$

is satisfied. In this case we have

$$\dot{\boldsymbol{\eta}}_0 = [D\mathbf{f}(\mathbf{s})]\boldsymbol{\eta}_0, \tag{7.10}$$

which is identical to Eq. (6.4) and represents the equation used to compute the Lyapunov exponents of a single oscillator along the trajectory $\mathbf{s}(t)$.

Let us now introduce the time evolution operator $\Lambda_0(t)$ for the variable $\boldsymbol{\eta}_0$, so that its dynamics can be formally expressed as

$$\begin{aligned} \boldsymbol{\eta}_0(t) &= \Lambda_0(t)\boldsymbol{\eta}_0(0) \\ \Lambda_0(0) &= 1. \end{aligned} \tag{7.11}$$

This evolution operator is

$$\Lambda_0(t) = \mathcal{T} \exp\left[\int_0^t \mathbf{Df}(\mathbf{s}(t'))dt'\right], \tag{7.12}$$

where $\mathcal{T}$ is the time-ordering operator. It might be clarifying to compare this equation with Eq. (6.6). The n eigenvalues of $\Lambda_0(t)$ form a set $\mu_0^i(t)$, $i = 1, \ldots, n$, and from those the Lyapunov exponents corresponding to the dynamics (7.10) are obtained as

$$\lambda_0^i = \text{Re}\left(\lim_{t\to\infty} \frac{1}{t} \ln \mu_0^i(t)\right), i = 1, \ldots, n. \tag{7.13}$$

The evolution equations for the transversal perturbations satisfy

$$\frac{d\boldsymbol{\eta}_k}{dt} = \left(\sum_{j=0}^{N-1} \mathbf{H}_j e^{2\pi ijk/N}\right) \boldsymbol{\eta}_k, k = 1, \ldots, N-1. \tag{7.14}$$

Denoting as $\Lambda_k(t)$ the time evolution operator for the $k-$th transverse variation and as $\mu_k^i(t)$ the corresponding eigenvalues, the transverse Lyapunov exponents are finally

$$\lambda_k^i = \mathrm{Re}\left(\lim_{t\to\infty}\frac{1}{t}\ln\mu_k^i(t)\right), k = 1, \ldots, N-1. \tag{7.15}$$

Further results cannot be derived without specifying the coupling functions.

This section is closed by considering an explicit example of global coupling among the oscillators. Using the procedure described above, transverse Lyapunov exponents can be expressed in this case as as functions of the exponents along the synchronous manifold. Consider the following global coupling,

$$\mathbf{F}_i(\mathbf{r}_0, \mathbf{r}_1, \ldots, \mathbf{r}_{N-1}) = \frac{1}{N}\sum_{k=0}^{N-1}\tilde{\mathbf{F}}(\mathbf{r}_k - \mathbf{r}_i), \tag{7.16}$$

which yields the matrix sequence

$$\{\mathbf{H}_i\}_{i=0}^{N-1} = \left\{\mathbf{Df}(\mathbf{s}) - \frac{K(N-1)}{N}\mathbf{D\tilde{F}}(\mathbf{0}), \frac{K}{N}\mathbf{D\tilde{F}}(\mathbf{0}), \frac{K}{N}\mathbf{D\tilde{F}}(\mathbf{0}), \ldots, \frac{K}{N}\mathbf{D\tilde{F}}(\mathbf{0})\right\}. \tag{7.17}$$

The equations for transversal perturbations take now the form

$$\frac{d\eta_k}{dt} = \left(\mathbf{Df}(\mathbf{s}) - K\mathbf{D\tilde{F}}(\mathbf{0}) + \frac{K}{N}\mathbf{D\tilde{F}}(\mathbf{0})\sum_{j=0}^{N-1}e^{2\pi ijk/N}\right)\eta_k. \tag{7.18}$$

The sum in the right-hand side of the latter equation vanishes for any $k \neq 0$, leaving the simple relation

$$\dot{\eta}_k = \left(\mathbf{Df}(\mathbf{s}) - K\mathbf{D\tilde{F}}(\mathbf{0})\right)\eta_k. \tag{7.19}$$

Assume finally that the coupling function is proportional to the identity matrix, $\mathbf{D\tilde{F}}(\mathbf{0}) \propto \mathbf{I}$, (with a proportionality constant that can be absorbed into K), and compare this evolution equation with (7.10). It turns out that the time evolution operator factors in two terms (since the Jacobian matrix commutes with the identity matrix $\mathbf{I}$ at all times) and can be written as

$$\Lambda_k(t) = \Lambda_0(t)\exp(-Kt). \tag{7.20}$$

The transverse Lyapunov exponents turn out to be

$$\lambda_k^i = \lambda_0^i - K\,, \tag{7.21}$$

a relation that was already among the first analytical results regarding the stability of synchronous, chaotic motion [Fujisaka and Yamada (1983); Yamada and Fujisaka (1983)].

The latter result clearly reveals how the stability of the fully synchronous state in coupled chaotic systems is the outcome of two counter-acting effects: the instability due to chaotic dynamics and the attraction due to coupling. When the effect of coupling is strong enough to overcome the intrinsic instability of the dynamics, the fully synchronous state becomes stable. Note, however, that for less symmetrical global coupling schemes, which will be considered below, too intensive coupling can also destabilize synchronization.

7.1.1 *Master stability functions*

The introduction of master stability functions [Pecora and Carroll (1998); Fink *et al.* (2000)] represents a further step toward the derivation of general stability criteria for different connection topologies of linearly coupled, identical oscillators. In this method, the nN evolution equations for small perturbations are expressed in a block-diagonal form, with the blocks having a common structure. The knowledge of the stability domains for some general evolution equation, corresponding to an elementary $n \times n$ block, permits to predict the stability of the fully synchronous state under an arbitrary coupling scheme.

The generic dynamical system that will be investigated is described by

$$\dot{\mathbf{r}}_T = \mathbf{I}_N \otimes \mathbf{f}(\mathbf{r}) + K(\mathbf{G} \otimes \mathbf{E})\mathbf{r}_T, \tag{7.22}$$

where $\mathbf{I}_N$ is the $N \times N$ identity matrix, $\mathbf{f}(\mathbf{r})$ specifies the dynamics of a single uncoupled oscillator, $\mathbf{G}$ is an $N \times N$ matrix of coupling coefficients which contains the topology of the couplings, $\mathbf{E}$ is an $n \times n$ matrix which contains the information on the variables which are coupled, and, finally, $\otimes$ indicates the external product of the two matrices. A large number of systems can be expressed in the form (7.22).

Let us consider as an example the case of a ring of N Rössler oscillators coupled in their variables x, y and z. The components of the nN−dimensional vector $\mathbf{r}_T$ are the variables for the N oscillators,

$\mathbf{r}_T = (x_1(t), y_1(t), z_1(t), x_2(t), \ldots, x_N(t), y_N(t), z_N(t))$. The first term on the right-hand-side contains the dynamical equations of the N identical, uncoupled oscillators, where $\mathbf{f}(\mathbf{r})$ are the functions corresponding to the three variables describing a single Rössler system, as in Eq. 6.3. Suppose that each oscillator is coupled to its two nearest neighbors in each of the variables, such that the oscillator i in the ring is described by

$$\begin{aligned} \dot{x}_i &= -\nu y_i - z_i + K(x_{i-1} - 2x_i + x_{i+1}) \\ \dot{y}_i &= \nu x_i + a y_i + K(y_{i-1} - 2y_i + y_{i+1}) \\ \dot{z}_i &= b + z_i(x_i - c) + K(z_{i-1} - 2z_i + z_{i+1}) \end{aligned} \tag{7.23}$$

If the ring has periodic boundary conditions, the matrices $\mathbf{G}$ and $\mathbf{E}$ are

$$\mathbf{G} = \begin{pmatrix} -2 & 1 & 0 & \ldots & 0 & 1 \\ 1 & -2 & 1 & \ldots & 0 & 0 \\ \vdots & \vdots & \vdots & \ddots & \vdots & \vdots \\ 1 & 0 & 0 & \ldots & 1 & -2 \end{pmatrix}, \quad \mathbf{E} = \begin{pmatrix} 1 & 0 & 0 \\ 0 & 1 & 0 \\ 0 & 0 & 1 \end{pmatrix}. \tag{7.24}$$

When the external product between these two matrices is performed, each element of $\mathbf{G}$ is substituted by an $n \times n$ sub-matrix which is the product of that coefficient by the matrix $\mathbf{E}$. Thus, the resulting matrix has constant coefficients and dimension $nN \times nN$. This matrix, when multiplied by the vector $\mathbf{r}_T$, yields all the linear coupling terms in the dynamical system.

The matrix $\mathbf{G}$ specifies the set of couplings in the oscillator ensemble. For an homogeneous, global coupling, the elements of $\mathbf{G}$ take value $1/N$ except in the diagonal, where $G_{ii} = -1 + 1/N$. The matrix $\mathbf{E}$ contains information about the variables which are coupled. For example, if coupling is introduced only trough the variable y, then the coefficients E_{11} and E_{33} are set to zero. Many other situations are thus implemented as simple modifications of the matrices $\mathbf{G}$ and $\mathbf{E}$.

The equation for the evolution of small perturbations corresponding to (7.22) is

$$\dot{\boldsymbol{\zeta}} = (\mathbf{I}_N \otimes \mathbf{Df} + K\mathbf{G} \otimes \mathbf{E})\,\boldsymbol{\zeta}\,. \tag{7.25}$$

Now the analysis proceeds as in the previous section. The essential step is the diagonalization of the matrix $\mathbf{G}$, which should be performed depending on each particular problem. Note however that this diagonalization does not affect the first term, since it includes only the identity matrix. Once the diagonalization is performed, Eq. (7.25) can be written in the form

$$\dot{\boldsymbol{\eta}}_k = (\mathbf{Df} + K\gamma_k \mathbf{E})\, \boldsymbol{\eta}_k, \tag{7.26}$$

where $k = 0, 1, \ldots, N-1$ and γ_k is the set of eigenvalues of $\mathbf{G}$. Again, $k = 0$ gives the evolution equation for perturbations along the synchronous trajectory, and all the rest represent transverse perturbations.

In general, the quantities $K\gamma_k$ are complex numbers which can be written in the form $K\gamma_k = \alpha + i\beta$. Hence, one can now forget that the set γ_k results from a particular network of couplings (the matrix $\mathbf{G}$) and consider the general dynamical system

$$\dot{\boldsymbol{\eta}}'_k = (\mathbf{Df} + (\alpha + i\beta)\mathbf{E})\, \boldsymbol{\eta}'_k. \tag{7.27}$$

The maximal Lyapunov exponent λ corresponding to this equation depends on α and β. The function $\lambda = \lambda(\alpha, \beta)$ is the *master stability function* of system (7.27). It defines a surface over the complex plane with certain domains for α and β where the Lyapunov exponent is negative, $\lambda < 0$.

Let us now return to the original problem, take the matrix $\mathbf{G}$ and calculate its eigenvalues. By using the information derived for the generic system (7.27) one can determine the sign of the Lyapunov exponent for each of the quantities $K\gamma_k$, which map on a pair (α, β). If all of the Lyapunov exponents are negative, then the coupling scheme given by $\mathbf{G}$ with a strength K produces a stable synchronous state. The main result of the master stability function approach is that the stability of other systems coupled through different matrices $\mathbf{G}$ (which yield other sets of eigenvalues γ_k) can be inferred from the knowledge of the $n-$dimensional system (7.27).

If the matrix $\mathbf{G}$ is symmetric, its eigenvalues are all real. The corresponding master stability function $\lambda(\alpha)$ has the typical form shown in Fig 7.1, right plot. In the most general case there are two eigenvalues α_1 and α_2 bounding the stability window. Whenever $\mathbf{E}$ is the $n \times n$ identity matrix $\mathbf{I}$, the coupling between the oscillators is called *vector coupling*. In this case, a direct relationship between the Lyapunov exponents of the single oscillator and those of the perturbation problem (7.27) exists, and the master function has only one value α_1 signaling the onset of stable full synchronization, with $\alpha_2 \to \infty$. If the coupling is not symmetric in all variables (i.e., $\mathbf{E} \neq \mathbf{I}$) then a non-trivial interplay between the coupling and

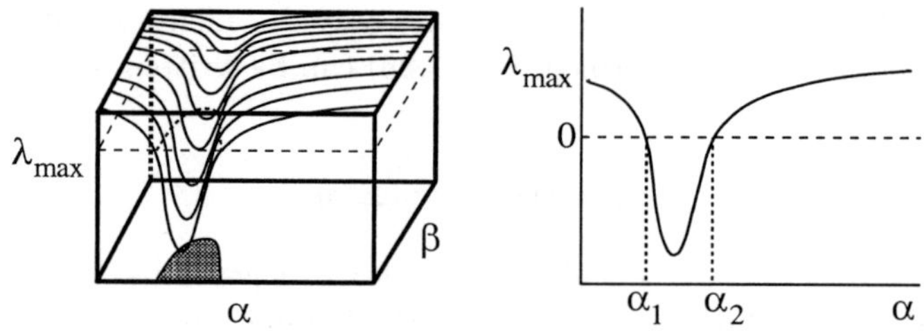

Fig. 7.1 Schematic representation of the master stability function in the (α, β) plane. The values $\lambda(\alpha, \beta) < 0$ (area in gray on the left) determine the synchronizability domain of the generic system (7.27). On the right, the typical shape of the master stability function for $\beta = 0$ is shown. This relevant case corresponds to symmetric matrices $\mathbf{G}$. The interval of stability is bounded by α_1 and α_2.

the dynamics occurs. As a result, the perturbation problem (7.27) cannot be further reduced, the correspondence with the Lyapunov exponent of the single oscillator does not exist, and the transversal exponents have to be calculated anew. In that case, K plays a role similar to any other parameter in a dynamical system, and a destabilization transition can occur when it overcomes a finite value, corresponding to the existence of a finite α_2.

This has further consequences regarding the synchronizability of a system with an arbitrary coupling. For each of the $k = 1, \ldots, N$ modes independently, it holds that a large enough K can make them stable. However, if for large enough K the corresponding exponent λ_k becomes again positive, it might occur that the minimal value of K required to ensure stability of mode $k = 1$ is already too large, such that the mode corresponding to the shortest wavelength has become unstable. As a consequence, it is possible that no domain of stability exists for such a system [Heagy *et al.* (1995); Pecora *et al.* (1997); Pecora (1998b)].

As an illustration of the method described above, let us analyze the stability of an array of N oscillators placed on a ring and diffusively coupled through all of their coordinates. The matrices $\mathbf{G}$ and $\mathbf{E}$ characterizing the evolution of small perturbations in Eqs. (7.25) are given in Eq. (7.24). Now, the use of Eqs. (7.5) and (7.6) gives a explicit form of the equations for small variations,

$$\dot{\boldsymbol{\eta}}_k = \left(\mathbf{Df} - 4K\sin^2(\pi k/N)\mathbf{I}\right)\boldsymbol{\eta}_k, \tag{7.28}$$

so that in this case the eigenvalues of matrix $\mathbf{G}$ are $\gamma_k = -4\sin^2(\pi k/N)$. This was one of the first analytical results on the stability of chaotic oscillator arrays [Fujisaka and Yamada (1983)]. In this case, the evolution operator for each transverse perturbation is given by

$$\Lambda_k(t) = \Lambda_0(t)\exp(-K\gamma_k t). \tag{7.29}$$

Thus, the relationship between the transverse Lyapunov exponents and those of the single oscillator is

$$\lambda_k = \lambda_0 - 4K\sin^2(\pi k/N). \tag{7.30}$$

In order to have stable synchronization in this system, all $N-1$ transversal perturbation modes must be damped, that is, $K\gamma_k > \lambda_0$ for all k. Since $k = 1$ corresponds to the maximal eigenvalue, this array of oscillators will be stable for any K larger than

$$K_s^R = \frac{\lambda_0}{4}\left[\sin^2\left(\frac{\pi}{N}\right)\right]^{-1}. \tag{7.31}$$

Therefore, higher values of K_s are required to synchronize increasingly large systems. The values of the Lyapunov exponent λ_0 corresponding to the single oscillator depend on each system and on the parameters chosen.

All the results discussed above can be immediately extended to coupled maps [Chen *et al.* (2003)]. Suppose that our dynamical system is now the logistic map with $a = 2$, for which $Df(x) = -4x$ and $\lambda = \ln 2$. As above, we consider a ring of N coupled logistic maps with the matrix $\mathbf{G}$ given in (7.24). Since in this case $n = 1$ the matrix $\mathbf{E}$ becomes simply unity. We can directly use the previously derived results and find that the fully synchronous state of the ring is stable for any coupling K larger than $K_s^{LM} = \ln 2/(4\gamma_1)$, where γ_1 is the ($N-$dependent) largest eigenvalue of $\mathbf{G}$. Note that for coupled logistic maps the coupling strength K should not exceed unity, in order to avoid dynamical instabilities. Taking into account the above results, this imposes a limit on the number of maps in the ring for which full synchronization can still be achieved,

$$N_{\max} = \pi\left[\arcsin\left(\frac{\ln 2}{2}\right)\right]^{-1} \simeq 8.88. \tag{7.32}$$

Since $N_{\max}$ can only take integer values, no more than eight maps can thus be fully synchronized for $a = 2$. For smaller values $a < 2$ of this parameter, synchronization can be achieved in larger systems. Note that the diagram in Fig. 6.1 contains the values of λ for each a, so that by inserting them into the previous equation the corresponding values of $N_{\max}$ can easily be found. Close to the transition to chaos, for $a = a_\infty + \epsilon$, the value of the Lyapunov exponent λ is almost zero, and consequently very large arrays can be fully synchronized for ϵ small enough.

7.1.2 *Synchronizability of arbitrary connection topologies*

The master stability function method allows to undertake the systematic comparison of ensembles of identical oscillators coupled through different topologies. Particularly, it permits to find the interval of coupling strengths where the synchronous state is stable, and to determine the topology for which synchronization would first occur if the number of connections is fixed. The onset of full synchronization and its robustness under changes in the network topology can be analyzed.

In this section, we compare ensembles of identical Rössler oscillators coupled through various topologies [Barahona and Pecora (2002)]. The coupling will always be through the variable x, so that

$$\mathbf{E} = \begin{pmatrix} 1 & 0 & 0 \\ 0 & 0 & 0 \\ 0 & 0 & 0 \end{pmatrix}. \tag{7.33}$$

For such coupling, $\alpha_2 < \infty$ and a desynchronization transition occurs for K large enough. We restrict the analysis to symmetric connection matrices such that the eigenvalues γ_k are all real.

Generally, a network with symmetric connections specified by a matrix $\mathbf{G}$ will by synchronizable if

$$\frac{\gamma_{\max}}{\gamma_1} < \frac{\alpha_2}{\alpha_1} \equiv \rho, \tag{7.34}$$

where γ_1 is the first non-zero eigenvalue of $\mathbf{G}$, and $\gamma_{\max}$ is its largest eigenvalue. The quantities $\alpha_{1,2}$ bound the domain of the master stability function where the largest Lyapunov exponent λ corresponding to (7.27) is negative (see Fig. 7.1). Note that ρ only depends on the dynamics of the single oscillator and on the matrix $\mathbf{E}$. If the ratio r $\gamma_{\max}/\gamma_1$ is well below ρ there is a wide interval of coupling strengths K where the system can be synchronized. The closer to ρ is this ratio, the less robust will be the

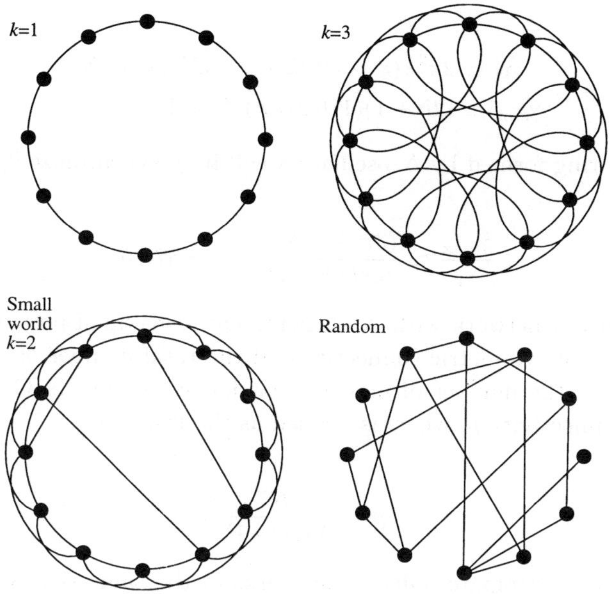

Fig. 7.2 Different connection topologies for the systems of Rössler oscillators described and compared in the text. Regular arrays with $k = 1$ (simple ring) and $k = 3$ are shown. A typical small-world network with an underlying regular structure with $k = 2$ has a small number of non-local connections. A random graph does not display any local order.

synchronous state. In our discussion, we fix the parameters as $a = 0.2$, $b = 0.2$, and $c = 2.5$, so that $\rho = 37.85$.

Let us consider a ring formed by N Rössler oscillators where each element is coupled to its $2k$ nearest neighbors (see Fig. 7.2). The respective matrix $\mathbf{G}$ has elements

$$G_{ij} = \begin{cases} 2k, & i = j \\ -1, & 1 \leq |i - j| \leq k \\ 0, & \text{otherwise} \end{cases} \tag{7.35}$$

The extremal (i.e., the lowest and the largest) eigenvalues of this matrix for $1 \ll k \ll N$ are

$$\gamma_1 \simeq 2\pi^2 k(k+1)(2k+1)/3N^2, k \ll N,$$
$$\gamma_{\max} \simeq (2k+1)(1+2/3\pi), k \gg 1, \tag{7.36}$$

and thus a ring formed by N oscillators will fully synchronize if

$$\frac{\gamma_{\max}}{\gamma_1} \simeq \frac{(3\pi+2)N^2}{2\pi^3 k(k+1)} \leq \rho \simeq 37.85. \tag{7.37}$$

Generally, a network with N elements has a maximal possible number $N(N-1)/2$ of symmetric connections. If the total number of actual connections, i.e. the number of non-zero entries in the matrix $\mathbf{G}$ is N^*, the network connectivity p, which is defined as the fraction

$$p = \frac{2N^*}{N(N-1)} \tag{7.38}$$

can be used to compare different networks of the same size N. For each considered network topology, there is a critical value p_c of the fraction of connections p above which the system becomes synchronized, for each of the topologies studied. The dependence of p_c on N for several different topologies in networks formed by Rössler oscillators is presented in Fig. 7.3. Further examples can be found in [Barahona and Pecora (2002)].

The calculation of the threshold p_c can be carried out analytically under certain approximations. For regular arrays (rings) with k neighbors, Eq. (7.37) can be used to estimate the minimum value of k for which the system will synchronize. This value turns out to be $k_{\min} \simeq N\rho^{-1/2}\sqrt{(3\pi+2)/(2\pi^3)}$. Considering now that a regular ring of size N with k neighbors per element has $N^* = Nk$ connections, we obtain that

$$p_c^{Reg} = \rho^{-1/2}\sqrt{\frac{6\pi+4}{\pi^3}} \simeq 0.1395\ldots. \tag{7.39}$$

Thus, synchronization in regular arrays with large N is only possible if their connectivity exceeds a certain threshold.

Next we consider random graphs where a fraction p of all possible edges is picked at random. It can be shown [Barahona and Pecora (2002)] that such random topologies are characterized by the following ratio of the two extreme eigenvalues

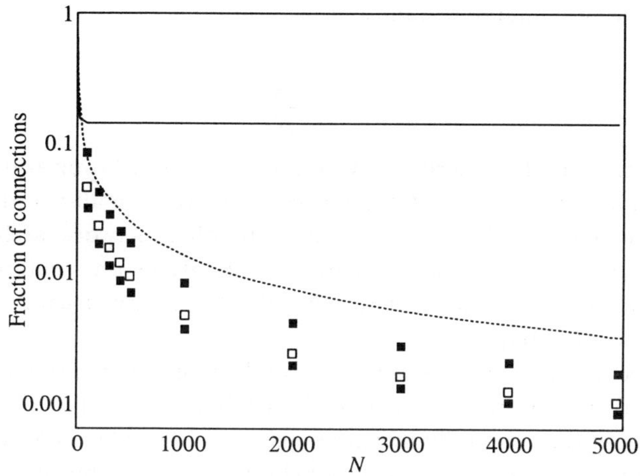

Fig. 7.3 Fraction of connections required in different topologies to achieve full synchronization. The continuous line corresponds to regular arrays, the dotted line to random graphs, and the squares to different small-world networks, with $k = 1, 2,$ and 3. Adapted from [Barahona and Pecora (2002)].

$$\frac{\gamma_{\max}}{\gamma_1} \simeq \frac{Np + \sqrt{2p(1-p)N\ln N}}{Np - \sqrt{2p(1-p)N\ln N}} \tag{7.40}$$

When random graphs are considered, one additional issue should be taken into account. For low values of the connectivity p, the network may split into disconnected subsets of elements, thus making impossible the synchronization of the whole ensemble. The typical number of connections required for a random graph with N nodes to be connected is proportional to $N \ln N$. It turns out that random graphs become almost surely synchronized just after they become connected at

$$p_c^{Ran} \simeq \frac{2\ln N}{N + 2\ln N}. \tag{7.41}$$

The threshold p_c^{SW} can also be considered for small-world networks that are constructed by adding a small amount of random connections to a regular array with k neighbors [Barahona and Pecora (2002)]. For such systems, a perturbation analysis applied to the matrix $\mathbf{G}$ (which is divided into a regular part plus a perturbation corresponding to the random connections) permits to approximately estimate it as

$$p_c^{SW} \simeq \frac{2k}{N}. \tag{7.42}$$

The analytical and numerical investigations of synchronization properties of networks with different topologies reveal that small-world networks synchronize more easily as compared with other network architectures. This was already noticed in the early investigations on small-world networks [Watts and Strogatz (1998)]. See also the recent discussion [Latora and Marchiori (2001)].

We have discussed in this section several examples of ensembles formed by identical oscillators. Such ensembles admit an analytical treatment which ensures that the fully synchronous state is stable. However, it was shown some time ago that chaotic synchronization can be realized experimentally [Anishchenko *et al.* (1992)] as well, even if the oscillators are then necessarily different. Recently, the coherent behavior of groups of chaotic oscillators has been applied to biological systems [Blasius *et al.* (1999)]. Other issues of interest in this broad field are the evolution of structurally different oscillators, the response of chaotic systems to external driving forces, or the enhancement of synchronicity due to the introduction of noise [Pecora *et al.* (1997); Boccaletti *et al.* (2002)]. Some of these subjects are analyzed in detail in the rest of this chapter and the next one.

7.2 Partial Entrainment in Rössler Oscillators

So far we have considered the existence and the stability of the fully synchronous state in ensembles of identical elements coupled through various topologies. Our discussion has been focused on the analytical techniques allowing to understand the synchronization mechanisms. Now we would like to address some problems related to the transition to full synchronization and analyze other dynamical regimes where, despite a high degree of collective coherence, full synchronization is absent. Such regimes are possible in ensembles of identical oscillators, as well as in inhomogeneous populations. In the latter case, there have been some analytical investigations on the onset of synchronization and the stability of the synchronous state [Ott *et al.* (2002)], though the theory is limited to globally coupled systems. Most of the results related to heterogeneous oscillator populations are obtained from numerical simulations and by using phenomenological approaches.

When the coupled chaotic elements of a system are not identical, full synchronization in the sense of the *exact* coincidence of the states of all the oscillators is no longer possible. At sufficiently strong coupling, all elements would usually move coherently, though there would be still a certain (small) dispersion in their positions. A system is $\delta-$synchronized (i.e., almost fully synchronized) if the distance between any pair of elements d_{ij} defined (for oscillators with three variables) as

$$d_{ij} = \sqrt{(x_i - x_j)^2 + (y_i - y_j)^2 + (z_i - z_j)^2} \tag{7.43}$$

is contained within an interval δ at any time t and for all pairs i and j, with δ taking values well below the amplitudes of individual oscillations.

In heterogeneous ensembles of Rössler oscillators, the trajectories of the individual elements show a transition to synchronization as the coupling strength is increased. In the parameter region preceding $\delta-$synchronization arises, a fraction of the elements becomes entrained while the rest of them evolves almost independently.

Before presenting a model where this transition has been described, we introduce the following ensemble of identical, coupled oscillators

$$\dot{\mathbf{r}}_i = \mathbf{f}(\mathbf{r}_i) + K\mathbf{A}(\langle \mathbf{r} \rangle - \mathbf{r}_i) + K'[\mathbf{f}(\langle \mathbf{r} \rangle) - \mathbf{f}(\mathbf{r}_i)], \tag{7.44}$$

where $i = 1, \ldots, N$, $\langle \mathbf{r} \rangle = N^{-1} \sum_i \mathbf{r}_i$, $\mathbf{A}$ is a constant matrix, and the coefficients K and K' specify the intensities of coupling given by the two last terms. When $K' = 1$, the dynamics is reduced to $\dot{\mathbf{r}}_i = K\mathbf{A}(\langle \mathbf{r} \rangle - \mathbf{r}_i) + \mathbf{f}(\langle \mathbf{r} \rangle)$, and the variations from the synchronous state are governed by an exact set of linear equations,

$$\dot{\boldsymbol{\eta}}_i = -K\mathbf{A}\boldsymbol{\eta}_i. \tag{7.45}$$

If all eigenvalues of the matrix A have positive real parts and $K > 0$, the global stability of the fully synchronous state is guaranteed. Using continuity arguments, one can expect that this situation would also hold in an interval of values $K' < 1$. Numerical simulations show that the system indeed undergoes robust synchronization under this coupling, even for relatively small values of K and K'.

Let us now fix $K = K'$ and choose the following form for the matrix $\mathbf{A}$:

$$\mathbf{A} = \begin{pmatrix} 0 & 1 & 1 \\ -1 & 1-a & 0 \\ 0 & 0 & c \end{pmatrix} . \tag{7.46}$$

To introduce heterogeneity, we allow the parameters a and c to take different values for each of the oscillators. The dynamical system that we consider is finally [Zanette and Mikhailov (1998a)]

$$\begin{aligned} \dot{x}_i &= -y_i - z_i \\ \dot{y}_i &= x_i + a_i y_i + K(\langle y \rangle - y_i) \\ \dot{z}_i &= 0.4 + z_i(x_i - c_i) + K(\langle x \rangle \langle z \rangle - x_i z_i). \end{aligned} \tag{7.47}$$

In order to characterize the synchronization transition, two order parameters can be defined. The first of them is the fraction $p(\delta)$ of pairs of elements which are found at a distance $d_{ij} < \delta$ (see Eq. (7.43)),

$$p(\delta) = \frac{1}{N(N-1)} \left\langle \sum_{i=1}^{N} \sum_{j=1}^{N} \Theta[\delta - d_{ij}(t)] \right\rangle , \tag{7.48}$$

where $\Theta(x)$ is the step function, defined as $\Theta(x) = 0$ for $x < 0$ and $\Theta(x) = 1$ otherwise. The brackets indicate that the enclosed quantity is averaged in time. The second order parameter is the fraction $w(\delta)$ of elements which have at least one other oscillator at a distance smaller than δ,

$$w(\delta) = 1 - \frac{1}{N} \left\langle \sum_{i=1}^{N} \prod_{j=1, j\neq i}^{N} \Theta[\delta - d_{ij}(t)] \right\rangle . \tag{7.49}$$

Values of $p(\delta)$ near zero indicate that the elements follow mainly independent orbits. When full synchronization is approached, $p(\delta)$ tends to unity because all elements eventually join the single coherent group. In contrast to this, the parameter $w(\delta)$ can take large values even when $p(\delta)$ is small, in a situation corresponding to the formation of different groups which are separated dynamically but within which a subset of elements follow close trajectories. Dynamical regimes with such cluster organization will be discussed in detail in the next chapter.

The two order parameters are presented in Fig. 7.4 as functions of the coupling strength K. The shown results correspond to identical oscillators,

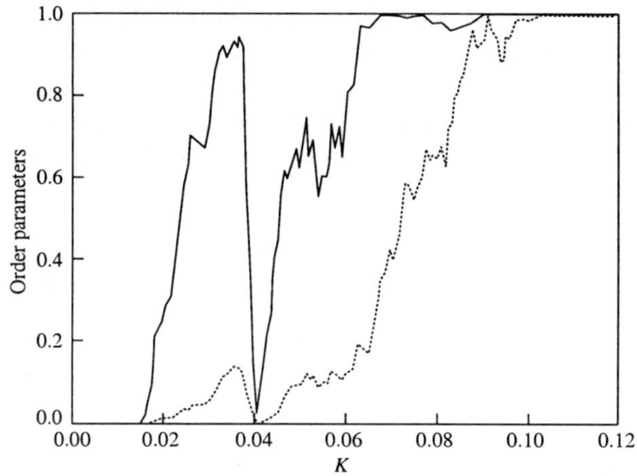

Fig. 7.4 The two order parameters $w(\delta)$ (solid line) and $p(\delta)$ (dotted line) for increasing values of the coupling K in system (7.47). Full synchronization sets in for coupling intensities approximately above 0.1. Organization of the elements into distinct clusters corresponds to large values of $w(\delta)$ and values of $p(\delta)$ close to zero. Adapted from [Zanette and Mikhailov (1998)].

with $c_i = 8.5$ and $a_i = 0.15$, for all i. In this case full synchronization takes place, and the order parameters have been estimated for $\delta = 0$.

The situation changes if heterogeneous ensembles are considered. If oscillators are not all identical, the synchronization among them is not complete, and a dispersion in their states persists in time. Consider the system defined by Eq. (7.47) where the parameter c_i is independently drawn for each oscillator from a flat distribution, $c_i \in [c-\sigma_c, c+\sigma_c]$, with average value $c = 8.5$. The typical radius $\langle R \rangle$ of the cluster formed by the positions of N oscillators can be calculated as

$$R^2(t) = \frac{1}{N} \sum_{i=1}^{N} \left[\Delta x_i^2(t) + \Delta y_i^2(t) + \Delta z_i^2(t) \right], \tag{7.50}$$

where $\Delta x_i(t) = x_i(t) - \langle x \rangle$, and similarly for y and z. The time average $\langle R \rangle$ for a fixed coupling intensity K provides a measure of the balance between the competing effects of dispersion in the parameters and the attraction due to coupling.

This quantity is plotted in Fig. 7.5 as a function of the dispersion σ_c of parameter c for a coupling strength $K = 0.2$, which corresponds to full synchronization when all oscillators are identical. The radius $\langle R \rangle$ grows proportionally to the heterogeneity strength. Similar results are obtained if a heterogeneity with respect to parameter a_i is considered. These numerical results are in agreement with the theoretical dependence expected in the limit $K \to 1$ when the heterogeneity is weak [Zanette and Mikhailov (1998a)].

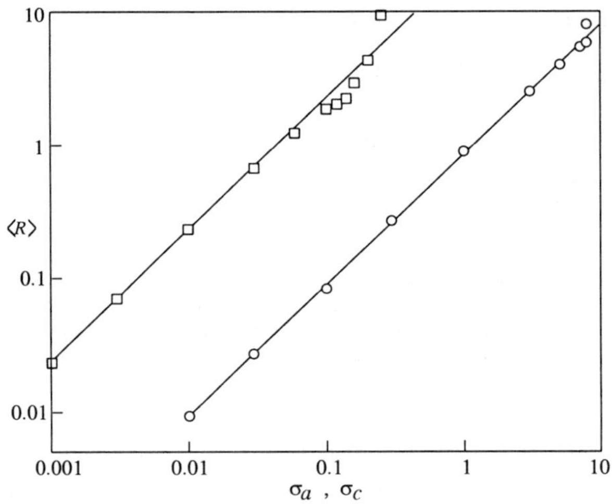

Fig. 7.5 Typical size of the cloud formed by the trajectories of $N = 10^3$ heterogeneous Rössler oscillators as a function of the dispersion σ_a (squares) and σ_c (circles) of the parameters a and c, respectively. Adapted from [Zanette and Mikhailov (1998)].

The addition of noise to a system of chaotic oscillators leads to an effect which is similar to that of quenched disorder (such as the heterogeneity in the parameters a_i and c_i in the previous example). The action of noise has been studied for a system of Rössler oscillators with vector coupling, with an additional noisy term applied to the variable x [Zanette and Mikhailov (2000)]:

$$\dot{\mathbf{r}}_i = \mathbf{f}(\mathbf{r}_i) + K(\langle \mathbf{r} \rangle - \mathbf{r}_i) + \mathbf{M}\xi_i, \tag{7.51}$$

where $\mathbf{f}(\mathbf{r})$ stands for the usual Rössler dynamics, and where the coefficients of the matrix $\mathbf{M}$ are all zero except for $M_{xx} = 1$. The noise has zero average, $\langle \xi_i \rangle = 0$ and is delta-correlated, $\langle \xi_i(t)\xi_j(t') \rangle = 2S\delta(t-t')\delta_{ij}$. Note that, if noise is absent, this global coupling is of the shift-invariant type. When written in the generic form (7.22) the corresponding matrices are

$$\mathbf{G} = \frac{1}{N}\begin{pmatrix} -N+1 & 1 & 1 \dots & 1 \\ 1 & -N+1 & 1 \dots & 1 \\ \vdots & \vdots & \vdots \ddots & \vdots \\ 1 & 1 & 1 \dots & -N+1 \end{pmatrix}, \quad \mathbf{E} = \begin{pmatrix} 1\,0\,0 \\ 0\,1\,0 \\ 0\,0\,1 \end{pmatrix}. \qquad (7.52)$$

Thus, there is a coupling strength K above which the deterministic system exhibits full synchronization.

Figure 7.6 shows the effect of very weak noise with $S = 10^{-10}$ on the order parameters w and p defined above. In numerical simulations, noise values were randomly chosen from a flat distribution. Here, the parameter δ is fixed to 10^{-3}, much smaller than the typical values of the variables and above the characteristic scale of the noise. If the noise intensity is increased, it is possible to observe a number of fuzzy groups of partly entrained elements. Each oscillator, however, can occasionally leave its group, wander for a while, and later join a different cluster. Finally, if the amplitude of noise is too large, any coherent behavior is destroyed.

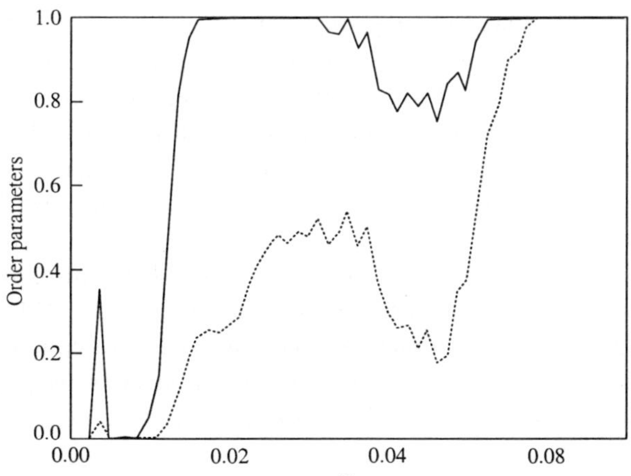

Fig. 7.6 Order parameters $p(\delta)$ (dashed line) and $w(\delta)$ (solid line), averaged over 20 independent realizations, as a function of the coupling intensity K for an ensemble of Rössler oscillators under the action of noise. The parameters are $a = b = 0.2$, $c = 4.5$. Adapted from [Zanette and Mikhailov (2000)].

7.2.1 *Phase synchronization*

In the previous chapter we have discussed the properties of phase synchronization, a form of entrainment where the frequencies of oscillations become locked while the amplitudes of the chaotic trajectories evolve weakly correlated. This effect was first studied for two oscillators, but it is also observed in heterogeneous ensembles of globally coupled Rössler oscillators [Pikovsky *et al.* (1996); Blasius *et al.* (2003)].

Let us consider a large number N of coupled Rössler oscillators,

$$\begin{aligned} \dot{x}_i &= -\nu_i y_i - z_i + K\langle x\rangle \\ \dot{y}_i &= \nu_i x_i + a y_i \\ \dot{z}_i &= b + z_i(x_i - c), \end{aligned} \tag{7.53}$$

where

$$\langle x\rangle = \frac{1}{N}\sum_{i=1}^{N} x_i(t) \tag{7.54}$$

Heterogeneity is introduced here through a distribution of values for the ν_i parameters. The natural frequencies of the oscillators are thus modified and take values ω_i proportional, but not equal to the parameters ν_i. In the Rössler system, the natural frequency ω is larger than the parameter ν and, in addition, the introduction of coupling in the form shown in (7.53) produces an increase in the effective frequencies ω_i' as larger values of the coupling strength K are used. The frequency of synchronization Ω grows with K as well.

Consider as an example the case of a Gaussian distribution for the values of ν_i with average $\nu_0 = 1$ and standard deviation $S = 0.01$. The set of natural frequencies ω_i is numerically computed before coupling is turned on, and is later compared with the new set of effective frequencies ω_i' for the system coupled with strength K. In practice, ω_i is defined as

$$\omega_i = 2\pi \frac{n_i^{PS}}{T_i}, \tag{7.55}$$

where n_i^{PS} is the number of times that the trajectory of oscillator i has crossed a suitably chosen Poincaré section, and T_i is the total time that was needed to complete the n_i^{PS} turns. Note that the frequencies ω_i can be thus determined only if a well-defined phase exists in the system, which however is true for Rössler oscillators (see Sec. 6.1.2 and 6.3).

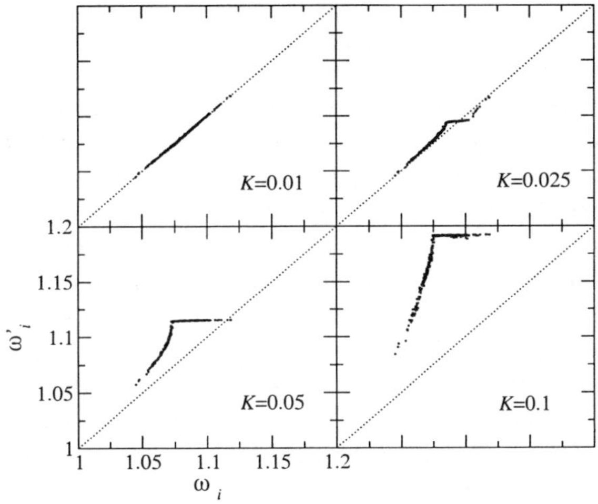

Fig. 7.7 Dependence between the natural frequencies ω_i and the effective frequencies ω_i' in an ensemble of globally coupled Rössler oscillators. For increasing coupling strength K a larger number of oscillators becomes entrained. The common frequency Ω is a function of K.

Figure 7.7 shows natural frequencies ω_i and the corresponding effective frequencies for an ensemble of $N = 500$ oscillators with parameters $a = 0.2$, $b = 0.2$, and $c = 4.5$. As coupling increases, locking of frequencies for a subset of the oscillators is observed. Typically, the entrainment frequency Ω grows with K, and oscillators with the lowest natural frequencies are more difficult to synchronize. However, when K overcomes a threshold (which in the present case is about 0.14) an instability occurs and the cluster of entrained oscillators melts away. Only partial phase synchronization can be observed for this choice of parameters. Other choices (for example $a = 15$, $b = 0.4$ and $c = 8.5$) yield collective behavior which is closer to full phase locking for all of the oscillators [Pikovsky *et al.* (1996)].

The instabilization and loss of synchrony above a certain threshold could have been expected if we recall the results derived in Sec. 7.1.1 for systems of identically coupled oscillators. If the coupling does not fulfill certain symmetry properties, it might happen than the system becomes unstable for a too large coupling.

For comparison, let us briefly analyze the synchronization properties of a different ensemble of heterogeneous Rössler oscillators with vector coupling of the form

$$\dot{\mathbf{r}}_i = \nu_i \mathbf{f}(\mathbf{r}_i) + K(\langle \mathbf{r} \rangle - \mathbf{r}_i), \tag{7.56}$$

where the values of $\langle y \rangle$ and $\langle z \rangle$ are defined analogously to (7.54), and the values of the parameters ν_i are randomly drawn from a Gaussian distribution with average unity and dispersion $S = 0.01$, as in the previous example. Note that now the parameters ν_i directly control time scales in each individual oscillator. Figure 7.8 shows again the dependence between natural ω_i and effective ω_i' frequencies. With this form of vector coupling, the effective frequency is systematically lower than the natural frequency. Full synchronization is obtained for high enough coupling intensity, and the disruption of this state is not observed under further increase of K. Full synchronization takes place despite the heterogeneity in the natural frequencies, in agreement with the theoretical results derived for this type of coupling in systems of identical oscillators.

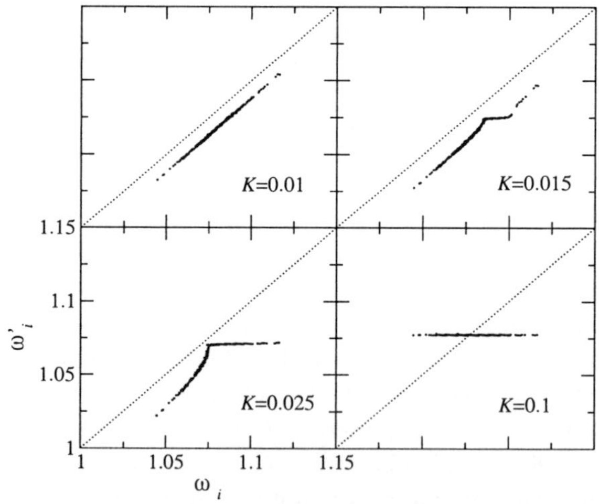

Fig. 7.8 Dependence between natural and effective frequencies in an ensemble of vector-coupled Rössler oscillators, Eq. (7.56).

In contrast to the coupling introduced in (7.53), the frequency Ω at which oscillators synchronize is almost independent on the coupling K and is determined by the initial distribution of ν_i values. This partly explains the stability of the system under increasing coupling strength. In the system Eq. (7.56), synchronization is achieved with lower values of the coupling strength K. This is simply due to the fact that coupling is introduced in all three variables instead of a single one, and its effect is consequently stronger.

Finally, note that in the previous case larger K values tended to increase the effective frequencies ω_i' as compared to the natural frequencies ω_i. In contrast to this, in the case of vector coupling there is a slight decrease in Ω for small values of K (see Figures 7.9 and 7.10) and, eventually, the entrained frequency stabilizes at the average of the natural frequencies, $\Omega \simeq N^{-1}\sum_i \omega_i$. This is observed also when only the coordinate y is coupled through $K(\langle y\rangle - y_i)$ [Montbrió and Blasius (2003)].

The transition to phase synchronization can be alternatively visualized by plotting the effective frequencies ω_i' of the oscillators in the ensemble as a function of the coupling strength K. The corresponding order parameter that signals the transition is the dispersion σ_ω in the values ω_i'. The condensation of the frequencies of oscillators as K increases is represented in Fig. 7.9 for simulations using Eqs. (7.53) and (7.56). Fig. 7.10 shows the dependence of the average effective frequency Ω and the dispersion in the values of ω_i' as K increases.

In the system Eq. (7.53), full synchronization cannot be achieved (for the given parameter values) at any coupling strength K. In an interval $0.02 < K < 0.07$ of coupling strengths, there is partial condensation of the trajectories, though those oscillators with the lowest natural frequencies cannot join the main group. The system enters an unstable regime for K approximately above 0.07. In the system Eq. (7.56), all effective frequencies coincide at sufficiently strong coupling.

The two discussed examples clearly show the sensitivity of the collective behavior to the way in which coupling is introduced. There are many different schemes to establish a functional relationship between the elements of an ensemble. The result can be highly dependent on the chosen coupling, and thus it is advisable to investigate different possibilities to extract those features which are characteristic for a class of systems and those which are highly model-dependent. Additionally, one has to keep in mind that very large dispersions in certain parameters, or very large values for the coupling K can bring about a qualitative change in the topological properties of the attractors corresponding to different oscillators. For example, a very large dispersion in the values of ν_i can result in the appearance of attractors with different structural properties, so that oscillators with periodic and chaotic dynamics will be present in the ensemble.

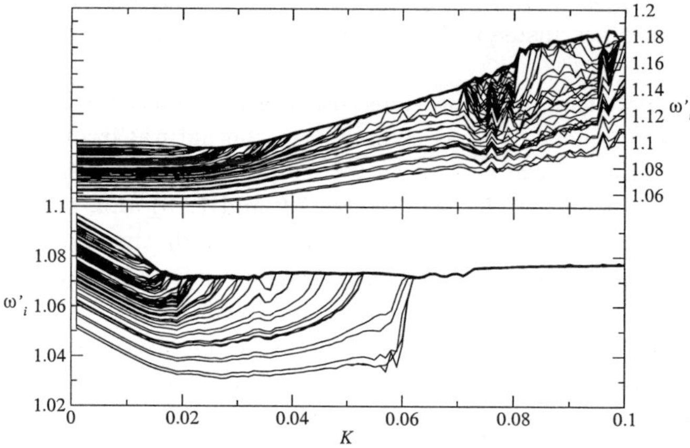

Fig. 7.9 Transition to phase synchronization in ensembles of $N = 100$ Rössler oscillators described by Eqs. (7.53) and (7.56). Effective frequencies are plotted as functions of the coupling intensity.

Phase synchronization of Rössler oscillators bears a strong similarity with frequency synchronization of periodic oscillators [Pikovsky *et al.* (2000)]. This is largely due to the isochronous behavior of the rotations around the origin in the Rössler system, that is, to the existence of a well-defined single peak in the power spectrum. Some important differences can be found, however. First, phase synchronization in chaotic oscillators refers to the coincidence of the frequencies while the amplitudes remain highly uncorrelated. There is no variable equivalent to the amplitude in periodic oscillators. Second, in frequency synchronization the entrained variable is the average frequency of the oscillators, irrespectively of their instantaneous phase values. In phase synchronization, instantaneous phases are locked.

Synchronization in phase has been also observed in heterogeneous arrays of diffusively coupled Rössler oscillators. We finish this section with the study of one-dimensional arrays [Osipov *et al.* (1997); Liu *et al.* (2001); Blasius *et al.* (2003); Montbrió and Blasius (2003)]. Two dimensional arrays [Davidsen and Kapral (2002)], as well as the effect of noise in those systems [Zhou and Kurths (2001)] have been studied.

Consider the following chain of diffusively coupled, non-identical Rössler oscillators [Osipov *et al.* (1997)]:

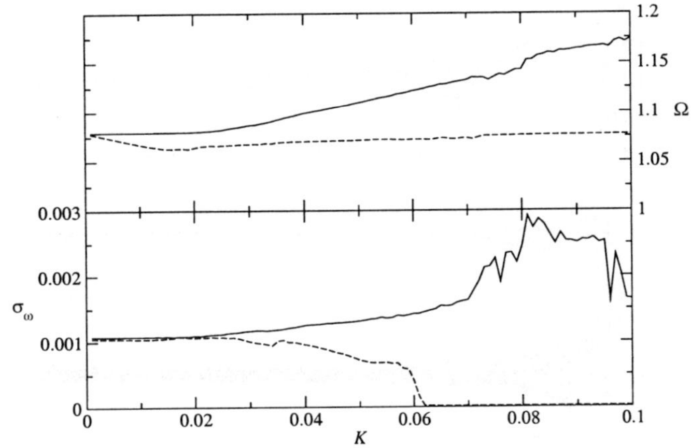

Fig. 7.10 Average frequency Ω and dispersion in the frequencies ω_i' for the two systems of Fig. 7.9 as functions of the coupling intensity K. The coupling used in Eq. (7.53) results in an entrainment frequency $\Omega(K)$ growing with the coupling strength (solid line). The respective dependences for the vector coupling (7.56) are shown by dashed lines. Note that in this case Ω is approximately equal to $\langle w_i' \rangle$.

$$\begin{aligned}\dot{x}_i &= -\nu_i y_i - z_i \\ \dot{y}_i &= \nu_i x_i + a y_i + K(y_{i+1} - 2y_i + y_{i-1}) \\ \dot{z}_i &= b + (x_i - c) z_i\end{aligned} \tag{7.57}$$

If the initial distribution of parameters ν_i is narrow enough, phase synchronization of the whole array is possible. However, if the dispersion in that distribution is large, the first effect observed as K increases is the suppression of chaos, which precedes the onset of the synchronous state. Another effect that can be illustrated with this system is the impossibility of synchronizing all the oscillators for too large rings. Although the results derived in Sec. 7.1.2 corresponded to full synchronization in identical systems, it is reasonable to expect that even when heterogeneity is present there will exist a maximum size of diffusively coupled oscillators above which synchronization (and phase synchronization as well) is not possible. This is shown in Fig. 7.11, where rings of different sizes are compared. While an array of size $N = 20$ becomes phase synchronized above $K \simeq 0.04$, doubling its size prevents synchronization.

The spatial ordering of the oscillators favors the synchronization of neighboring elements and the formation of locally coherent groups. In

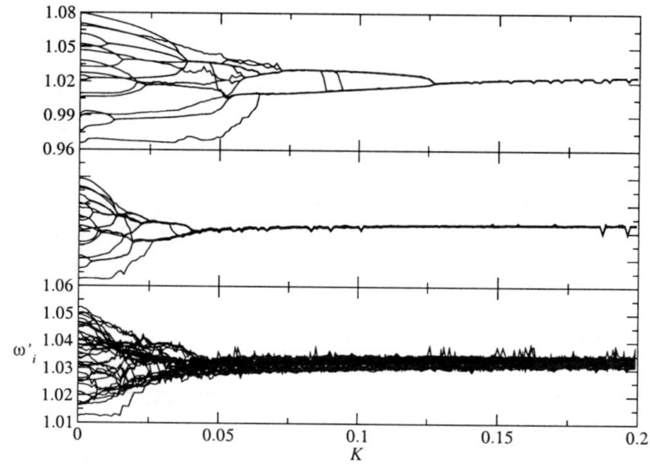

Fig. 7.11 Transition to phase synchronization in an array of diffusively coupled Rössler oscillators. The upper plot corresponds to a ring of $N = 20$ oscillators with a Gaussian distribution of dispersion $S = 0.03$ for the initial values of ν_i. The two lower plots have a lower initial dispersion $S = 0.01$ and correspond to a system of size $N = 20$ (middle) and $N = 40$.

the shown examples, there is a wide domain of K values for which two branches with two different synchronization frequencies are found. In large rings where entrainment is far from complete, the formation of spatiotemporal patterns is often observed. We show an example of this behavior in Fig. 7.12, where the evolution of the amplitude and the phases in a ring formed by $N = 300$ oscillators is presented. Parameters are as in Fig 7.11, with $S = 0.01$. The phases are defined here as

$$\sin \phi_i = \frac{y_i}{A_i}. \tag{7.58}$$

The amplitude $A_i = \sqrt{x_i^2 + y_i^2}$ is displayed in the upper plot of Fig. 7.12, and $\sin \phi_i$ in the lower plot. Despite the absence of a fully synchronous state or even of a clear condensation of the individual trajectories, it can be seen that local coupling leads to the appearance of correlations between neighboring oscillators.

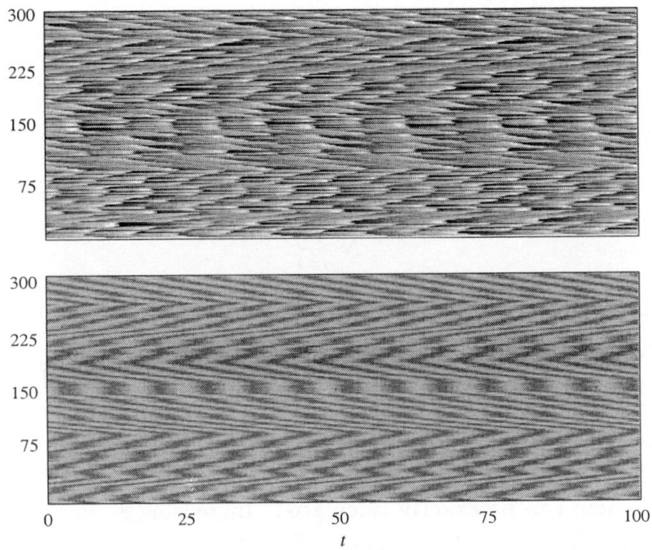

Fig. 7.12 Spatiotemporal evolution of the phase and amplitude of oscillators in a large ring of $N = 300$ elements. Though synchronization of all of the elements cannot be achieved, the emergence of local order can be clearly seen.

7.3 Logistic Maps

Large ensembles of coupled logistic maps display synchronization transitions similar to those observed for continuous-time systems. However, synchronization of different elements proceeds here through the entrainment of only one variable determining their states.

As in other considered systems, various forms of disorder hinder full synchronization. Noise, inhomogeneous couplings, or complex connection topologies make the elements non-identical and result in partial synchronization.

7.3.1 *Globally coupled logistic maps*

Globally coupled logistic maps were originally introduced as a mean-field approximation to coupled map lattices [Kaneko (1989); Kaneko (1990a)]. The collective behavior of globally coupled logistic maps is extremely rich, and has been investigated in great detail by different research groups. The basic equations describing the dynamics of the coupled ensemble are

$$x_i(t+1) = f(x_i(t)) + K\left[\langle f(x)\rangle - f(x_i(t))\right], \tag{7.59}$$

where the coupling is introduced through the average

$$\langle f(x)\rangle = \frac{1}{N}\sum_{j=1}^{N} f(x_j(t)). \tag{7.60}$$

In this case the coupling contains the transformed variables $f(x)$, and is such that it vanishes when full synchronization is attained. For logistic maps, $f(x) = 1 - ax^2$. To begin the analysis of globally coupled maps, note that some of the exact results derived for systems of identical oscillators can be directly applied. First, the exact results derived in Sec. 7.1 reveal that the fully synchronous state is stable for coupling strengths above a threshold which can be exactly calculated. Increasing K further does not destabilize it. Similar to other systems displaying full synchronization, it is expected that that (almost) full synchronization can be reached when the ensemble is only weakly heterogeneous.

The stability of the fully synchronous attractor is determined by the eigenvalues of the matrix

$$\mathbf{J} = f'(s(t))\left((1-K)\mathbf{I} + \frac{K}{N}\mathbf{G}\right), \tag{7.61}$$

where $\mathbf{I}$ is the identity matrix and $\mathbf{G}$ is a matrix whose elements take all value unity [Kaneko (1990a)]. Perturbations along the synchronous state $s(t)$ grow according to the largest Lyapunov exponent λ (larger than zero if the dynamics is chaotic), while there are $N-1$ identical exponents determining the transversal stability of the fully synchronous state. Due to the high degree of symmetry in the system, all the transversal directions become simultaneously unstable. The Lyapunov exponent for transversal perturbations can be simply obtained in this case,

$$\lambda_{\perp} = \lambda + \ln(1-K), \tag{7.62}$$

and is identical to the exponent characterizing the stability of the synchronous state for two coupled logistic maps. Note that the global coupling in Eq. (7.59) is the immediate generalization of that introduced in Sec. 6.2.

For values of K below the full synchronization transition, a plethora of different partially ordered states has been identified in this system. We discuss them in detail in the forthcoming two chapters. In the remaining of this section, instead, we describe the effect of heterogeneity in full synchronization of globally coupled logistic maps.

7.3.2 *Heterogeneous ensembles*

Even if any element in a system responds only to the global signal coming from the rest of the elements, its individual sensitivity to such a signal may vary. If this happens, each of the elements is slightly different regarding its response to the global signal, and this situation can be modeled, for example, by introducing a distribution of coupling strengths K [Kaneko (1994a)]. The corresponding evolution equation is

$$x_i(t+1) = (1 - K_i)f(x_i(t)) + \frac{K_i}{N}\sum_{j=1}^{N} f[x_j(t)], \tag{7.63}$$

where the random numbers K_i are drawn from some distribution with average K_0. Two different distributions have been tested: a flat distribution [Kaneko (1994a)], and a truncated Gaussian [Zanette (1999)]. In both cases, the dynamics of elements in the ensemble affected by different values of K_i was modified qualitatively, and, for instance, some oscillators were stabilized close to periodic trajectories. An example of such loss of structural stability is shown in Fig. 7.13. For that example, the distribution $P(K)$ of coupling strengths is

$$P(K) = \begin{cases} \exp[-(K - K_0)^2/2S^2], & \text{for } 0 \leq K \leq 1 \\ 0, & \text{otherwise} \end{cases} \tag{7.64}$$

The top panel in this figure represents a single snapshot of the states of $N = 10^4$ oscillators as a function of their coupling strength, whereas the bottom panel shows eight superimposed consecutive snapshots, illustrating the highly non-trivial evolution of the ensemble. Note that around $K_i \simeq 0.2$ chaotic and periodic dynamics (for different oscillators) seem to coexist. Interestingly, in this case the behavior of the ensemble is almost insensitive to the distribution $P(K)$, and a uniform distribution in the interval $(K_0 + 0.3, K_0 - 0.3)$ yields a snapshot which is almost indistinguishable from that shown in Fig. 7.13. This implies that the individual dynamics is much more dependent on the particular value of K_i assigned to an element than on the average $\langle x \rangle$.

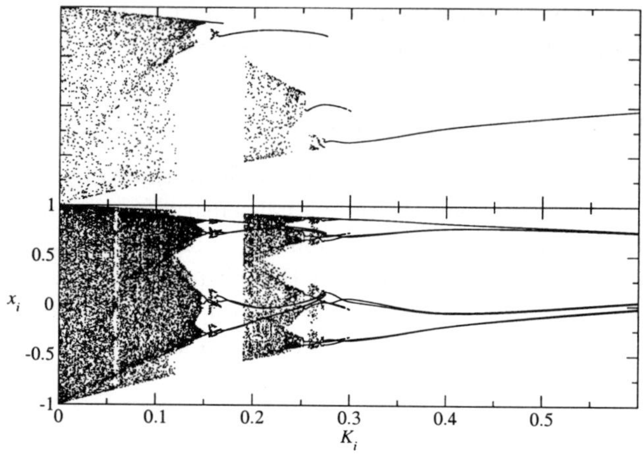

Fig. 7.13 Snapshot of a population of $N = 10^4$ heterogeneous, globally coupled logistic maps. The larger K_i, the more regular the dynamics of the elements become. Here $K_0 = 0.3$, and there is a high dispersion in the Gaussian distribution, $S = 0.5$. The upper plot shows the state x_i of the elements at a fixed time t. The lower plot displays the same quantity for eight consecutive time steps. Elements following periodic orbits coexist with others moving along chaotic trajectories.

The appearance of periodic motion in coupled systems of logistic maps is extremely common, even if the initial parameter values for each individual oscillator correspond to chaotic behavior. This fact has been related to the dense presence of windows of periodic behavior in the parameter domain of chaotic dynamics. The introduction of coupling effectively modifies the parameters, and this can push the oscillators close to periodic windows. Sometimes, stabilization of periodic motion as a result of coupling can occur.

The existence of a stable fully synchronous state depends on the mean coupling K_0 as well as on the statistical dispersion S of the distribution $P(K)$. For too small average coupling strength and for too high dispersion S, the system is not synchronizable, though a broad area where full synchronization takes place may still exist. The desynchronization boundary can be estimated numerically by performing simulations of the ensemble at a fixed a, varying the average value of the coupling strength K_0 and its dispersion S, and determining when full synchronization is lost. Alterna-

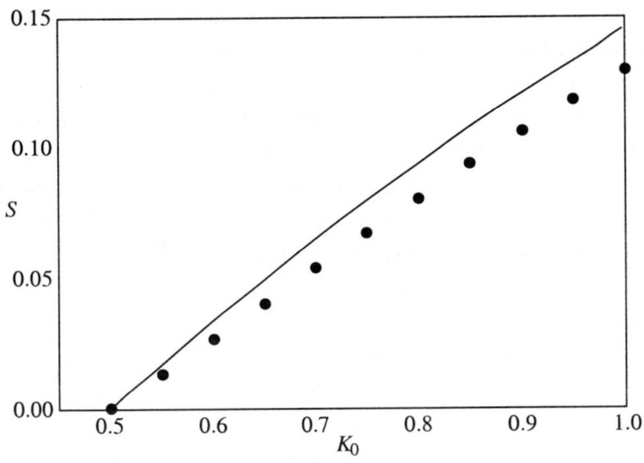

Fig. 7.14 Full synchronization of a heterogeneous ensemble of globally coupled logistic maps is possible for relatively small dispersions S in the K_i parameters and large enough average values. Above the solid line the fully synchronous state has never been observed in numerical simulations of system (7.63) with $N = 10^4$ elements and $a = 2$. Adapted from [Zanette (1999)].

tively, one can argue that the fully synchronous state will be destabilized if at least one element has a coupling strength below the stability threshold Eq. (7.62). As an example let us consider the case $a = 2$ and apply this criterion, which implies that the stability boundary (dependent on N) is obtained from the identity

$$N^{-1} = \int_0^{1/2} P(K)dK. \tag{7.65}$$

Figure 7.14 displays the stability boundaries obtained in numerical simulations (solid line) and determined by this equation (circles), and a reasonable agreement between the two estimated quantities can be seen. Note that there is a maximal dispersion $S \simeq 0.14$ above which stable full synchronization cannot be achieved. This situation is reminiscent of that observed in ensembles of Rössler oscillators, where it is also the dispersion in the natural frequencies which sets a limit to synchronizability.

An alternative way of introducing heterogeneities in such systems is to assume that an oscillator can receive information not from all of the elements in the ensemble, but only from a subset of them. In this case, the system represents a network with some connection topology. Above in this chapter, we have explored the effect of (mainly) local coupling in the synchronization properties of Rössler oscillators.

The behavior of randomly coupled networks of logistic maps has been studied [Manrubia and Mikhailov (1999)]. Even if a large fraction of connections (compared to the globally coupled case) is deleted, synchronization is still possible in such networks. In this sense, it is interesting to recall that synchronizability of randomly coupled Rössler oscillators is very robust against edge removal. Indeed, we have seen that the latter network becomes fully synchronized shortly after the set of nodes gets connected in a single group (see Sec. 7.1.2).

A randomly coupled ensemble is characterized by a fixed matrix of connections $\mathbf{G}$ whose elements $\{G_{ij}\}$ take value one if the oscillators i and j are connected, and zero otherwise. All the diagonal elements are set to zero. The average connectivity κ of such a network is defined as

$$\kappa = \frac{1}{N(N-1)} \sum_{i,j=1}^{N} G_{ij}. \tag{7.66}$$

It is convenient to modify the evolution equations for the elements in the ensemble and write them in the form

$$x_i(t+1) = \left(1 - \frac{K}{\kappa(N-1)} \sum_{j=1}^{N} G_{ij}\right) f[x_i(t)] + \frac{K}{\kappa(N-1)} \sum_{j=1}^{N} G_{ij} f[x_j(t)]. \tag{7.67}$$

Disorder introduced in the form of a fixed random matrix of connections $\mathbf{G}$ bears strong similarity with the case of heterogeneous couplings discussed in the previous section. Since an element is being connected only to a subset of the whole ensemble, its coupling strength becomes effectively changed. For large enough systems, the average over the states of the subset of oscillators to which i is connected can be approximated by an average over the whole ensemble. As a result, the term $\sum_j G_{ij} f[x_j(t)]$ in Eq. (7.67) approaches $\langle f[x_j(t)]\rangle \sum_j G_{ij}$. Then, the effective coupling can be defined as

$$K_i \simeq \frac{K}{\kappa(N-1)} \sum_{j=1}^{N} \xi_j, \tag{7.68}$$

where the random variable ξ_j would take value 1 with probability κ and 0 with probability $1-\kappa$. Therefore, the values of the coupling strength K_i in the limit of a large system turn out to obey a Gaussian distribution with mean value K and dispersion

$$S = \sqrt{\frac{\kappa(1-\kappa)}{N}}. \tag{7.69}$$

Thus, the effective dispersion in the random couplings decreases with increasing system size and the heterogeneity of randomly coupled networks is suppressed in the limit $N \to \infty$. In this limit, randomly coupled maps converge to an ensemble of homogeneous, globally coupled maps with effective coupling identical to the average value K. This is further related to the observation that the critical coupling intensity K_c above which a randomly connected system of size N synchronizes fulfills

$$K_c - K_{GC} \simeq \frac{1}{\sqrt{N}}, \tag{7.70}$$

where K_{GC} is the synchronization threshold of the globally coupled system formed by identical elements. Finally, this implies that for any value of κ such that the elements form a single connected group there is a size $N(\kappa)$ above which the system synchronizes. This size is however very large for relatively small values of κ, so that systems with several hundreds of elements still show a behavior remarkably different from that of the globally coupled system.

For fixed values of the system size N and the connectivity κ, states with different degrees of order are observed as the coupling strength K increases. The transition to full synchronization can be quantified by using the two order parameters w (Eq. (7.49)) and p (Eq. (7.48)) introduced in Sec. 7.2.

The appearance of partially ordered states in finite ensembles of randomly coupled maps for intermediate values of the coupling strength is less pronounced than in globally coupled maps, as the results portrayed in Fig. 7.15 show. The maximal distance between a pair of elements to contribute to the order parameter was $\delta = 10^{-3}$ (see Eq. (7.43) and Sec. 7.2). For a domain of coupling values around $K/\kappa \simeq 0.4$, a remarkable increase in the number of elements with at least another oscillator at a distance smaller than δ is observed, as quantified by w. This is in contrast with the low values taken by p in this interval, meaning that it is mainly isolated

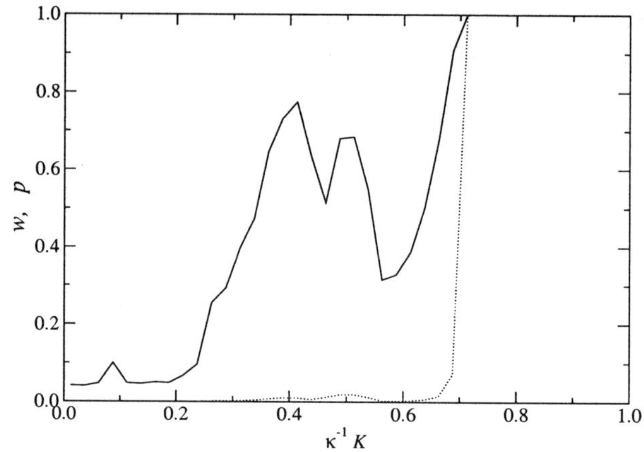

Fig. 7.15 Transition to synchronization in a system of $N = 250$ randomly coupled maps with connectivity $\kappa = 0.8$. The order parameter p corresponds to the dashed line, while w is represented through the solid curve. The parameter of the logistic map is $a = 2$, and $\delta = 10^{-3}$ in this example. Adapted from [Manrubia and Mikhailov (1999)].

pairs that approach each other, and large coherent groups are not formed. As long as fluctuations due to the finite size of the system are relevant, this behavior is analogous to what was observed in the system of globally coupled maps with heterogeneous couplings [Zanette (1999)], where no clustering phase preceding the onset of full synchronization was detected. However, as shown in the next chapter, some forms of hierarchical ordering similar to clustering can still be found for finite systems of randomly coupled maps.

In domains of parameters where globally coupled maps show partial ordering, the addition of either multiplicative or additive external noise of low intensity has an effect very similar to the introduction of disorder in the form of random connections. This equivalence is, however, only qualitative, since the mechanisms by which the dynamics is modified in the case of quenched disorder (random links) or noise are clearly different. These two situations can be compared by plotting the distribution of pair distances between the elements. Note that full synchronization corresponds to a single peak at $d = 0$, while the formation of groups corresponds to a number of isolated peaks. If entrainment is partial but the system tends to be synchronized, then a broad distribution with a maximum at the origin is expected. Fig. 7.16 compares those histograms for a globally coupled system with white noise of amplitude 10^{-3} added and for randomly coupled

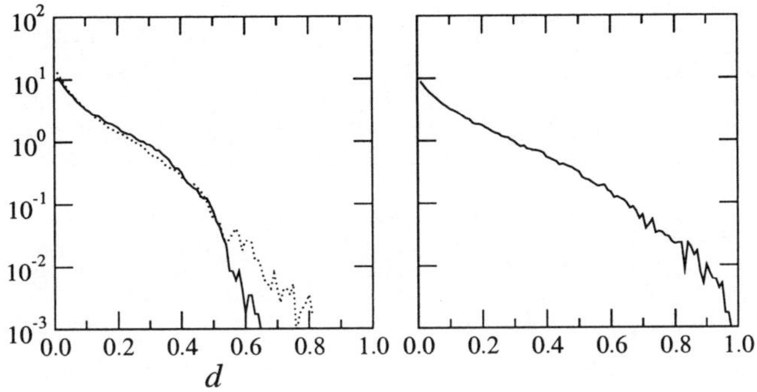

Fig. 7.16 Histograms of pair distances for globally coupled logistic maps with added noise (left, the continuous line is for multiplicative noise, the dotted line for additive) and for randomly coupled logistic maps (right). The histograms have been averaged over time. Adapted from [Manrubia and Mikhailov (1999)].

maps with parameters $a = 2$, $K = 0.45$, and $\kappa = 0.8$. If neither noise nor disorder in the links are present, the histograms have a much more irregular structure, with many peaks that are maintained even if time averaging is performed.

Such approximate equivalence in the dynamical behavior between the disorder due to external noise and that produced by random connections is observed only in the regime where coupled logistic maps present "intermittent" dynamics. In the intermittent phase it is common that many clusters of low stability are formed, and that the final attractor for the dynamics is strongly dependent on the initial conditions.

7.3.3 *Coupled map lattices*

We close this chapter with a brief presentation of coupled map lattices, which are one of the most paradigmatic examples of spatiotemporal chaos. They were first introduced in 1984 with the aim of systematically investigating high-dimensional, chaotic extended systems within a simple framework [Kaneko (1984)]. The simplicity of the model is, however, only apparent. It displays various complex collective behaviors and can be used as a toy model for such classical problems as fully-developed turbulence, pattern formation, and phase transitions in spatial structures.

The equations of motion for a one-dimensional coupled map lattice with diffusive coupling are

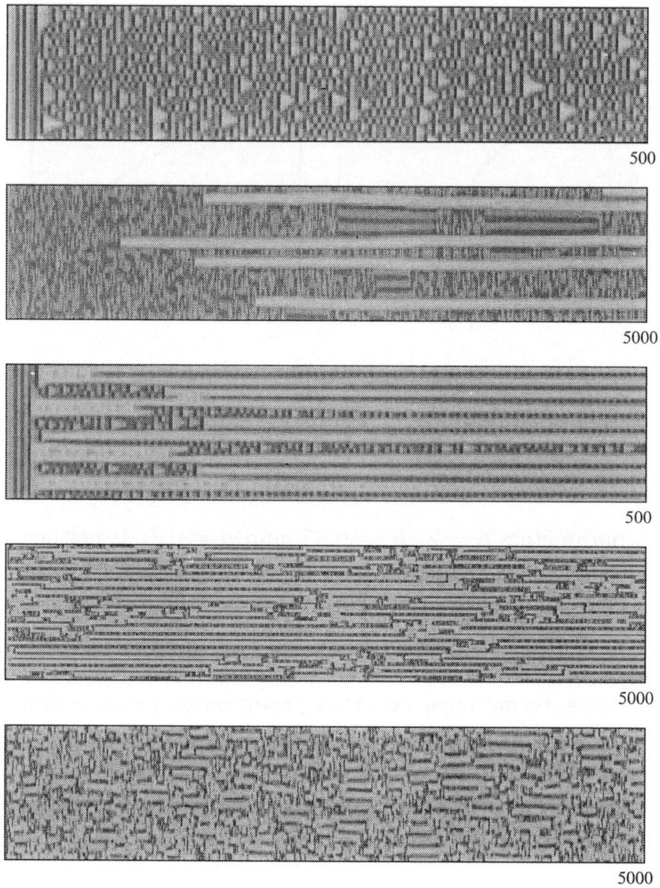

Fig. 7.17 Spatiotemporal patterns in arrays of coupled logistic maps formed by $N = 200$ elements. From top to bottom the parameters are: $a = 1.56$, $K = 0.45$, $a = 1.56$, $K = 0.45$ (but the time scale has been increased tenfold); $a = 1.63$, $K = 0.3$, $a = 1.7$, $K = 0.12$, $a = 1.8$, $K = 0.42$. The states of the maps are displayed at every second step, and the total evolution time is specified on the lower right corner below each plot.

$$x_i(t+1) = f[x_i(t)] + K\left\{f[x_{i-1}(t)] - 2f[x_i(t)] + f[x_{i+1}(t)]\right\}, \qquad (7.71)$$

where periodic boundary conditions are usually considered. We know already that there exists a maximal size for a ring of coupled logistic maps above which full synchronization is not stable. But, similarly to all other systems studied, even if the complete coherence of all the states of the oscillators in the system cannot be achieved, the attractive effect of coupling induces sometimes an intermittent behavior and often causes the emergence of local order [Kaneko (1985)].

Figure 7.17 presents some characteristic patterns that can be found in a one-dimensional array of logistic maps under variation of the parameters a and K. The two upper plots correspond to the same parameter values, $a = 1.56$, $K = 0.45$, and show the formation of patterns on different time scales. On the scale of ten time steps and with at most some tens of oscillators participating, a behavior similar to that observed in some of Wolfram's cellular automata [Wolfram (1983)] takes place. On a longer time scale, the coherent local bursts do not die out but become stabilized. The coexistence of ordered regions of a characteristic size and adjacent domains with disordered dynamics is observed. Sometimes periodic motions are developing at the boundaries of coherent domains. As a result, these domains are able to drift through the lattice. The evolution of the array shown in the middle plot, for $a = 1.63$ and $K = 0.3$, settles faster and apparently more regularly to a stripped pattern with alternating phases for the oscillators. As a is further increased (second panel from below) intermittency can be observed. Long regions of laminar behavior are interrupted by disordered bursts. For still larger a, even if the coupling strength is large, the "turbulent" behavior is dominant, and the laminar regions become much smaller.

In this chapter, we have explored different systems which, despite their different definitions, behave similarly. It is of great interest to disentangle the properties which are common to all of them, or the different types of universal behavior they display, as opposed to other features which might depend on particular characteristics of the model under study. Often, systems which differ in their microscopic formulation display the same qualitative (and even quantitative) behavior when they are close to critical points of phase transitions. Frequently, the presence of long-range correlations between the states of the elements, despite the connectivity being local, is a signature of such transitions. Moreover, close to instability points (e.g. a Hopf bifurcation), different dynamical systems behave analogously because, near the instability, they are structurally similar.

Indeed, certain patterns and dynamical behaviors recur in many of the systems explored. This qualitative observation points to the plausibility of a statistical theory of collective behavior in extended systems [Chaté and Manneville (1992)]. For example, spatiotemporal intermittency has been observed in a large number of systems and different studies have tried to determine some of its potentially universal properties. A typical quantity that can be compared is the distribution of laminar lengths —that we have described for coupled Lorenz oscillators (Sec. 6.3.3)— which in particular has been analyzed for inhomogeneous arrays of logistic maps [Sharma and Gupte (2002)]. Still, the question about the universal characteristics of different spatiotemporal behaviors remains open and constitutes an active field of research. Taking a broader viewpoint, different synchronization transitions in extended systems can be described within a common framework, and two clearly different types of transition —in the universality class of directed percolation and in that of the Kardar–Parisi–Zhang equation— have been characterized by means of a general Langevin equation [Muñoz and Pastor-Satorras (2003)]. Within this formulation, all the different types of spatiotemporal behavior described in coupled maps can be recovered.

Chapter 8

Clustering

Among the plethora of various collective behaviors displayed by ensembles of chaotic elements, clustering plays a prominent role. The formation of a number of separated dynamical groups in a system is a highly unpredictable phenomenon which, nonetheless, is characterized by a remarkable degree of order. Clustering is an emergent property resulting from a non-trivial feedback between the individual and the global dynamics. When it takes place, the initial symmetry inherent to the system's equations of motion is spontaneously broken.

In globally coupled ensembles formed by identical elements, all oscillators within each cluster are fully synchronized. In heterogeneous ensembles, a weaker form of clustering is also possible. Similar to full synchronization in heterogeneous systems, the dynamical states of elements in a cluster are not exactly the same, and clustering holds only up to a certain precision. The phenomenon of transient dynamical clustering, where elements wander between clusters or whole clusters merge and split in the course of evolution, is observed.

A special form of collective behavior closely related to clustering is the development of hidden order. This refers to the appearance of correlated dynamical states though, apparently, the individual oscillators follow independent trajectories and the system is in a turbulent state.

In this chapter, we consider clustering and other kinds of collective dynamics which are different from full synchronization, and analyze the relationship between the properties of the individual oscillators and various collective behaviors in their ensembles.

8.1 Dynamical Phases of Globally Coupled Logistic Maps

We have already discussed in previous chapters some of the rich phenomenology exhibited by coupled logistic maps. The logistic map is a model system representative of a broad class of equations of motion. Most collective phenomena encountered in logistic maps are also found in other dynamical systems where the individual chaotic map (or oscillator) has windows of periodic behavior. Similar collective phenomena are found, for example, in Rössler oscillators.

It is hardly possible to give a complete description of all the dynamical regimes of globally coupled maps for all values of the parameters a and K. Except for a few cases, there is no predictable relationship between the values of these two parameters and the collective dynamics. When a is varied, the logistic map undergoes structural changes, so that chaotic and periodic dynamics intermingle. Hence, minimal changes in a can induce a transition from one behavior to another. Moreover, the addition of coupling can effectively modify the nature of the orbits. And finally, if the system is run starting with different initial conditions, different attractors may be chosen.

Let us recall the equation defining the dynamics of globally coupled maps,

$$x_i(t+1) = (1-K)f[x_i(t)] + K\frac{1}{N}\sum_{j=1}^{N} f[x_j(t)], \tag{8.1}$$

where K is the coupling constant and $f(x) = 1 - ax^2$ is the logistic map. We have seen that, for any a, there is a sufficiently large value of K such that full synchronization is achieved. As discussed in Sec. 7.3.1, the condition for stable full synchronization is $\lambda + \ln(1-K) \leq 0$. The dependence of the Lyapunov exponent λ on the parameter a of the logistic map is highly irregular due to the frequent appearance of periodic windows in the chaotic domain (see Fig. 6.1). The curve $K = 1 - \exp(-\lambda)$ defines the boundary of stable full synchronization. In the clustering phase where the fully synchronous state is stable, it coexists with other attractors.

For low values of K and relatively large a, the global behavior of the system is apparently uncorrelated. No obvious dependence between their states is found, and the ensemble is in the turbulent phase. Still, there are some correlations in the collective dynamics of the system, which are however more difficult to unveil than the obvious coincidence of the states of the elements.

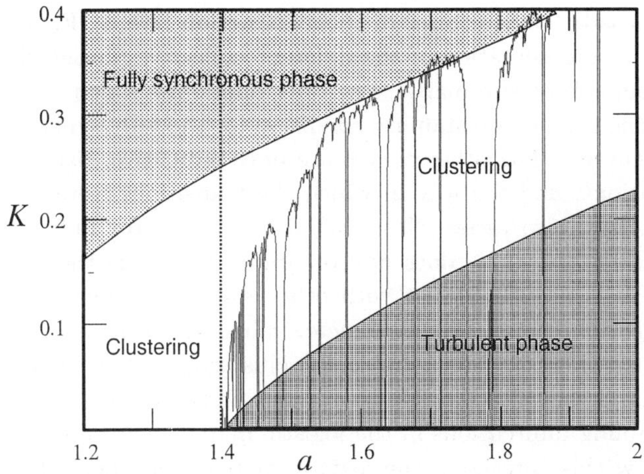

Fig. 8.1 Phase diagram of globally coupled logistic maps. The main three phases of the system, separated by smooth curves, are turbulence, clustering, and full synchronization. The irregular line represents the exact boundary where the fully synchronous state becomes unstable, $K = 1 - \exp(-\lambda)$. Adapted from [Kaneko (1990)].

Between the turbulent and the fully synchronous phase, a broad band with complex collective behavior is found [Kaneko (1989); Kaneko (1990a); Balmforth *et al.* (1999); Popovych *et al.* (2001); Manrubia and Mikhailov (2001)]. Here, the symmetry of the initial ensemble is broken as time elapses. In the phase with clustering, the ensemble splits into a number M of groups, each with N_i elements, for $i = 1, \ldots, M$. Inside each of the clusters, full synchronization occurs. Usually, different clusters tend to occupy different areas of phase space or display anti-phase oscillations. Stabilization of periodic orbits is typically observed when only a few clusters are present.

When the system is attracted to a state with many different clusters, it might also happen that some elements remain non-entrained or join one or another cluster, in an itinerant fashion. Sometimes, long transients are required before all of the elements eventually fall on the final attractor. As a consequence, it may be difficult to completely characterize the nature and properties of the final set of attractors and their basins of attraction.

In the whole domain where the formation of clusters occurs (between the turbulent and fully synchronous phases), the finally chosen attractor depends on the initial conditions. Usually, many different attractors coexist, their number depending on the system size N for fixed a and K, and with broadly varying sizes of their attraction basins. Figure 8.1 is an approximate representation of the different collective phases displayed by globally coupled logistic maps. The only boundary which is exact corresponds to the transversal stability condition for the fully synchronous state (irregular curve). The other curves are only indicative, based on numerical simulations, and the way in which the transitions from one phase to another proceed also varies. Note that the fact that the fully synchronous state exists and is stable does not yet imply that it is the only possible attractor. It may well coexist with other solutions which can have even larger attraction basins. The parameter values for which such situation is found are classified as belonging to the clustering region in the phase diagram. The dotted vertical line at a_∞ signals the accumulation point for period doubling bifurcations in the logistic map. For values of the parameter a below this threshold, the logistic map is periodic. This does not, however, imply that the synchronization properties of the system become then trivial. Even though the fully synchronous state is always linearly stable in that whole domain, it cannot be reached starting from random initial conditions if coupling is too weak.

8.1.1 *Two-cluster solutions*

The simplest form of clustering in an ensemble consists in a partition of the system into two subgroups of sizes N_1 and N_2 [Xie and Hu (1997); Balmforth *et al.* (1999); Popovych *et al.* (2001)]. Suppose that two groups are formed in a system with $N = N_1 + N_2$ elements. All of the oscillators within each group follow exactly the same dynamics, and the set of N original equations can be reduced to two effective evolution equations,

$$\begin{aligned} x_1(t+1) &= (1-K)f[x_1(t)] + K\left(pf[x_1(t)] + (1-p)f[x_2(t)]\right) \\ x_2(t+1) &= (1-K)f[x_2(t)] + K\left(pf[x_1(t)] + (1-p)f[x_2(t)]\right), \end{aligned} \tag{8.2}$$

where

$$p = \frac{N_1}{N}, 1 - p = \frac{N_2}{N}. \tag{8.3}$$

For convenience, the elements in the system will be labeled in such a way that numbers 1 to N_1 correspond to elements in the first cluster, and N_1+1 to N_1+N_2 correspond to those in the second cluster. In addition, clusters are ranged according to their sizes, so in this example we would have $N_1 \geq N_2$.

Despite the formal similarity of Eq. (8.2) with the case of two coupled maps described in Chapter 6, they have intrinsic differences. Each of the "effective" oscillators in Eq. (8.2) is actually a cluster composed of many elements. Each group can be understood as a fully synchronized subset and immediately the question arises, whether each of the groups will be stable under perturbations of the states of its constituent elements.

In a system where clustering is present, each cluster has an associated transversal exponent which determines its stability. In the case of two clusters, the linearized equations for perturbations of the two trajectories $s_1(t)$ and $s_2(t)$ are

$$\begin{aligned} \boldsymbol{\eta}_1^j(t+1) &= (1-K)f'[s_1(t)]\boldsymbol{\eta}_1^j + Km(t) \\ \boldsymbol{\eta}_2^k(t+1) &= (1-K)f'[s_2(t)]\boldsymbol{\eta}_2^k + Km(t) \end{aligned} \tag{8.4}$$

where $j = 1, \ldots, N_1$ and $k = 1, \ldots, N_2$ label all possible directions along which the orbits of oscillators in clusters 1 and 2 can be perturbed, respectively [Balmforth *et al.* (1999)]. The linearized coupling field $m(t)$ is

$$m(t) = pf'[s_1(t)]\sum_{j=1}^{N_1}\boldsymbol{\eta}_1^j + (1-p)f'[s_2(t)]\sum_{k=1}^{N_2}\boldsymbol{\eta}_2^k. \tag{8.5}$$

Perturbations along the orbits s_1 and s_2 are quantified through the usual Lyapunov exponents for the two-dimensional system (8.2). There are two types of perturbations that are transverse to the two-cluster manifold and that determine their stability:

$$\boldsymbol{\eta}_1^j(t) = \mathbf{A}(t)\mathbf{u}(j), \quad \boldsymbol{\eta}_2^k(t) = \mathbf{B}(t)\mathbf{v}(k), \tag{8.6}$$

These two matrices are mutually orthogonal and independently probe the stability of one or the other cluster. The vectors $\mathbf{u}$ and $\mathbf{v}$ satisfy

$$\sum_{j=1}^{N_1} \mathbf{u}_j = 0, \sum_{k=1}^{N_2} \mathbf{v}_k = 0, \tag{8.7}$$

so that the coupling field vanishes in both cases and the orientation of the two perturbations remains fixed. The linearized equations can be written as

$$\begin{aligned} \mathbf{A}(t+1) &= (1-K)f'[s_1(t)]\mathbf{A}(t) \\ \mathbf{B}(t+1) &= (1-K)f'[s_2(t)]\mathbf{B}(t), \end{aligned} \tag{8.8}$$

yielding finally for the transversal exponents

$$\begin{aligned} \lambda_\perp^1 &= \ln(1-K) + \lim_{t\to\infty} \tfrac{1}{t}\textstyle\sum_{i=1}^t \ln|f'[s_1(t)]| \\ \lambda_\perp^2 &= \ln(1-K) + \lim_{t\to\infty} \tfrac{1}{t}\textstyle\sum_{i=1}^t \ln|f'[s_2(t)]| \end{aligned} \tag{8.9}$$

with an $(N_1 - 1)$–fold degeneracy for the first one and an $(N_2 - 1)$–fold degeneracy for the second one. The above exponents are called *splitting exponents.* Two elements inside a cluster would separate (split) at a rate depending on the transversal exponent along their common orbit [Kaneko (1994b)]. Note moreover that the value of the exponent does not explicitly depend on the size of the cluster (though each of the synchronous trajectories do).

The reduced system (8.2) permits a systematic exploration of the stability associated with partitions into two clusters without explicitly carrying out numerical simulations for the whole ensemble. Note that p can be viewed as an external parameter which can be independently varied. The condition for a given partition to be stable is that all the transversal exponents corresponding to each of the clusters are negative. The transversal exponents take different values for each cluster unless they are identical, that is, formed by the same number of elements.

A systematic study of two-cluster solutions has shown that there is a strong correlation between the transversal stability of the clusters and their dynamics. When the dynamics is periodic, the typical value of the transversal exponent associated turns out to be much lower than when the dynamics is chaotic [Balmforth *et al.* (1999)]. In fact, a close correspondence between the change of sign of the Lyapunov exponent and the transversal exponent is found in many cases. When large ensembles are considered, the appearance of clustered states is typically accompanied by the stabilization of periodic dynamics.

The stability domains in the plane (p, K) are very irregular. Not only some large areas where the system is stable exist, but also very small stability islands scattered all over the plane are present. Figure 8.2 shows a representative example for $a = 1.9$. For values of K slightly larger than 0.4 the stable situation is that of full synchronization.

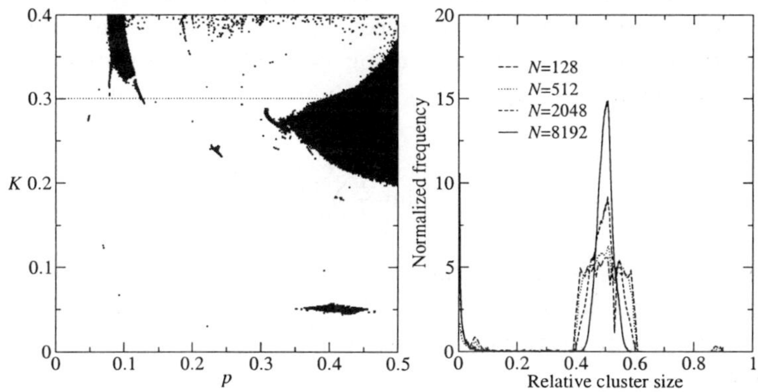

Fig. 8.2 Left: Black areas are those domains of parameters for which the partition in two clusters represented as two coupled oscillators (8.2) is stable. The dotted line indicates the coupling strength $K = 0.3$ used in the simulations yielding the distributions shown in the next plot. Right: Distribution of cluster sizes for increasingly large systems. The partition chosen by the whole ensemble falls on the stability area.

An actual ensemble of globally coupled logistic maps does not have the effective parameter p as a real degree of freedom. However, it is often observed that the ensemble eventually falls into the attraction basin of a stable partition, so not all values of p have the same probability. In Fig. 8.2 we can see an example of the partition chosen by several large ensembles with the same parameter values. The partition is quantified through the normalized distribution of relative cluster sizes p_k, which is defined as

$$p_k = \frac{N_k}{N}, \tag{8.10}$$

for $k = 1, \ldots, M$, thus allowing to compare systems with different total number of elements. In Fig. 8.2 the parameters are chosen as $a = 1.9$ and $K = 0.3$. As the size of the system increases, the partition corresponding to two identical clusters gains weight [Manrubia *et al.* (2001)]. Note that this particular partition belongs to a large stability domain in the diagram displayed on the left side of Fig. 8.2.

8.1.2 *Clustering phase of globally coupled logistic maps*

The stability of a two-cluster partition is characterized by two different transversal Lyapunov exponents given by Eq. (8.9). This result can be generalized to an arbitrary number of clusters. A partition $\{N_1, N_2, \ldots, N_M\}$ of an ensemble of N globally coupled logistic maps will be stable if all transversal exponents calculated along each of the corresponding synchronous trajectories $s_1, s_2, \ldots, s_M$ are simultaneously negative,

$$\lambda_{\perp}^{i} = \ln(1-K) + \lim_{t\to\infty} \frac{1}{t} \sum_{i=1}^{t} \ln |f'[s_i(t)]| < 0, \tag{8.11}$$

where $i = 1, \ldots, M$ is the total number of clusters. Non-entrained elements (that is, "clusters" formed by a single element) can have positive transversal exponents without destroying the stability of a given partition [Kaneko (1994b)].

An attractor of the dynamical system (8.1) is completely specified by the set of numbers $\{M; N_1, N_2, \ldots, N_M\}$. Since under global coupling all elements are equal, the partition is not sensitive to the particular labeling of the maps. We assume that, for each partition, cluster sizes are ordered in such a way that $N_1 \geq N_2 \geq \cdots \geq N_M$. There is a one-to-one correspondence between the partition characterizing the attractor and its dynamics. This is important because, then, a study of the dynamical attractors for systems of globally coupled, identical maps, can be carried out through the investigation of final partitions. Note that this is no longer true if the elements are not identical and the equations of motion are not invariant under an arbitrary permutation of the elements. Such asymmetries among the elements naturally arise in connection topologies different from global coupling.

A stable partition, once reached, is invariant under the dynamics. However, the typical time to fall into a given attractor might vary broadly depending on the parameters chosen [Manrubia and Mikhailov (2000b); Abramson (2000)]. As a result, a system eventually collapsing into a small number of periodic clusters can be mistaken for an attractor with many clusters, though, actually, it is only a transient state. Since most investigations of large systems rely on numerical simulations, it is important to be careful and assure that (i) the system has reached the final attractor, and (ii) the corresponding partition is stable. This latter property is checked through the computation of the transversal exponent corresponding to each cluster.

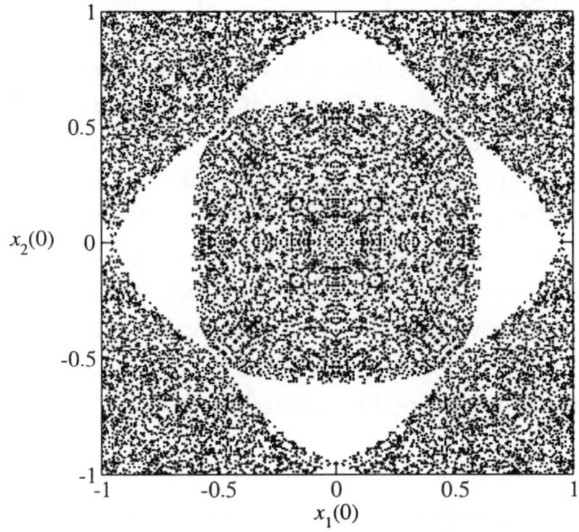

Fig. 8.3 Initial conditions for two oscillators out of a system with four leading to the fully synchronous state. This attractor with chaotic dynamics coexists with a two-cluster attractor where two pairs of oscillators follow periodic orbits.

Let us now give an example of coexistence of two attractors in a small system formed by $N = 4$ globally coupled maps. We have chosen the parameters inside the clustering phase, but so that the fully synchronous state is linearly stable: $a = 1.5$, $K = 0.235$, see Fig. 8.1. The system is prepared as follows: the initial conditions for two of the oscillators are kept fixed, while the initial values for the states of the other two are systematically varied to explore the entire parameter plane. Black dots in Fig. 8.3 indicate all pairs of initial values $x_1(0)$ and $x_2(0)$ leading to the fully synchronous state, for the other two initial conditions fixed at $x_3(0) = -0.9$, $x_4(0) = -0.2$. In this case, the attraction basin of the synchronous state is such that arbitrarily near every point in this basin there are other initial conditions leading a different dynamical attractor. When this happens, we say that the basin of attraction of the full synchronous state is *globally riddled* with respect to the two-clusters partition [Ott *et al.* (1993); Popovych *et al.* (2001)].

For a large system there are thus regions in the parameter space where many different attractors coexist. The cluster distribution function $Q(M)$ weights the fraction of initial conditions leading to a given attractor, and thus the attraction basin of a stable partition into M clusters. It is defined as the ratio between the number of initial conditions leading to a particular attractor divided by the total number of initial conditions explored [Kaneko (1990a)]. The collective behaviors corresponding to different parameters can be classified in terms of the relative abundance of partitions with a certain number of clusters. In most of the clustering phase, partitions with $M = 2$ are dominant. As the coupling strength K decreases and the turbulent phase is approached, also cases with $M = 3, 4$, and higher can be found. Figure 8.4 shows how the cluster distribution functions $Q(M)$ change when a is varied from 1.4 to 2 with coupling $K = 0.1$. The displayed curves reflect changes of dominance of partitions with an increasing number of clusters. For a above 1.62, approximately, only the turbulent phase is found, and all the partitions identified have more than $M = N/2$ clusters. This calculation was performed in a system with $N = 200$ elements [Kaneko (1990a)].

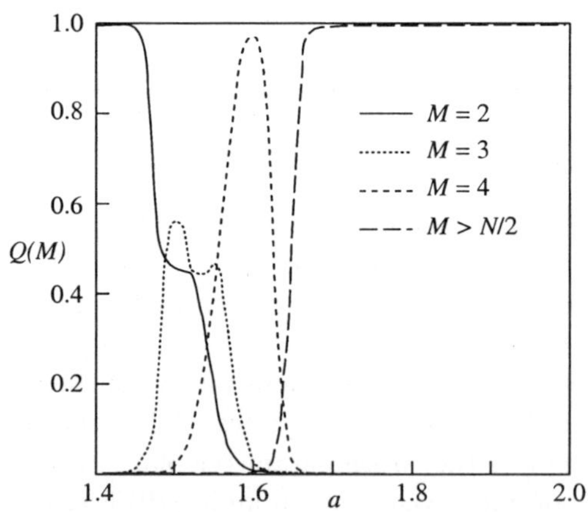

Fig. 8.4 Fraction of initial conditions leading to partitions with $M = 2, 3$, and 4 clusters. Also the curve for the fraction of partitions with more than $N/2$ clusters is shown. It reaches unity approximately when the system enters the turbulent phase. Adapted from [Kaneko (1990)].

The total number of different attractors can be huge even if they correspond only to two or three clusters. Consider for example the situation where two groups of similar size are formed, and the particular case shown in Fig. 8.2, where the region of stability for $K = 0.3$ starts around $p_s = 0.4$. A system with N elements forming 2−cluster partitions would have then around $(1/2 - p_s)N$ possible stable attractors, a number that grows proportionally to N. If the restriction of a fixed number of clusters is eliminated, the maximal number of attractors $A(N)$ for a system formed by a large number of elements ($N \gg 1$) is asymptotically given by the expression [Goldberg *et al.* (1970); Crisanti *et al.* (1996)]

$$A(N) \simeq \frac{e^{\pi\sqrt{2N/3}}}{4N\sqrt{3}}. \tag{8.12}$$

Considering that the realized attractors are limited by two conditions, stability and accessibility, this number will be much smaller in practice, but still extremely large for parameters lying in the clustering phase.

The coexistence of attractors also permits that, responding to an external perturbation, an element belonging to a cluster leaves it and joins another group. This allows the exploration of the set of attractors available for given parameters, and accessing of states which might not be spontaneously achieved by a large ensemble starting with random initial conditions (recall the example shown in Fig. 8.2). Since clusters of an actual attractor are all stable by definition, there is a minimum perturbation strength required for this reorganization to happen. Switching of elements between clusters in response to an external input at a single site has been considered in the first study of globally coupled logistic maps [Kaneko (1989); Kaneko (1990a)]. A change in the number of elements composing each cluster might also imply a qualitative modification of the dynamics of the ensemble, for example the periodicity can increase or decrease through cascade bifurcations or the current partition might become unstable.

In summary, the defining property of the clustering phase is the coexistence of a large number of different attractors depending on the initial conditions. This permits to establish a parallelism between multi-attractor dynamical systems and the so-called glassy systems, a topic that we treat in detail in the next chapter.

8.1.3 *Turbulent phase*

There is a broad range of values of a and K where a large ensemble of globally coupled logistic maps seems to evolve in an uncorrelated fashion, in the sense that no two elements share their dynamical state. This domain is called *turbulent* because of its qualitative similarity with the turbulent state observed in hydrodynamics. However, as well as in hydrodynamical turbulence, some degree of order persists at low values of the coupling intensity, even when clustering is not present. Such *hidden order* can be disentangled by using a suitable measure of the global dynamics: the dynamical properties of the coupling field [Shibata and Kaneko (1997); Shibata *et al.* (1999); Cencini *et al.* (1999)].

The behavior of a large ensemble of globally coupled maps is intrinsically dependent on the collective dynamics of all its elements, while at the same time it is the influence of the rest of the ensemble on each element that determines its evolution. The collective action of the population is represented by the coupling field

$$m(t) = \frac{1}{N} \sum_{i=1}^{N} f[x_i(t)]. \tag{8.13}$$

If the coupling field shows chaotic dynamics, so do also the elements of the ensemble. The periodicity of cluster dynamics is reflected in the regular dynamics of $m(t)$ as well. Figure 8.5 displays two time-series of $m(t)$ for a system with $N = 100$ elements. Different initial conditions have been used, so that two different final attractors are finally reached. Above, the fully synchronous state corresponds to a coupling field with chaotic motion whose dynamics is identical to that of the single oscillator; below, the coupling field reflects the period-two dynamics corresponding to an attractor consisting of two clusters with sizes $N_1 = 65$ and $N_2 = 35$.

The coupling field is also a suitable quantity to detect the presence of more subtle forms of order in the system, even when no clustering is present and when all the elements seem to evolve independently. In the turbulent phase, it could be naively expected that the coupling field behaves as the sum of N independent random variables (with a distribution given by the corresponding invariant measure of the logistic map). If this is true, for increasing N the coupling field $m(t)$ should converge to a Gaussian distribution with mean square deviation σ^2 inversely proportional to N,

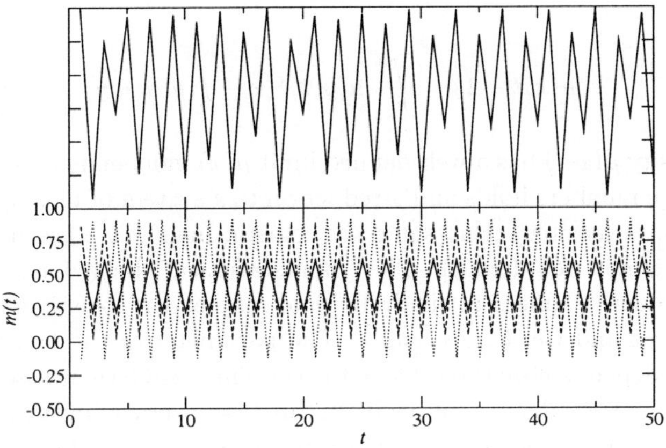

Fig. 8.5 Dynamics of the coupling field for a system with 100 globally coupled logistic maps. Two different attractors have been reached: a fully synchronous state (above) and two clusters of different size (below). The coupling field $m(t)$ is represented with a solid line, the two trajectories $s_1(t)$ and $s_2(t)$ are displayed with dotted and dashed lines, respectively. In this example $a = 1.43$, $K = 0.17$.

$$\sigma^2 = \langle m^2(t) \rangle - \langle m(t) \rangle^2 \propto \frac{1}{N}. \tag{8.14}$$

If this holds, then, in the limit $N \to \infty$, it would be possible to reduce the coupled system to a set of uncoupled logistic maps of the form

$$x_i(t+1) = (1-K) f[x_i(t)] + K \langle m \rangle \tag{8.15}$$

where $\langle m \rangle$ is the constant average value of the coupling field. However, it turns out that, though the distribution of coupling-field values first converges to a Gaussian function, the convergence is terminated at a finite value of N and after that the mean square deviation becomes constant independently of N [Kaneko (1990b)]. Such irregular behavior of large systems results from the fact that the distribution of the states of the elements is time-dependent [Kaneko (1992); Pikovsky and Kurths (1994)]. At a given time t, one can calculate the probability $\mu(x; t)$ that an element in the ensemble takes value x. The evolution of this density is governed by the mapping

$$\mu(x; t+1) = \int_{-1}^{1} \mu(y; t)\delta(x - f(y, a))dy, \tag{8.16}$$

If the density $\mu(x;t)$ has a well-defined limit $\mu(x)$ independent of t, then the law of large numbers holds and a reduction to a system of uncoupled, noisy oscillators, is feasible. But since the logistic map $f(y, a)$ is non-mixing, this limit is not well-defined. This "non-statistical" behavior is related to the non-smooth changes in the individual map as the logistic parameter a varies [Pérez and Cerdeira (1992)]. The parameter of the logistic map is effectively dependent on the value of the coupling field (see next section) and thus becomes time-dependent itself. In this case, averages over the ensemble at a fixed time are not equivalent to averages over time for the coupling field and, as a consequence, deviations from the law of large numbers are observed.

Again, it is the dense presence of periodic windows that qualitatively modifies the collective behavior of globally coupled logistic maps. It has been shown in Sec. 7.3.2 how the introduction of disorder in the form of inhomogeneous coupling translates into large structural changes in the trajectories of individual maps, thus preventing the appearance of clustering. In the current case, the heterogeneity in the logistic parameter a appears in time and is internally generated by the ensemble, eventually modifying the structural properties of the attractor.

The addition of noise to the globally coupled system (8.1) permits to recover the law of large numbers for the fluctuations in the coupling field [Kaneko (1994b)]. Some other forms of "noise" also lead to the expected scaling $\sigma^2 \propto N^{-1}$. One possibility is to abandon the classical synchronous update of the ensemble of logistic maps and use instead an asynchronous updating scheme [Abramson and Zanette (1998)].

Asynchronous updating amounts to updating one randomly chosen element at a time, recalculating the coupling field with the updated element, and then proceeding with the next randomly chosen element. Such stochastic procedure introduces an effective noise into the system and leads to collective states of high order. With this scheme it was found that, at any fixed a, the dynamics of any element in the ensemble exhibits an inverse bifurcation cascade which terminates with a stable fixed point for the dynamics as the coupling K increases. For example, for $a = 2$ chaos is completely suppressed (except for some noise in the orbits) for any K above 0.2. Now it is possible to use the approximate form (8.15) to calculate some features of the phase diagram characterizing an ensemble of globally coupled logistic maps with asynchronous updating.

Let us consider the most ordered situation in which the motion goes to a stable fixed point. In the case where the dynamics is described by the effective equation (8.15), the measure $\mu(x)$ equals a delta function at $x = x_0$, where x_0 is the fixed point defined by

$$x_0 = (1 - K)f(x_0) + Kf(x_0). \tag{8.17}$$

This fixed point is stable if the coupling intensity satisfies $K > 1 - |f'(x_0)|^{-1}$, a condition that is obtained from the linearized form of the effective evolution equation. The stability boundary obtained from this condition, as well as the corresponding curve for the period-2 solution fully agree with the numerical results, thus supporting the validity of the approximation (8.15). Moreover, this is an indirect proof that the fluctuations of the coupling field follow the law of the large numbers in this case. Now, the addition of noise just enhances the degree of independence among the elements and outweighs the ordering effect of the coupling, therefore playing a role equivalent to that of external noise.

Still, the effect of noise in dynamical systems can be subtle and highly unpredictable. Generally, noise neither necessarily increases the disorder that might be present in the system due to other factors nor enhances the independence of different elements.

The addition of noise of small amplitude to globally coupled logistic maps in their turbulent phase is an example of a situation where noise enhances the internal coherence of the system. Other forms of disorder, for example a distribution of parameters a for the maps in the ensemble, are also able to induce a high degree of order [Shibata and Kaneko (1997)].

It has been demonstrated that if additive Gaussian noise of amplitude S is included in the dynamical equations of an ensemble of globally coupled logistic maps, the macroscopic behavior of the ensemble becomes simplified as S increases [Shibata *et al.* (1999)]. The emergence of collective order despite the chaotic and incoherent individual dynamics is shown in Fig. 8.6, where the return map of $m(t)$ is shown. The *return map* for the coupling field is constructed by representing the pair of quantities $(m(t), m(t+1))$ on a plane. If it consists of a single point, then the coupling field takes a single value along the dynamics, and the law of large numbers holds. The appearance of more complex patterns, as a closed curve indicative of quasiperiodic motion for $m(t)$, is related to hidden order. In Fig. 8.6,

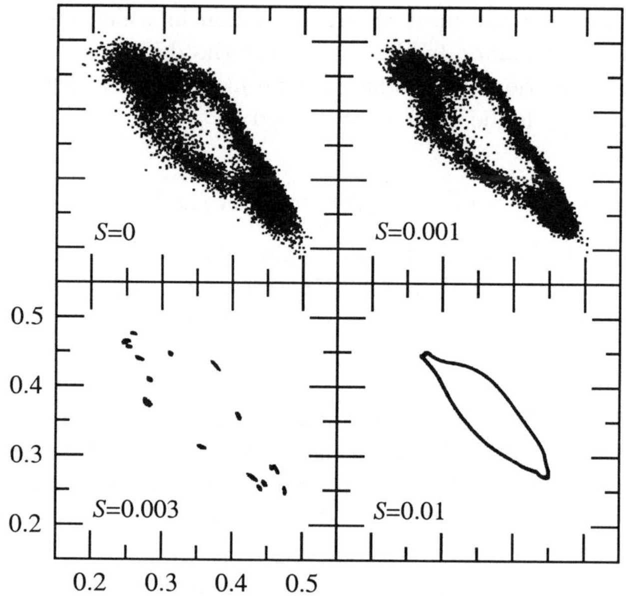

Fig. 8.6 Return map of the coupling field $m(t)$ for an ensemble of logistic maps. The collective motion becomes more ordered as the noise intensity increases. The variance S of the Gaussian noise is indicated in each plot. The logistic parameter is $a = 1.86$.

the return map of the coupling field $m(t)$ corresponding to a large system formed by $N = 10^6$ logistic maps is presented in the cases where noise is absent (upper left) and when Gaussian noise of zero average and dispersion S is added. As the intensity of noise increases, the behavior of the coupling field becomes more deterministic, as indicated by the disappearance of scattered points around the quasiperiodic dynamics of the return map. This counterintuitive effect of noise clearly exemplifies the non-trivial interplay between individual and collective dynamics. The addition of noise of small amplitude sharpens the collective dynamics. Eventually, for a high enough intensity, the return map of the coupling field collapses into a single point (except for fluctuations which now decay as $\sigma \propto 1/\sqrt{N}$), the law of large numbers is recovered, and an effective separation of the dynamical system in the form shown in Eq. (8.15) becomes possible.

8.2 Universality Classes and Collective Behavior in Chaotic Maps

Full synchronization, clustering, and collective motion in the turbulent phase (hidden order) are three types of emergent behavior observed in globally coupled logistic maps. As has been discussed at large in the previous chapter, full synchronization can be present in many different systems, composed of oscillators with different individual characteristics, and coupled under different schemes. But, how widespread is the presence of clustering? Is collective behavior universal or system-dependent? In this section we analyze the relationship between the dynamical properties of the individual oscillator and the different classes of collective phenomena possible in large ensembles of identical chaotic maps that are globally coupled in the form (8.1). No theory completely explaining this relationship is yet available. However, a number of numerically and partly analytically worked out examples provides us with a relatively complete picture of the kinds of systems able to display the different types of cooperative phenomena which we have already described.

The universality class represented by the logistic map is characterized by the dense presence of periodic windows as the parameter a is varied and by the map $f(x)$ being contracting for some values of x. This last property means that two oscillators with values of x in the appropriate range would tend to approach each other, at least for some time, since there are values of x where the map has a derivative smaller than 1 in absolute value,

$$|f'(x)| < 1. \tag{8.18}$$

At the turbulent phase, the presence of periodic orbits results in deviations from the law of large numbers. Even if no cluster formation is observed, individual elements tend to "cooperate" producing correlations in their individual trajectories which are reflected in the properties of the coupling-field values. If the fluctuations in the mean coupling field $\langle m \rangle$ do not vanish in the limit $N \to \infty$ limit, coupling cannot be treated as external noise, and it is not possible to effectively write down the dynamics as an individual oscillator plus a constant term. As evidenced by the numerical results that have been obtained for different systems, it seems that hidden order, remnant fluctuations of the coupling field (implying non reducibility of the coupled system in the limit $N \to \infty$), and time-dependence of the density $\mu(x;t)$, are always simultaneously observed.

There is a simple method allowing to get insight into how the dynamics of a single map is modified when it is coupled to other oscillators. It consists in using a single driven map as reduced analogue of the full system [Parravano and Cosenza (1998); Cosenza and Parravano (2003)],

$$x_i(t+1) = f[x_i(t)] + ch_t. \tag{8.19}$$

The driving term ch_t is intended to take into account, in an approximate manner, the influence of the other maps (compare this equation with Eq. (8.15)). Generally, the driving is time-dependent, but its value might be bound to a relatively narrow interval. In the simplest case, h_t can be considered as constant. Note that even small values of the driving term can change the qualitative nature of the dynamics and stabilize periodic windows which might be nearby in the parameter space. This was indeed proposed as a mechanism for synchronization in coupled maps [Shinbrot (1994)]. Though it is clearly not necessary for synchronization, this seems to be the way in which clustering originates. In fact, when stable clusters with two or more elements exist, their dynamics is in most cases periodic.

Consider a globally coupled dynamical system as that of Eq. (8.1), where the individual oscillator is now described by a tent map

$$f(x) = a(0.5 - |x - 0.5|). \tag{8.20}$$

This map belongs to a universality class different from the logistic map. In particular, stable periodic windows are not present, and the dynamical behavior of the individual oscillator changes smoothly as a varies. For $a < 1$, this function has a stable fixed point at $x = 0$. At $a = 1$ a line of marginally stable fixed points appears, and for $a > 1$ full chaos develops. The dynamics corresponds to two-band chaos until $a = \sqrt{2}$, where the two bands merge into a single one. The Lyapunov exponent corresponding to the tent map is $\lambda = \ln a$.

Full synchronization in an ensemble of globally coupled tent maps can be attained for values of the coupling $K > K_s = 1 - 1/a$ (see the condition in Eq. (7.62)). For $a > 1$, where the tent map is not contracting anywhere, two maps never approach each other if $K < K_s$. Extensive numerical simulations of globally coupled ensembles of tent maps have never found synchronization of a pair of elements below K_s [Kaneko (1995)].

As can be seen in Fig. 8.7, periodic windows are absent from the bi-

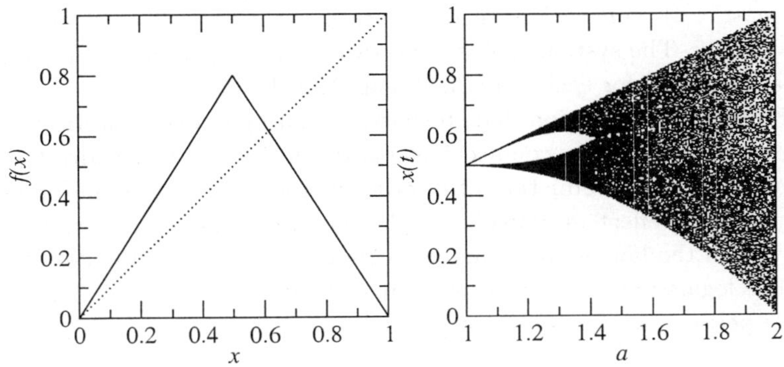

Fig. 8.7 Plot of a tent map with $a = 1.6$. On the right, the corresponding bifurcation diagram of the single tent map is shown.

furcation diagram of the tent map. Clustering is not observed in coupled tent maps, in agreement with the conjecture that clustering is only found when the map is locally contracting and stable periodic windows are present [Cosenza and González (1998); Cosenza and Parravano (2003)].

A system of globally coupled tent maps shows however hidden coherence for all values of a. The coupling field has non-vanishing fluctuations in the limit $N \to \infty$, though they have an amplitude much smaller than that characteristic for logistic maps. Examining the typical values taken by the coupling field in logistic maps (Fig. 8.6), we notice that they are only several times smaller as compared with the amplitude of the map dynamics. The same analysis for coupled tent maps in the domain $a > \sqrt{2}$ reveals that the coupling field shows here quasiperiodic dynamics with a magnitude in a range of $10^{-5} - 10^{-10}$ [Nakagawa and Komatsu (1998)]. Such small fluctuations are only detectable when they exceed the fluctuations due to finite-size effects, and hence can be observed only for very large values of N. As a consequence, the fluctuations in the coupling field follow the law of large numbers down to very large system sizes [Pérez and Cerdeira (1992)]. In tent maps, the presence of finite fluctuations in the limit $N \to \infty$ has been ascribed to the existence of "golden values" for the parameter a. When a equals one of these golden values, a trajectory of the individual map that begins at the peak of the map returns to it after a small number of time steps [Nakagawa and Komatsu (1998)]. In other words, there is an unstable periodic orbit. This produces a non-monotonous dependence of the amplitude of the collective motion on a and is the reason for the appearance of weak cooperative behavior. In the two-band domain (for

$1 < a < \sqrt{2}$), correlations arise naturally, since the orbits change band periodically. The system "splits" into two groups depending simply on the initial condition for each individual map [Kaneko (1995)].

Finally, let us mention that stochastic updating is able to destroy collective motion in globally coupled tent maps [Morita and Chawanya (2002)]. Recall that the introduction of disorder in the form of random updating had the same effect in globally coupled logistic maps, since it caused the recovery of the law of large numbers [Abramson and Zanette (1998)].

The logarithmic map is a function without maxima or minima, where stable periodic windows are not present. It is defined as

$$f(x) = b + \ln|x|. \tag{8.21}$$

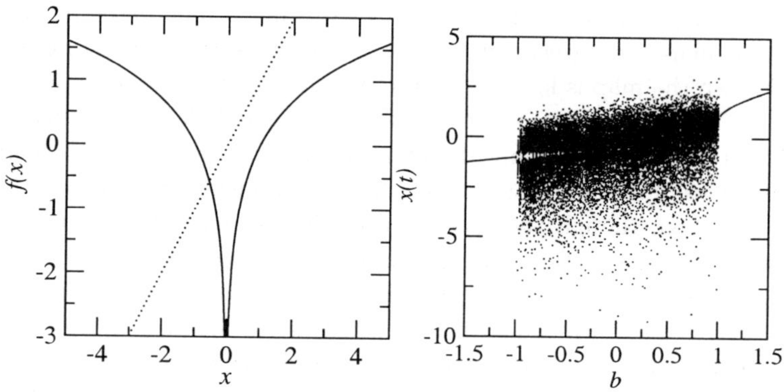

Fig. 8.8 Plot of a logarithmic map with $b = 0$. On the right side, the bifurcation diagram of the single map is represented.

The logarithmic map has chaotic dynamics in the interval $b \in (-1, 1)$. Within this range, there exist several unstable periodic orbits. Ensembles of logarithmic maps can fully synchronize [Cosenza and González (1998)] when a coupling of the form given in Eq. (8.1) is used. The stability condition for the fully synchronous state is equivalent to the one calculated for an ensemble of logistic maps (Eq. (7.62)),

$$\left|(1-K)\mathrm{e}^{\lambda}\right| < 1, \tag{8.22}$$

where λ is now the Lyapunov exponent of the single logarithmic map.

Note that despite the fact that the logarithmic map has locally contracting pieces where its derivative is smaller than unity in absolute value, no clustering has been numerically observed in this system. When the fully synchronous state looses stability, the system enters a turbulent regime.

Different collective states have been observed in the desynchronized phase of globally coupled logarithmic maps. For a fixed value of the coupling, and as b increases, the coupling field $\langle m \rangle$ might undergo a series of inverse bifurcations, until eventually a single point appears. An example of such behavior is shown in Fig. 8.9. The trajectories of the oscillators cluster alternatively around each value of the coupling field, but they do not synchronize within each cloud. Whenever the coupling field takes two or more different values, the law of large numbers does not hold, as expected, and non-trivial collective behavior arises. It is only when the values of the coupling field coincide with its average value that the fluctuations become Gaussian and the law of large numbers is recovered [Cosenza and González (1998)].

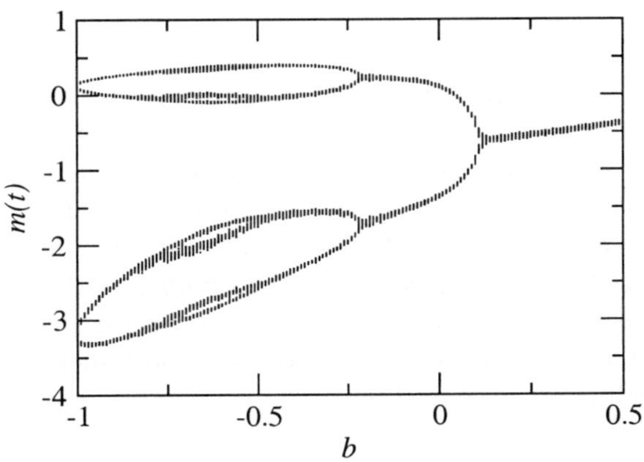

Fig. 8.9 Bifurcation diagram for the coupling field $\langle m \rangle$ corresponding to an ensemble of $N = 10^5$ logarithmic maps. For values of $b < 0.100...$, non-trivial collective behavior is observed. For larger values of b, the statistical behavior of the coupling-field fluctuations is recovered. The coupling intensity is fixed to $K = 0.25$.

Summarizing the discussed examples, it seems that the law of large numbers becomes violated whenever the attracting set of a considered map changes its topological structure as the parameters are varied. Coupling modifies the effective parameter experienced by each individual oscillator, and inhomogeneities in the measure $\mu(x;t)$ unavoidably arise.

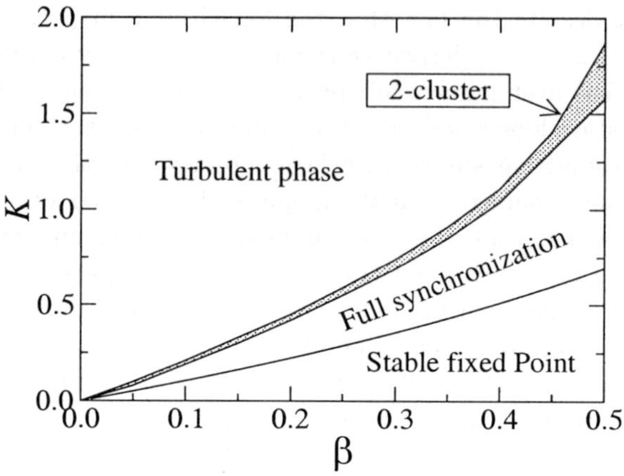

Fig. 8.10 Phase diagram of globally coupled, nonlinear saw-tooth maps described by Eq. (8.23). Note that the domains of the different phases shrink for $\beta \to 0$ and only the turbulent phase is found at $\beta = 0$. Then, the dynamical behavior is qualitatively equivalent to that of globally coupled tent maps. Adapted from [Manrubia and Mikhailov (2000)].

Individual maps have to be locally contracting for clustering to occur, but this alone is not sufficient. Usually, the presence of periodic windows is essential to get clustering. There is however an example of a system where a simple kind of clustering has been observed in absence of periodic windows. This example is provided by globally coupled nonlinear saw-tooth maps introduced in Sec. 6.2. Note that the coupling in this system is such that, when full synchronization is achieved, the dynamics does not coincide with that of the single element. Nonlinear saw-tooth maps with this form of global coupling are described by

$$x_i(t+1) = \{F(\{x_i\})x_i(t)(1-\beta x_i(t))\}, \tag{8.23}$$

where

$$F(\{x_i\}) = \exp\{K\langle x\rangle\}, \langle x\rangle = \frac{1}{N}\sum_{i=1}^{N} x_i(t), \tag{8.24}$$

and $\{\cdot\}$ denotes the operation of taking the fractional part of the argument. Under this coupling, the mapping becomes locally contracting in a domain of values of K and β, and two identical clusters are formed [Manrubia and Mikhailov (2000a)]. Note that a single map, even for positive values of β, never possesses a derivative lower that one in absolute value (see Fig. 6.3), and it is the coupling that induces it. The phase diagram of (8.23) is depicted in Fig. 8.10. The domains of full synchronization and desynchronization are separated by a narrow band where two clusters are observed. In the two-cluster phase, the system shows a transient during which contraction of the trajectories takes place. Interestingly, this transient always leads to the formation of two identical clusters, each with $N/2$ elements. Once such clusters are formed, the contraction of the trajectories disappears and further synchronization of the two clusters is not possible. The asymptotic dynamics is then equivalent to that of the antiphase state for two coupled maps discussed in Sec. 6.2 and shown in Fig. 6.4.

It should however be noted that the formation of two clusters in the latter ensemble of saw-tooth maps is atypical. In the majority of studied systems, complex clustering (i.e., coexistence of a large number of different attractors) is observed in globally coupled identical chaotic maps only if the individual map is locally contracting and has stable periodic windows.

8.3 Randomly Coupled Logistic Maps

The introduction of quenched disorder in coupled logistic maps does not prevent the appearance of clustering. In this section we analyze an ensemble of logistic maps coupled through a random network of connections. Instead of global connectivity, links between maps will be chosen at random with a fixed probability. This leads to a new dynamical phenomenon that can be described as fuzzy clustering. Under fuzzy clustering, only partial entrainment of the elements trajectories, not full synchronization, takes place within each cluster. The situation thus differs from that of the globally coupled logistic maps, where the synchronization among the elements forming a cluster is complete, and external perturbations are needed for a change of partition to occur (see Sec. 8.1.2). In the case of randomly coupled logistic maps [Manrubia and Mikhailov (1999)], we speak of clustering to the precision δ, meaning that two elements with states at a distance smaller than δ are considered to belong to the same cluster. A similar notion has been already discussed in the context of synchronization in heterogeneous systems (Secs. 7.2 and 7.3.2).

When frozen disorder is present, it is possible that elements become

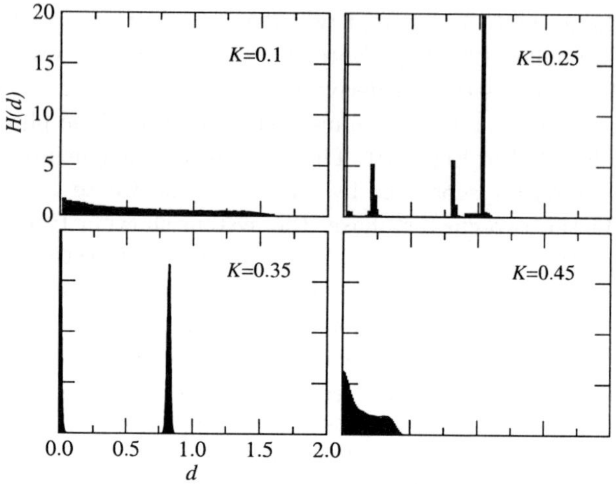

Fig. 8.11 Distributions over pair distances for an ensemble of $N = 10^3$ randomly coupled logistic maps, with average connectivity $\kappa = 0.8$ and different coupling intensities. Adapted from [Manrubia and Mikhailov (1999)].

itinerant, and that they occasionally change affiliation, i.e., the cluster to which they belong. An additional result of partial entrainment is that clustering is *hierarchical*, that is, as δ increases the elements merge into increasingly large clusters. Let us illustrate these behaviors in the system of randomly coupled logistic maps (Eq. (7.67)), that is

$$x_i(t+1) = \left(1 - \frac{K}{\kappa(N-1)} \sum_{j=1}^{N} G_{ij}\right) f[x_i(t)] + \frac{K}{\kappa(N-1)} \sum_{j=1}^{N} G_{ij} f[x_j(t)], \tag{8.25}$$

where $\mathbf{G} = \{G_{ij}\}$ is the matrix of connections and κ is the average connectivity of the network, Eq. (7.66). For very low values of the connectivity or of the coupling K, the system has a turbulent phase similar to that of globally coupled logistic maps. As the coupling intensity increases, $\delta-$clusters start to form, although they might be short lived. The average lifetime of a given partition increases for increasing K. Eventually, it is possible that the system gets locked in attractors where all the elements follow periodic (though different) orbits.

Figures 8.11 and 8.12 show different quantities characterizing the dynamical phases of randomly coupled logistic maps. The distribution of pair

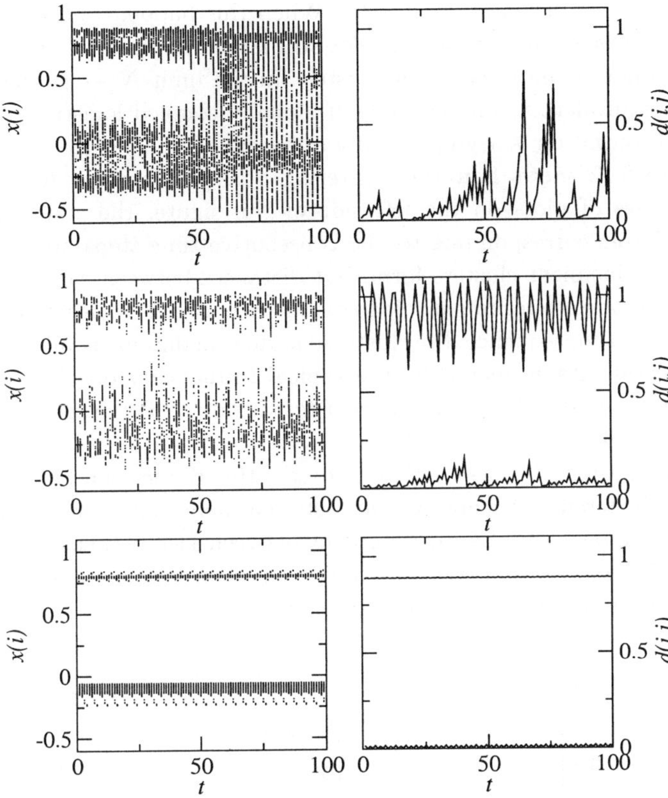

Fig. 8.12 Formation of broad clusters in randomly coupled logistic maps. For $K = 0.2$ (upper plot) two clusters exist until large fluctuations force them to merge at $t \simeq 60$. On the right, the distance between two elements which were in the same cluster before the merging is shown. In the middle plot ($K = 0.23$), clusters are better defined and the intracluster distance (right, lower curve) and intercluster distance (right, upper curve) are more separated. For K large enough ($K = 0.28$, bottom plots), the groups become well defined and itinerancy of elements finishes when locked states appear.

distances $H(d)$ between two elements i and j is represented in Fig. 8.11. This distribution characterizes the probability that two elements are found at a distance $d_{ij} = |x_i(t) - x_j(t)|$. For low values of the coupling the system is in the desynchronized phase, and the pair distribution function is almost flat. As K increases, clusters of varying width are formed. They are broad and highly unstable at the onset of the clustering phase, and gain stability as the coupling intensity grows. In this example, the system is locked into exactly periodic orbits at $K = 0.35$. When this happens, no exchange of elements between clusters occurs. For still larger values of K the elements tend to condense into one single cluster. In the limit $N \to \infty$ the system becomes equivalent to an ensemble of globally coupled logistic maps, and full synchronization is asymptotically attained.

Figure 8.12 shows how the clustering phase changes as K increases. In the series of plots on the left side of this figure, the states of $N = 200$ oscillators corresponding to 100 consecutive time steps are displayed. For low K, incipient clusters form, but distances between elements in the same cluster are often comparable to distances between elements in different clusters. As a consequence, the clustered state is maintained only for a finite time interval. As K increases, the typical intracluster distance becomes significantly smaller than the intercluster distance. For large enough K, locked states appear.

In the clustering phase with itinerancy, the system can explore many different clustered states and wander between them. Then, well-defined attractors disappear and chaotic itinerancy between a set of them sets in. Separation of the attraction basins is recovered when locked (periodic) states appear and itinerancy is no longer possible.

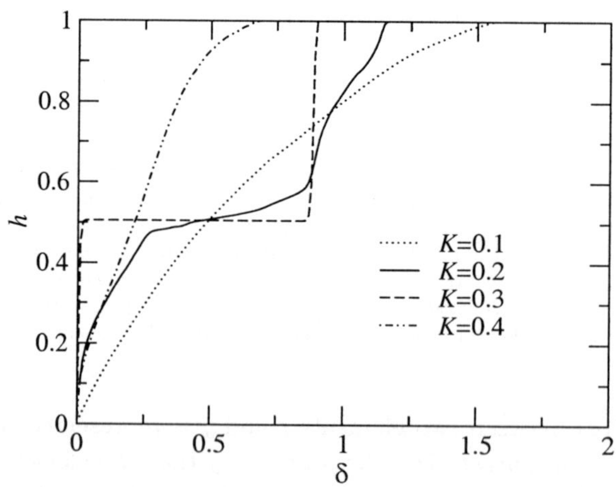

Fig. 8.13 Variation of the quantity $h(\delta)$ with the resolution δ in a system of $N = 10^3$ randomly coupled logistic maps. The curves correspond to four representative dynamical states: asynchronous phase ($K = 0.1$), clustering with itineracy for $K = 0.2$, locked periodic state for $K = 0.3$, and onset of condensation into a single cluster for $K = 0.4$. The logistic parameter is $a = 2$.

The hierarchical clustering can be characterized by calculating the fraction $h(\delta)$ of pairs of elements which are at a distance $d_{ij} < \delta$. A similar quantity, for δ fixed, was introduced as an order parameter to characterize the synchronization transition (Eq. (7.48)). In globally coupled logistic maps, $h(\delta)$ equals unity at full synchronization, for all values of δ. If two clusters of sizes N_1 and N_2 are formed, then $h(\delta) \approx (N_1^2 + N_2^2)/(N_1 + N_2)^2$ until δ reaches the typical separation between clusters, and then jumps to unity. When more than two clusters are present, $h(\delta)$ has a ladder-like shape with jumps at the typical separations between clusters. In randomly coupled logistic maps the increase in $h(\delta)$ is smoother, while locked states behave similarly to many-cluster attractors in the globally coupled case. Four representative situations are depicted in Fig. 8.13.

8.4 Clustering in the Rössler System

The collective behavior in ensembles of continuous-time oscillators with stable periodic windows is very similar to that described for logistic maps. Complex clustering is also observed in such systems. The introduction of noise causes the itinerancy of some elements and occasionally the coalescence of whole clusters. However, since most clustered states also correspond to periodic dynamics, weak noise does not destroy such stable clusters.

Consider a system of Rössler oscillators with the following vector coupling,

$$\begin{aligned} \dot{x}_i &= -y_i - z_i + K(\langle x \rangle - x_i) + \xi_i(t) \\ \dot{y}_i &= x_i + a y_i + K(\langle y \rangle - y_i) \\ \dot{z}_i &= b - c z_i + x_i z_i + K(\langle z \rangle - z_i). \end{aligned} \tag{8.26}$$

In numerical simulations of this system, the noise values $\xi_i(t)$ are drawn independently at each time step and for each oscillator from a uniform distribution, $\xi \in (-\eta_0, \eta_0)$. The dispersion of this distribution is $S = \eta_0^2/6\Delta t$, where Δt is the integration step used in the numerical simulations [Zanette and Mikhailov (2000)].

When the dispersion of the noise is small, the behavior of this system with respect to clustering is close to that of the noiseless system. The calculation of the pair distance distributions reveals that robust attractors are present. Figure 8.14 shows the distributions obtained for four different values of the coupling strength. We see that the formation of stable, well-defined clusters is preceded by a condensation process. In the final clustered phase, the low dispersion in the states of the elements within a cluster and the large inter-cluster separation are indicative of high stability.

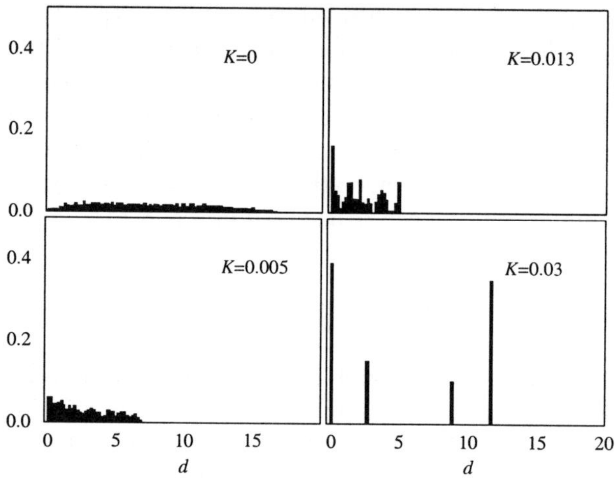

Fig. 8.14 Distribution of pair distances between elements in an ensemble of $N = 10^3$ globally coupled Rössler oscillators at a fixed time moment. The decoupled ensemble ($K = 0$) shows a flat distribution which becomes, however, modified even for very small coupling intensity. This is a sign of the appearance of correlations in the asynchronous phase. For sufficiently strong coupling, clustering appears. Adapted from [Zanette and Mikhailov (2000)].

This high stability is challenged in the presence of noise of large enough intensity. The situation is then similar to what happens in randomly coupled logistic maps, where the presence of (quenched) disorder makes clusters temporary entities. Then, the neighborhoods of many attractors of the noiseless system are explored, and clusters can in the course of time coalesce and then split again. This process can be once more visualized by using the pair distribution function. Figure 8.15 shows a dynamical behavior equivalent to that displayed in Fig. 8.12 for low coupling, where the population switches intermittently between a single-cluster and a two-cluster state. Such dynamical coexistence of attractors takes place for intermediate intensities of the noise strength, $2 \times 10^{-6} < S < 2 \times 10^{-4}$. A larger noise strength causes the elements to wander around a single (broad) cluster. The effect of increasing the noise strength is illustrated in the lower plot of Fig. 8.15, where the fluctuations in the density of pairs of elements at a distance d as a function of time are represented (compare this behavior with Fig. 8.12).

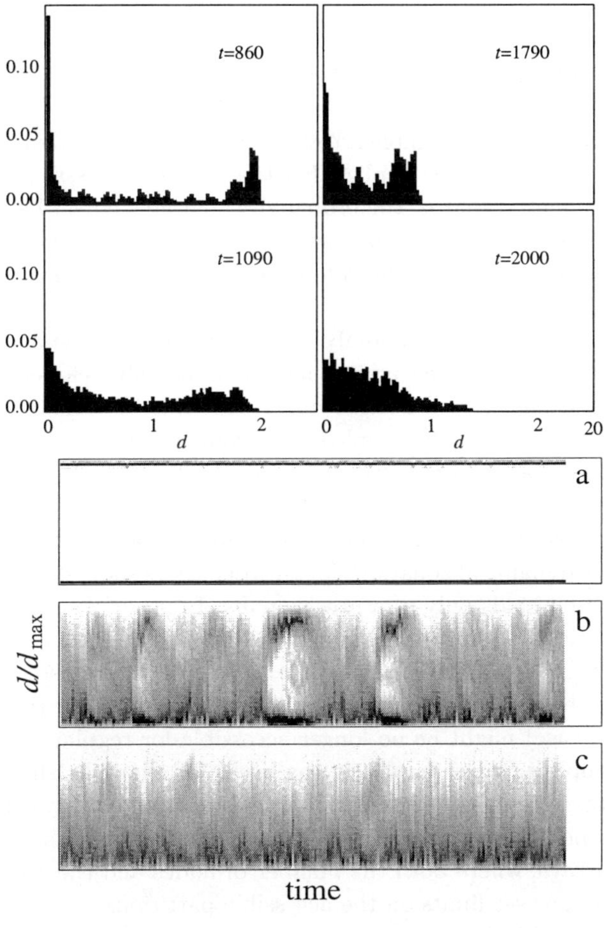

Fig. 8.15 Above: Distributions of pair distances between elements in an ensemble of $N = 10^3$ globally coupled Rössler oscillators at subsequent time moments for a fixed coupling intensity, $K = 0.4$. The dispersion of noise is $S = 5 \times 10^{-6}$. Below: Density plots of histograms over normalized pair distances. An interval corresponding to 3000 time steps is displayed. The maximal distance is $d_{\max} = 2$. Three different values of the noise intensity are used: (a) $S = 5 \times 10^{-7}$, (b) $S = 5 \times 10^{-5}$, and (c) $S = 5 \times 10^{-4}$. From [Zanette and Mikhailov (2000)].

Clustering has been observed also in Rössler oscillators under different coupling schemes [Zanette and Mikhailov (1998a)]. This gives further support to the conjecture that it is the presence of stable periodic windows, and not other factors, which are essential to the appearance of clustered states in ensembles of identical, globally coupled chaotic elements.

8.5 Local Coupling

When oscillators in an ensemble are globally coupled, their dynamical equations are all identical. This implies that the system possesses complete permutation symmetry, that is, the probability that an element belongs to one or another cluster, or synchronizes with one or another oscillator, is *a priori* the same for each element. This additionally permits that synchronization be full within each cluster.

In large ensembles of randomly coupled maps, this symmetry is only weakly broken as long as the average number of neighbors for each oscillator takes values around a well defined mean with small fluctuations. Only in this limit is the randomly coupled case comparable to a noisy globally coupled system, as we have just seen. Still, it is observed that elements within the same dynamical cluster tend to be more strongly connected than elements belonging to different clusters [Manrubia and Mikhailov (1999)], and since the dynamical equations for the oscillators show slight differences, synchronization holds only to a precision $\delta < 1$.

Network architectures departing from the two previous cases can however significantly constrain the possible partitions into clusters. Some attractors which are present in a state with high symmetry (like the globally coupled case) might be no longer accessible for regular networks with neighbor coupling, for example. It has been, for example, shown that full synchronization cannot be attained in one-dimensional arrays of oscillators if too many nodes are present (recall Sec. 7.1.1). A similar situation arises in regular arrays, where both the number of nodes and the architecture of their connections set limits on the accessible partitions.

The relationship between the symmetry of the connection network and the possible clusters can be clearly illustrated by using small systems. The ways in which elements can cluster depends on their relative positions in the system. Whenever a given network is invariant under certain permutation of the elements (partial symmetry is present) synchronization within clusters can be full. As an example, let us consider the cases presented in Fig. 8.16. Above, the four different possible configurations of a network with $N = 5$ elements and connectivity $\kappa = 0.6$ are shown [Manrubia and Mikhailov (1999)]. Below, two chains with open boundary conditions (which break the translation symmetry along the chain) and $N = 6$ and $N = 7$ are depicted [Belykh *et al.* (2003)]. Due to the symmetries inherent to each system, it is observed that elements represented with the same symbol synchronize more easily than other pairs.

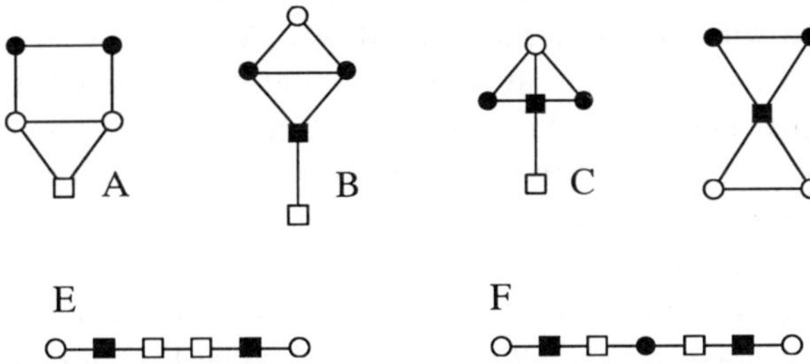

Fig. 8.16 Different possible network architectures of the connectivity network in small systems. Nodes represented with the same symbol tend to belong to the same cluster. Partitions into clusters that do not respect the permutation symmetries are strongly suppressed. Adapted from [Manrubia and Mikhailov (1999)] and [Belykh *et al.* (2003)].

The symmetry of the network is reflected in the symmetry of the equations of motion. Consider the set of equations defining the dynamics of network A (nodes 1 and 2 are full circles, 3 and 4 open circles, and the square is the node 5),

$$\begin{aligned} \dot{x}_1 &= f(x_1) + K\left[f(x_2) + f(x_3)\right]/2 \\ \dot{x}_2 &= f(x_2) + K\left[f(x_1) + f(x_4)\right]/2 \\ \dot{x}_3 &= f(x_3) + K\left[f(x_1) + f(x_4) + f(x_5)\right]/3 \\ \dot{x}_4 &= f(x_4) + K\left[f(x_2) + f(x_3) + f(x_5)\right]/3 \\ \dot{x}_5 &= f(x_5) + K\left[f(x_3) + f(x_4)\right]/2. \end{aligned} \tag{8.27}$$

These equations remain invariant if the permutation $\{1,2,3,4,5\} \rightarrow \{2,1,4,3,5\}$ is performed. This relabeling defines an invariant cluster synchronization manifold [Belykh *et al.* (2003)]. In other networks shown in Fig. 8.16, similar permutation symmetries also determine the most probable clustering regimes. Numerical simulations agree with the theoretical predictions. If $f(x)$ is the logistic map and the domain $a \in [1.42, 2]$ and $K \in [0.01, 0.6]$ is explored, the most probable clustering configurations as defined above are obtained in a 59% of the realizations for graph A, in a 75% for B, 17% for graph C (for which a 62% are desynchronized, however), and in a 57% for D. Moreover, certain attractors are never observed, for example in graph A the situation where 1 and 5 form one cluster, 2 and 3 a second one, and 4 remains non-entrained.

Clustering behavior which maintains the symmetries present in the network of connections has been observed in chaotic maps [Manrubia and Mikhailov (1999)], and for oscillator systems [Belykh *et al.* (2000)], all having regimes with stable periodic windows. Analysis of large systems, where the network symmetries are not exact and where some heterogeneity is present, is just beginning. Similar to random networks, where some degree of randomness does not prevent the appearance of clustering (up to a certain precision), the formation of coherent groups in complex networks of Rössler and Lorenz oscillators has also been observed [Belykh *et al.* (2003)]. It is interesting that clustering disappears in ensembles of saw-tooth maps of the kind discussed in Sec. 8.2 (see also 6.2) when they are not globally coupled, though such systems can still attain full synchronization.

The fact that clusters constituted by elements which are dynamically identical have a much larger probability to form can be used to design networks with *a priori* specified attractors. The number of stable attractors that a system of N elements can have is huge and largely exceeds the system size. In consequence, such systems might have applications in information storing and have a potential as classifier systems, since a correspondence between a set of initial conditions and a final attractor can be established.

Chapter 9

Dynamical Glasses

Statistical mechanics studies the collective behavior of systems formed by a large number of elements. Traditionally, it has been mainly concerned with the analysis of the properties of matter, but in the last decades its powerful techniques have been successfully applied to the description of other complex systems, for instance in the field of biological or social sciences.

In the framework of mechanics, one needs to specify the microscopic state of a physical system (positions and velocities of all the particles constituting it) as a function of time in order to completely characterize its state. The equilibrium statistical mechanical description of the same system is based on a few macroscopic observables, such as internal energy and density, averaged over a very long observation time. The fundamental assumption of equilibrium statistical mechanics, the ergodic hypothesis, states that these time-averages can be replaced by an average over a probability distribution in the space of the microscopic states accessible to the system. This probability distribution constitutes the statistical mechanical description, or *macroscopic state*, of the system. The macroscopic state of an isolated system that evolves in time maintaining its internal energy constant is a uniform probability distribution over the micro-states with constant internal energy. The ergodic hypothesis asserts in this case that the systems visits during its evolution all microscopic states compatible with its energy, without any preference for one or another. This simple behavior is however not universal. Many macroscopic systems violate the ergodic hypothesis, in the sense that some regions of their configuration space are mutually inaccessible.

In the last decades, macroscopic systems with complex statistical behavior have attracted great attention from theoretical and experimental physicists. The paradigm of such systems are spin glasses, magnetic alloys with complex magnetic ordering. In mathematical models of spin glasses, an exponentially large number of macroscopic states, each one concentrated in a different region of the configuration space of the system, can coexist. The phenomenology of spin glasses arises from the fact that there is a very large number of microscopic states which are local minima of the internal energy, many of them sharing a close value of the energy. During its evolution at low temperature, the system becomes trapped in the attraction basin of one of these minima, and the corresponding macroscopic state consists of a probability distribution strongly peaked around the energy minimum.

The existence of many energy minima is a consequence of the so-called *frustration* of the system. This means that there are constraints which prevent the system to attain the global minimum of energy. Finally, in the best known model for spin glasses, the infinite-range (mean-field) model proposed by Sherrington and Kirkpatrick, the energy minima are hierarchically organized in a very special way: the metric properties of the distance between attractors imply that the space of attractors is an *ultrametric* space.

We show in this chapter that the rich phenomenology of spin glasses finds a counterpart in the collective behavior of globally coupled logistic maps. This model is paradigmatic of a whole class of dynamical systems constituted by coupled chaotic oscillators with periodic windows. As we have seen, this is apparently the only kind of dynamical system able to undergo complex clustering. Periodic windows act as traps for individual trajectories and lead to frustration, and thus glassy behavior, in this dynamical system.

9.1 Introduction to Spin Glasses

Symmetry breaking in thermodynamical systems is a collective phenomenon. Once in the non-symmetric phase, but close to the transition separating it from the symmetric phase, fluctuations are enhanced and result in a macroscopic state where the symmetry of the microscopic system is lost. This behavior can be illustrated by means of a classical example: the Ising ferromagnet.

A ferromagnetic Ising system consists of a regular lattice, for instance simple cubic, in which each position i is occupied by a magnetic moment (spin) σ_i with two possible orientations, $\sigma_i = \{-1, 1\}$. Depending on the relative orientation of two neighboring spins, the energy of the system is lowered or increased by an amount proportional to $\sigma_i \sigma_j$, so that the total internal energy is given by the expression $\mathcal{H}(C) = -\sum_{ij} \sigma_i \sigma_j$. The sum runs over all nearest neighbor pairs, and $C = \{\sigma_i\}$ defines the microscopic state of the system, that is the set of values of spin variables at each position. According to statistical mechanics, the equilibrium state of a closed system that can exchange energy, but not particles, with an external reservoir at temperature T, is described by the canonical or Boltzmann distribution. In this state, a microscopic configuration C having internal energy $\mathcal{H}(C)$ has a probability

$$P(C) \propto \exp\left\{-\frac{\mathcal{H}(C)}{k_B T}\right\}, \tag{9.1}$$

where k_B is the Boltzmann constant. For a discrete configuration space, the Boltzmann distribution maximizes the entropy of the system for a fixed value of the average energy, and minimizes the free energy.

In the Ising ferromagnet, as in all thermodynamic systems, the equilibrium state results from a compromise between the trend to align the individual spins in the same direction (thus lowering the energy of the system) and the thermal disorder that "randomizes" the system, increasing its entropy. A useful order parameter to monitor the behavior of the system is the magnetization, defined as an average over the individual spins,

$$m = \frac{1}{N}\sum_i \sigma_i. \tag{9.2}$$

The thermodynamic average of this quantity in the Boltzmann state is zero, since the internal energy $\mathcal{H}(C)$ is symmetric with respect to the inversion of all spins and two micro-states with magnetization $+m$ and $-m$ have exactly the same Boltzmann weight. In fact, at high temperature the system is indeed in a paramagnetic phase where the average magnetization is zero when there is no external magnetic field applied. Nevertheless, experiments show that at low temperature the system has a non-vanishing magnetization even in the absence of an external field. How can one reconcile this result with the microscopic symmetry of the system?

At zero temperature, only the micro-states with minimum energy have a non-vanishing Boltzmann weight. These are the two states where all spins are aligned, either in the positive or in the negative direction. These two states, however, are separated, in the thermodynamic limit of infinitely large systems, by an infinite free-energy barrier. Therefore, the time to switch from one state to the other is also infinite, and the Boltzmann distribution where the two micro-states are equiprobable is never reached. Thus, the system is not ergodic. It gets trapped in one of the two pure-states corresponding to the two energy minima, and never escapes from it. This transition from a symmetric equilibrium state to two or more non-symmetric pure states is called *spontaneous symmetry breaking*. The pure state where the system ends up is not selected due to some external field or energetic preference, but is chosen due to microscopic fluctuations of the initial microstate, and the opposite state, *a priori* equally probable, is never visited.

Such situation can be represented graphically by computing the free energy as a function of the order parameter m, for instance using the mean-field approximation. Above the critical temperature, the minimum of the free energy is reached at $m = 0$. Below this temperature, two minima of the free energy appear, corresponding to the two pure states with spontaneous magnetization $+m$ and $-m$ (see Fig. 9.1). These two states are separated by a free-energy barrier that increases with the system size, so that a large system can never switch from one state to the other. To characterize the system below the critical temperature, one should consider different realizations of the system corresponding to different initial conditions (replicas). Each realization ends up in one of the two pure states, and the distribution of the time-averaged magnetization obtained with different replicas has a bimodal shape, approaching two delta functions centered at $\pm m$ in the infinite size limit, whereas the replica average of the magnetization is zero because of the microscopic symmetry.

The ferromagnetic Ising magnet is the simplest example of a system displaying spontaneous symmetry breaking, and having just two pure states. There is another class of systems where the number of states among which the system can choose is huge, since there is an extremely large number of energy minima where the system can be trapped at low temperature. The paradigmatic model of such systems is the mean-field model of spin glasses, proposed by Sherrington and Kirkpatrick to represent the behavior of complex magnetic alloys. These are described by the energy function

$$\mathcal{H} = -\frac{1}{2}\sum_{i\neq j}^{N} J_{ij}\sigma_i\sigma_j. \tag{9.3}$$

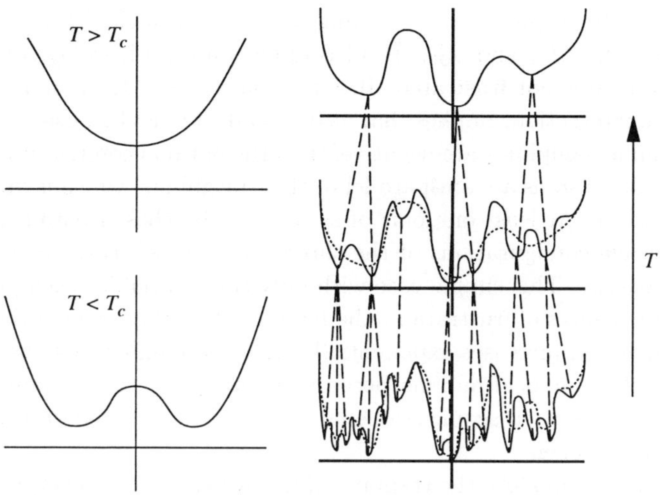

Fig. 9.1 Left: schematic representation of symmetry breaking in an Ising ferromagnet. The plots show the states of minimal energy. At a critical temperature T_c the system magnetizes spontaneously and its symmetry is broken. Right: we show the hierarchy of nested minima developing in a spin glass when the temperature is lowered.

Here the sum runs over all pairs of binary variables, $\sigma_i = \pm 1$. Effectively, the model is characterized by an infinite spatial range and belongs to the mean-field type, making it amenable to analytic treatment. Note that the corresponding model with local interactions in three dimensions can only be studied through numerical simulations. Though apparently simple, this mean-field model displays extremely complex collective properties.

The contribution of each pair of elements to the energy is $J_{ij}\sigma_i\sigma_j$. While in the Ising model the coefficient $J_{ij} = J$ is constant and positive for all pairs at nearest neighbor positions (and zero for all other pairs), in spin-glass models the coefficients J_{ij} form a random matrix, which is aimed at representing the local impurities in the complex magnetic alloy. The coefficients J_{ij} are quenched random variables, and are kept frozen throughout the evolution of the system. The aim of spin-glass theory is to calculate the thermodynamic variables for a fixed realization of the J_{ij} couplings (thermodynamic average) and then to average them over different realizations of the quenched disorder (quenched average).

A physical consequence of the Hamiltonian Eq. (9.3) is the presence of *frustration*. This concept refers to the impossibility of minimizing all local interactions simultaneously, along a loop of interacting variables. Let us consider three spin variables arranged along a triangle, interacting with coefficients J_{12}, J_{23} and J_{31}. If all couplings are positive (ferromagnetic case), the loop is not frustrated. If, for instance, $\sigma_1 = 1$, the minimization of all local interactions implies that also σ_2 and σ_3 must be positive. On the contrary, if all couplings are negative, then the optimal configuration would require that all spins are antiparallel to their neighbors, which is impossible to accomplish. At least one pair of spins must be chosen parallel. Of the eight possible configurations of the three variables, six correspond to this minimal energy. This simple case with only three variables exemplifies the deep consequences of frustration: the number of local minima of the energy is very high and increases exponentially with the number of spins N. The "valleys" of these local minima become attractors for the dynamics of the system, or pure macroscopic states, at sufficiently low temperature, where ergodicity is broken.

In spin-glass models, the trapping of the system close to energy minima, with ensuing ergodicity breaking, is associated with the breaking of *replica symmetry* at low temperature. To characterize this phase transition, one uses as order parameter the *overlap* $q_{\alpha\beta}$ between replicas. This quantity measures the similarity between the final states α and β, resulting from two independent trajectories (replicas) of the same system at fixed quenched disorder. It is defined as

$$q_{\alpha\beta} = \frac{1}{N}\sum_{i=1}^{N}\langle\sigma_i\rangle_\alpha\langle\sigma_i\rangle_\beta. \tag{9.4}$$

The quantity $\langle\sigma_i\rangle_\alpha$ represents the time average of the spin variable σ_i along a trajectory converging to the state α. For an Ising ferromagnet above the critical temperature, where the system is ergodic, this average is zero and $q_{\alpha\beta} = 0$ for all pairs of trajectories. Below the critical temperature, the situation changes, since the trajectories may converge to one of the two pure states with magnetization $\pm m$. The overlap thus takes two different values, $q = \pm m^2$. The situation changes qualitatively in spin glasses, where the overlap q takes values out of a continuum. The statistics of all possible overlaps in the space of states is described through the overlap distribution $P(q)$, defined as

$$P(q) = \sum_{\alpha\beta}\delta(q_{\alpha\beta} - q). \tag{9.5}$$

To measure $P(q)$ in computational studies, many different pairs of independent trajectories have to be run and compared at fixed values of J_{ij}. Some schematic examples of overlap distributions are shown in Fig. 9.2.

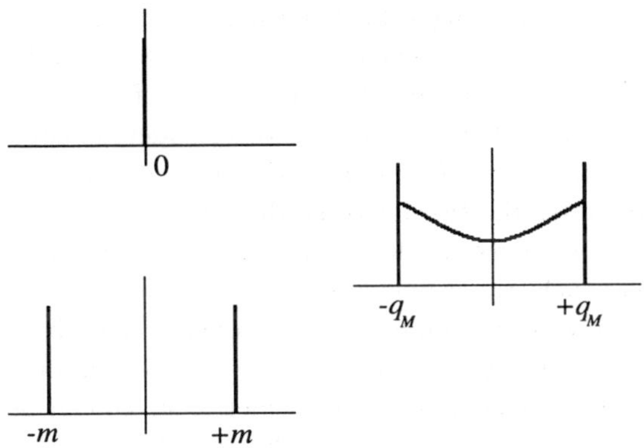

Fig. 9.2 Overlap distributions (left) for an Ising ferromagnet above and below the critical temperature, and (right) for a spin glass. In the latter case, the distribution of overlap values is a continuous function.

The overlap $q_{\alpha\beta}$ allows us to define a metric in the space of pure states. A distance $d_{\alpha\beta}$ between two states can be defined as

$$d_{\alpha\beta} = 1 - q_{\alpha\beta}. \tag{9.6}$$

The fact that, for spin glasses, $P(q)$ is a continuous function (in the limit of infinitely large system sizes) is related to the hierarchical organization of the space of pure states. The dependence of the overlap distribution on temperature, and the fact that the distribution of overlap values is a continuous function, suggest the following qualitative representation of the phase space of the system, illustrated schematically in Fig. 9.1. The valleys around local minima have a finer structure and are themselves formed by valleys. This process continues until the similarity between two states becomes arbitrarily close to the similarity of a state with itself (in the limit $T \to 0$). Broad valleys are less similar among themselves than smaller valleys contained inside them. At high temperature, thermal disorder allows the system to overcome energy barriers between small valleys, but not barriers between broad valleys, which are higher. If the temperature is lowered, the system becomes trapped into a narrower region of the configuration space. Therefore, the lower the temperature, the higher are the values of the overlap. The maximal overlap $q_{\alpha\alpha} = 1$ is only found in the limit $T \to 0$, where a valley shrinks to the corresponding energy minimum.

The above argument suggests that the states of the system are hierarchically organized. This can be indeed confirmed by a special investigation. Analytical studies of the mean-field model (9.3) of spin glasses, where distances $d_{\alpha\beta}$ between different states were considered, have demonstrated that these distances are *ultrametric* [Mézard and Virasoro (1985)]. For each triplet of states the two maximal distances in the set $\{d_{\alpha\beta}, d_{\alpha\gamma}, d_{\beta\gamma}\}$ are identical. Using a geometric interpretation, this would correspond to a situation where each triplet of elements can be placed on the vertices of an isosceles triangle with the two major edges of equal length. The ultrametric property is important because, as it can be shown, it implies a hierarchical organization. The elements of an ultrametric space can be mapped onto the end point of the branches in a tree-like graph.

There is a convenient way to check, through a computational study, whether a system is ultrametric. It consists in calculating the distribution $H(\Delta q)$ of the differences between the two minimal overlaps out of the three in a triplet,

$$\Delta q = |q_{\alpha\beta} - q_{\beta\gamma}|, \tag{9.7}$$

where we have arranged labels so that $q_{\alpha\gamma} > q_{\alpha\beta}, q_{\beta\gamma}$. If $H(\Delta q) \to \delta(\Delta q)$ in the limit $N \to \infty$, then the system is ultrametric. If the width of the distribution remains finite in this limit, the considered system departs from exact ultrametricity.

Another important quantity characterizing an attractor α is its weight W_α, defined as the probability to fall in the state α starting from a random initial condition. This quantity plays the same role as the size of the attraction basin for dynamical systems. In the case of a ferromagnet below the critical temperature, the two states have weights $W_\pm = 1/2$. Since in spin glasses the number of states is infinite in the infinite size limit, one may expect that all weights tend to zero in this limit. To check this, it is convenient to measure the average weight of a state, defined as

$$Y = \sum_{\alpha} W_{\alpha}^2. \tag{9.8}$$

If all states tend to have a vanishing weight, then Y tends to zero. Alternatively, if only one state dominates the phase space and all others have vanishing weight, Y tends to one. Intermediate values of Y indicate that there is a finite number of states with non-vanishing weights.

Up to now, we have been describing a fixed realization of the quenched disorder (the couplings J_{ij}). How do the properties of the system change when different realizations are considered? In usual thermodynamic systems, it is common that intensive variables (i.e., those not proportional to the size of the system) have vanishing fluctuations in the infinite-size limit. These variables are therefore said to be self-averaging. In disordered systems, however, some intensive variables may have non-vanishing fluctuations even in the infinite-size limit. The overlap distribution is one such variable, since different systems have overlap distributions with different moments. The weight distribution is also not self-averaging, and different realizations of a system have different values of Y, because the number of states with non-vanishing weights changes from one realization to the other.

The properties described in this section, i.e. frustration, multiple attractors with a non-trivial overlap distribution and a non-trivial weight distribution, and lack of self-averaging, are not unique to systems with quenched disorder, but mean-field spin glasses are the system where these properties are better understood analytically. As we will show in the following, some of these properties are shared by coupled dynamical systems, where disorder has a different origin.

9.2 Globally Coupled Logistic Maps as Dynamical Glasses

The first indication that there could exist a parallelism between globally coupled logistic maps and the behavior of glassy systems is the high multiplicity of clustered attractors. Even for a fixed number of clusters, many different partitions are possible and do indeed coexist [Kaneko (1991b)]. Let us recall the definition of partition into clusters. Starting with random initial conditions, we let the system evolve until a stable attractor is reached (see Sec. 8.1.2). In the clustering phase, M clusters with N_i elements, $i = 1, \ldots, M$, are formed. An attractor $\mathcal{A}$ can be coded through

$$\mathcal{A} = \{M; N_1, N_2, \ldots, N_M\}, \tag{9.9}$$

where the clusters will be ordered according to their sizes,

$$N_1 \geq N_2 \geq \cdots \geq N_M. \tag{9.10}$$

This also specifies the dynamical properties of the attractor, given K and a, since there is a one-to-one correspondence between a partition and the dynamics of the system. The variation of the cluster number M with the parameters of the system (coupling intensity K and logistic parameter a) were used to uncover some of the structure of the clustering phase. This phase was called glassy or partially ordered phase, depending on whether the values of M are proportional to the system size N or not, respectively [Kaneko (1989); Kaneko (1990a)]. However, since even for a fixed value of M the number of different partitions that the system can reach is *a priori* huge, the whole phase where clustering is present is potentially a glassy phase [Manrubia *et al.* (2001)].

The partition variety $\tilde{Y}$ was introduced as a measure of the non-uniformity of partitions [Kaneko (1991b)]. It is defined as

$$\tilde{Y} = \sum_{i=1}^{M} \left(\frac{N_i}{N} \right)^2 . \tag{9.11}$$

Consider a situation in which M clusters of almost equal size can be formed in a system. For $N_i = N/M$, for all i, the partition is $\tilde{Y} = 1/M$. If clusters vary in size around N/M, the distribution of $\tilde{Y}$ develops a broad peak slightly above $1/M$.

The quantity analogous to $\tilde{Y}$ in the spin glass and which inspired its introduction is the probability Y that two arbitrarily chosen initial conditions fall on the same metastable state, Eq. (9.8). However, there is a main difference between the two quantities, since $\tilde{Y}$ is calculated with a single realization of the system, while Y arises from the comparison of two different realizations. Hence, $\tilde{Y}$ is actually a measure of the complexity of individual partitions.

In the turbulent phase, the average value of $\tilde{Y}$ almost vanishes. Indeed, when no clusters are present, $N_i = 1$ for all i, and $\tilde{Y} = 1/N$. Hence, $\tilde{Y} \to 0$ for $N \to \infty$ in the desynchronized phase. When clustering starts, the values

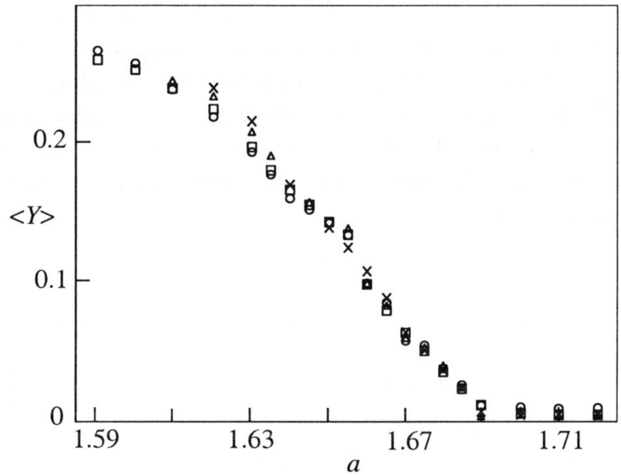

Fig. 9.3 Average value of the partition complexity $\langle\tilde{Y}\rangle$ as a function of the logistic parameter a for a fixed value of the interaction strength, $K = 0.1$. When $\langle\tilde{Y}\rangle \approx 0$ the system is in the turbulent phase, while finite values, independent of N, correspond to the glassy (clustering) phase. Symbols correspond to different systems sizes: $N = 100$ for circles, $N = 200$ for squares, $N = 400$ for triangles, and $N = 800$ for crosses. Adapted from [Kaneko (1991b)].

of $\langle\tilde{Y}\rangle$ —obtained by averaging over different initial conditions— remain finite even for very large systems, and are almost independent of N. This behavior is represented in Fig. 9.3.

A measure of the complexity of the space of attractors is the probability that a given partition $\mathcal{A}$ is observed in a system of size N for given parameter values [Crisanti *et al.* (1996)]. In order to obtain this quantity, a large number of random initial conditions has to be tested, and all the different asymptotic attractors have to be recorded. The entropy associated with the space of states is

$$S(N) = -\sum_{\mathcal{A}} P_N(\mathcal{A}) \ln P_N(\mathcal{A}), \tag{9.12}$$

from which the number N_W of macroscopic configurations with non-negligible probability (in the language of spin glasses, those having a non-vanishing weight W_α), can be extracted,

$$N_W(N) = \mathrm{e}^{S(N)}. \tag{9.13}$$

Heuristic arguments, together with numerical simulations, show that the squared entropy grows approximately proportional to N (see Fig. 9.4) [Crisanti *et al.* (1996)]. For a value of the coupling $K = 0.1$, and $a \in [1.5, 1.7]$, the system has a very large number of macroscopic states, while there are very few outside this range. In agreement with the results obtained through the calculation of $\tilde{Y}$, this signals again the domain where the glassy phase exists.

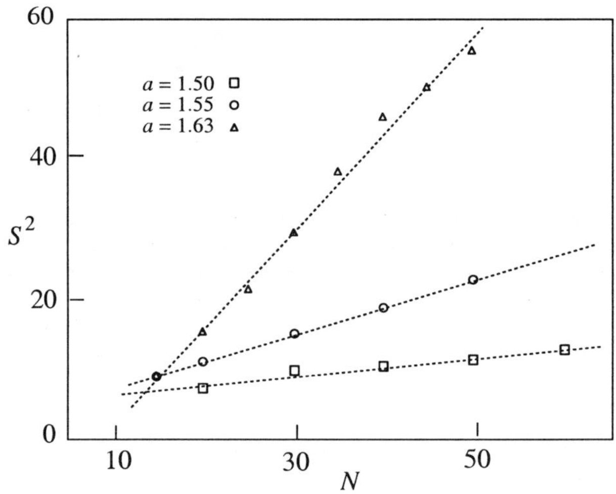

Fig. 9.4 Increase in the squared entropy $S^2(N)$ with the system size N for three different values of the logistic parameter a. The value $a = 1.63$ is located inside a periodic window for the individual oscillator. The coupling strength is $K = 0.1$, and a number of different initial conditions of the order of 10^4 was used. Adapted from [Crisanti *et al.* (1996)].

Finally, let us mention that the fine-scale structure of the space of states can also be probed with increasingly larger sets of initial conditions. If the hierarchy of valleys within valleys is present, then more of them should be visited as more initial conditions are assayed. The ones with the larger attraction basins will be detected fast, but some other might be rarely observed. Numerical simulations indicate that the number of attractors visited by the system grows as a power-law of the number of initial conditions probed, with an exponent which depends on the system size [Manrubia *et al.* (2001)].

9.3 Replicas and Overlaps in Logistic Maps

The similarity between the rich behavior of spin glasses and that of globally coupled logistic maps goes beyond the analogy between two multi-attractor systems described in the previous section. Indeed, replicas can be suitably defined for globally coupled logistic maps and the whole formalism developed for spin glasses can then be translated into the language of dynamical systems [Manrubia and Mikhailov (2001); Manrubia *et al.* (2001)].

The main problem with defining an overlap in globally coupled dynamical systems formed by identical elements comes precisely from their indistinguishability. For spin glasses, this problem does not exist, since the quenched disorder makes them different. Hence, the comparison between the state of the *same* element in two replicas, as characterized by the intensity of the interaction with its neighbors, can be immediately carried out. This is a problem that has to be sorted out for logistic maps. Notice that, even if the system reaches the same attractor in two different runs, the element i might have exchanged affiliation with the element j, and if their states would be directly compared, non-existing differences between the two attractors would appear.

The problem of the indistinguishability of identical, globally coupled elements, can be solved with an appropriate relabeling of the elements once a stable attractor has been reached. Since elements within each cluster are indeed identical, we assign labels in the following manner. The N_1 elements in the largest cluster (recall that clusters are ordered according to their sizes) get labels 1 to N_1. The N_2 elements in the second largest cluster are labeled from N_1+1 to N_1+N_2. This process continues down to the smallest cluster. Now, if two realizations α and β would lead to the same attractor, and the previous relabeling procedure would be performed, elements with the same label would have identical dynamics. This is the first step towards an operative definition of an overlap in dynamical systems.

The second main difference with respect to spin glasses is the number of scalar quantities defining the state of an element. While the time-averaged spin $\langle \sigma_i \rangle$ characterizes well the element i in a spin glass, this is not so in a dynamical system. For example, if a map follows a $p-$periodic orbit, p consecutive values of x_i are required to define the state. Performing time-averages destroys relevant dynamical information. Finally, one has to take into account that, due to periodicity in the dynamics, two different realizations falling into the same attractor might have a difference in phase which has to be eliminated in order to provide a suitable definition of overlap. With all these considerations in mind, a replica α in globally coupled logistic maps can defined as an orbit $\{x_i^\alpha(t)\}$ of the whole system, and different replicas correspond to different sets of initial conditions.

In order to give a suitable definition of overlap, it is convenient to transform the orbits into binary sequences $\{\sigma_i^\alpha(t)\}$. An element i gets a value $\sigma_i^\alpha(t) = 1$ (or $\sigma_i^\alpha(t) = -1$) at time t if its state is $x_i^\alpha(t) > x^*$ (or $x_i^\alpha(t) < x^*$), with

$$x^* = \frac{-1 + \sqrt{1 + 4a}}{2a}. \tag{9.14}$$

Here x^* is the fixed point of a single logistic map, such that the measure $\mu(x;t)$ has equal weights around this value (see Sec. 8.1.3 for the definition of $\mu(x;t)$). Due to this property, the variable $\sigma_i(t)$ can also be understood as well as the *phase* of the element i at time t.

Once all the previous operations have been performed, the two replicas can be compared. The dynamical overlap is defined as

$$q_{\alpha\beta} = \frac{1}{NT} \sum_{t=1}^{T} \sum_{i=1}^{N} \sigma_i^\alpha(t) \sigma_i^\beta(t). \tag{9.15}$$

This equation is almost identical to the definition given for spin glasses, Eq. (9.4), except for the way in which the time averages are performed. The average, which in the case of spin glasses is carried out over the states of each individual element with respect to time, is now taken as a sum along the global orbits of the two replicas. As it has been discussed, the deterministic trajectories of individual elements in the dynamical case contain relevant information that cannot be discarded, while in the case of the spin glass the possible changes in the spin state of an element are due to random, thermal fluctuations. Finally, the averaging time T for a dynamical system has to be the minimal common multiple of the two periods if both attractors are periodic, or a large enough interval if the trajectories are chaotic.

Thus, to compute overlaps in a dynamical system the following procedure can be applied:

(1) Perform simulations with two different initial conditions. These are the replicas α and β.

(2) Check, for both replicas, that the state corresponds to an attractor, that is, that the dynamical situation is stable.
(3) Eliminate degeneration due to arbitrary labeling of the elements.
(4) If feasible, define a simple quantity characterizing its dynamical evolution (similar to σ_i).
(5) Calculate the overlap between α and β according to Eq. (9.15). Repeat for all possible values of the phase difference between the replicas.
(6) The minimum overlap obtained in (5) is the overlap for that particular pair of realizations.

This yields a single value of q. The calculation of the distribution $P(q)$ requires the repetition of the previous steps until the shape of the distribution is well characterized statistically.

9.4 The Thermodynamic Limit

There is an additional issue of relevance when exploring the phase space of globally coupled systems, concerning the proper exploration of all possible initial conditions for the system.

In many systems, the value given to initial conditions is not very significant. Often, there are noisy processes which wash out the effect of initial conditions, the final state is unique, and thus, except for rare situations, every possible initial condition is equivalent to any other. By now, it is however clear that this is not the case in systems with glassy properties.

In most investigations on the collective behavior of globally coupled logistic maps, starting with a random initial condition means that, for each element i, a random number in $(-1, 1)$ is independently drawn from a uniform distribution and assigned to the initial state of that element. For systems formed by a small number of elements, this procedure tends to explore most of the attractor space. This is also the case in arrays of oscillators, where only few neighboring elements affect the dynamical evolution of each oscillator.

Large ensembles of globally coupled logistic maps have a high symmetry which prevents the efficient exploration of all the space of attractors. In the limit $N \to \infty$, the coupling field affecting each of the elements turns out to be the same,

$$\lim_{N\to\infty} m(0) = \lim_{N\to\infty} \frac{1}{N} \sum_{i=1}^{N} \left[1 - a[x_i(0)]^2\right] = 1 - \frac{a}{3}. \tag{9.16}$$

For finite N, this coupling field is the sum of N independent random variables and thus follows a Gaussian distribution with fluctuations of order $1/\sqrt{N}$. Moreover, in the limit $N \to \infty$ there would be a constant density of elements for each interval $\Delta x(0)$ of initial conditions. What is to be expected? Now, the initial condition preserves the symmetry implicit in the equations of motion, and it can be predicted that only one attractor will be reached in that limit. This situation holds strictly only for an infinitely large ensemble, since finite systems will deviate from the limiting behavior. It is one of the defining properties of deterministic chaos, namely the amplification of initially small differences, that masks this averaging effect and produces a seemingly large number of different attractors even when relatively large systems are used.

Thus, in the limit $N \to \infty$, there are no macroscopic differences in the initial state of the system corresponding to two replicas α and β, if the initial states of all elements are independently chosen. This has a number of relevant consequences. First, the distribution of cluster sizes becomes narrower as the system size increases. Second, the differences between the replicas become increasingly smaller, and the overlap distribution tends to a delta function at $q = 1$. Third, the transition between the different synchronous regimes of globally coupled logistic maps must be discontinuous, so that there is a one-to-one correspondence between the values of K and a and the attractor reached. These three properties have confirmed in numerical investigations [Manrubia *et al.* (2001)]. Figures 9.5, 9.6, and 9.7 illustrate quantitatively the described features.

Consider the fraction $p_k = N_k/N$ of elements belonging to a given cluster out of an ensemble with N elements. The distribution of relative cluster sizes $Q(p_k)$ averaged over many replicas is a measure of the variety of clusters in the attractors of the system. In a sense, it conveys an information complementary to the overlap distribution regarding the space of attractors. In Fig. 9.5 several distributions $Q(p_k)$ for increasingly large systems are represented. It is clear that the distribution depends strongly on N, and the peaks become narrower as N increases. In fact, in this case the system chooses a configuration where two clusters are formed. The largest one accumulates around 65% of the elements, while the remaining ones form

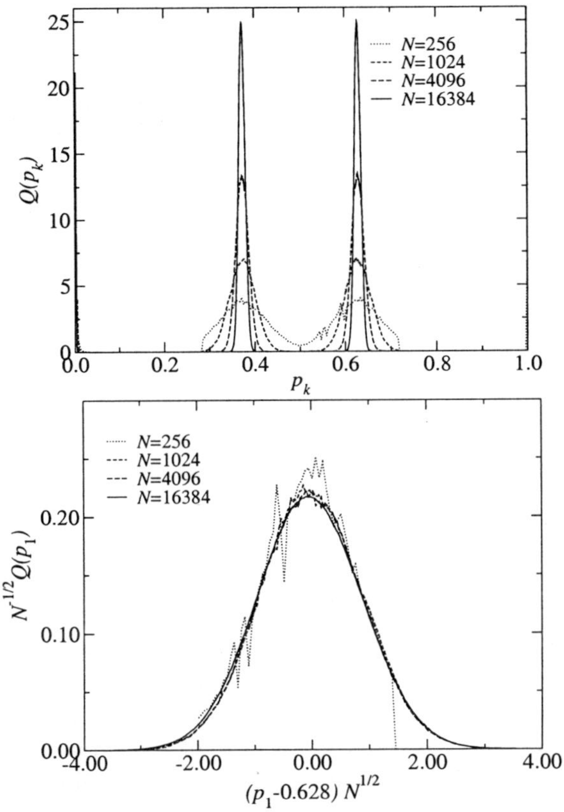

Fig. 9.5 Top: Distribution of clusters sizes for increasingly large ensembles of globally coupled logistic maps. Below: It can be seen that upon appropriate rescaling of the variables, the size of individual clusters follow a Gaussian distribution. The parameters are $K = 0.15$ and $a = 1.3$. The choice of initial conditions has been carried out independently for each map. Adapted from [Manrubia *et al.* (2001)].

the second cluster. Fluctuations of these quantities vanish as $1/\sqrt{N}$ and, upon appropriate rescaling of the distribution $Q(p_k)$, the functions collapse on a single Gaussian curve.

The situation is similar with the overlap distribution. An example is displayed in Fig. 9.6. The overlap tends to a delta function simply because there is no real diversity of attractors in the limit $N \to \infty$. Though the number of attractors is indeed diverging, they are all macroscopically identical.

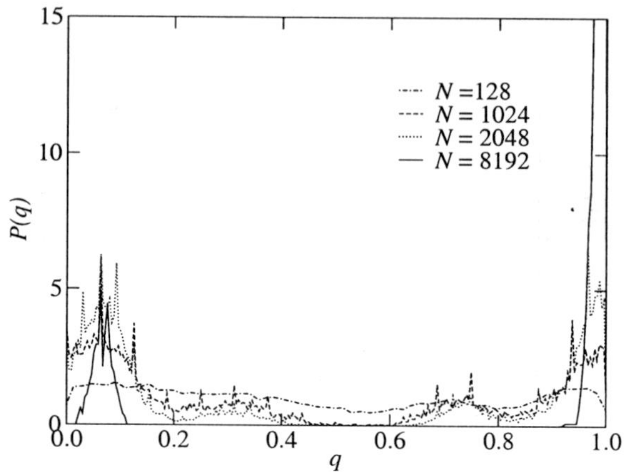

Fig. 9.6 Normalized overlap distributions $P(q)$ for increasingly large systems when the initial states of all elements are independently chosen. In the limit $N \to \infty$ all the replicas become identical. Parameters are $K = 0.1$ and $a = 1.55$. Adapted from [Manrubia *et al.* (2001)].

Finally, we have mentioned that the weight of attraction basins can be used as an order parameter to characterize the transition between the phases of globally coupled logistic maps. In particular, the transition to the fully synchronous phase has been numerically studied [Manrubia *et al.* (2001)]. Figure 9.7 represents the weight of the attraction basin corresponding to the fully synchronous state for different values of the coupling strength and for increasing system sizes. Once more, the coexistence of this attractor with others (which would yield a value of the attraction basin between zero and one) disappears as N increases. For small systems coexistence is observed, but for large enough systems either all of the initial conditions drive the system to full synchronization or none of them does. At $K \simeq 0.155$, in an infinite system, the space of attractors suddenly changes from one to the other situation.

Summarizing this section, we conclude that, if the set of initial conditions does not permit the appearance of macroscopic fluctuations in the coupling field from replica to replica, then the diversity of attractors detected in globally coupled logistic maps is a finite-size effect. This behavior is however a direct consequence of a particular procedure, used to generate initial conditions (when the initial states of all elements are independently and randomly chosen). A different procedure, described in the next section, should be employed to undertake an exhaustive exploration of the space of

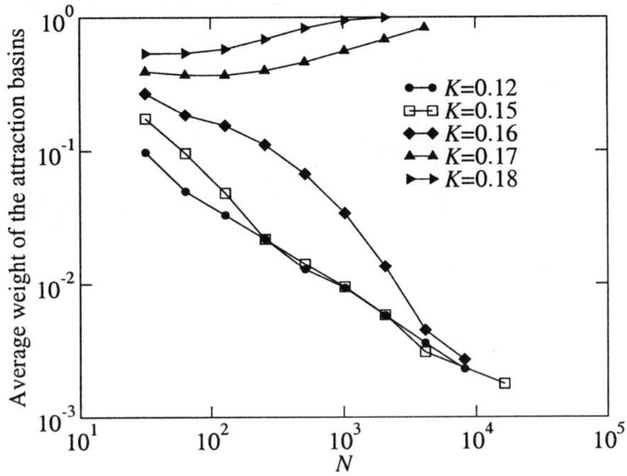

Fig. 9.7 Average weight of the largest attractor basin as a function of the system size for $a = 1.3$ and different values of the coupling K. Initial states of all elements are independently chosen. There is a well-defined coupling strength at which the transition from the fully synchronous state to the two-cluster state occurs. Adapted from [Manrubia and Mikhailov (2001)].

initial conditions.

9.5 Overlap Distributions and Ultrametricity

In order to have replica symmetry breaking in the thermodynamic limit when a system is globally coupled, it is necessary that the initial coupling field presents macroscopic variations from replica to replica. Different algorithms can fulfill this property. As a particular case, we consider the situation where the initial states of the elements are drawn from a uniform distribution in the interval $[-\xi^\alpha, \xi^\alpha]$. The quantity ξ^α is itself a random variable which is chosen anew for each replica α. This prescription maintains the average value of the set of initial conditions around the value $x = 0$, following a Gaussian distribution. But since the dominant term in the coupling is of the form $x^2(0)$, the initial coupling field is now affected by the value ξ^α, such that

$$\lim_{N\to\infty} m(0) = 1 - \frac{a}{3}\xi^\alpha, \tag{9.17}$$

and macroscopic differences thus arise.

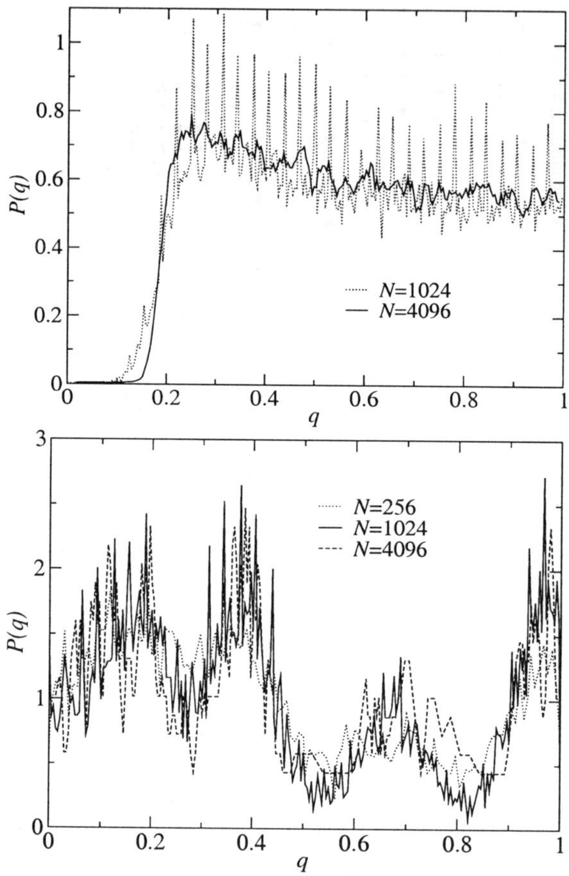

Fig. 9.8 Overlap distributions for ensembles of globally coupled logistic maps. The initial conditions for each replica α are chosen at random from a uniform distribution in $[-\xi^\alpha, \xi^\alpha]$, with a random choice of $\xi \in (0, 1]$. Many different attractors coexist for fixed values of K and a, and the distribution $P(q)$ occupies the whole domain of values of q also in the thermodynamic limit. Adapted from [Manrubia *et al.* (2001)].

Each value of ξ^α corresponds to a macroscopically different initial condition. However, there might be intervals of the value ξ^α that are mapped onto the same attractor. For example, if the values of $a = 1.3$ and $K = 0.15$ are selected, different attractors are explored as ξ^α varies. At $\xi^\alpha = 1$, the selected partition has two clusters, the largest one containing about 63% of the elements. For $\xi^\alpha = 0.944$ still the two-cluster partition is selected, but now its size is $0.72N$. When the value of ξ^α becomes smaller than 0.944, the

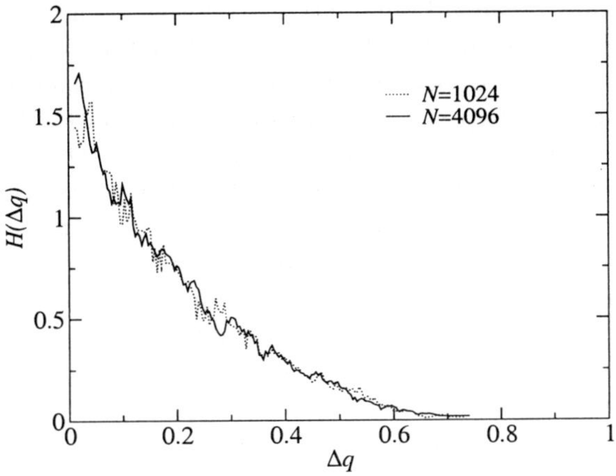

Fig. 9.9 Distribution of distances between the two smallest overlaps out of a triad. For the attractors to be ultrametrically organized, this function should have vanishing width in the thermodynamic limit. From [Manrubia *et al.* (2001)].

single synchronous state is selected. Hence, different attractors do coexist if the initial coupling field is different enough from replica to replica.

The overlap distributions obtained by using this macroscopically fluctuating coupling field occupy a broad, continuous interval. As N grows, they approach a fixed shape which depends only on the parameter values. Two examples are shown in Fig. 9.8 for $a = 1.3$ and $K = 0.15$, where the trajectories of the elements have period 2 (above), and $a = 1.55$ and $K = 0.1$, where more complex dynamical states take place (below). In the first case, there is a series of equally spaced peaks in the distribution. This is an interesting finite-size effect often observed in globally coupled logistic maps, where the system chooses partitions with a small number of clusters (usually two) and differing in a fixed amount of elements. The peaks occur in this case at intervals of size $s = 0.03125$. This indicates a preference to internally organize in groups of 16 elements, since the quantity $sN = 32$ corresponds to the number of out-of-phase elements between two replicas. Interestingly, the effect disappears for larger system sizes, and the overlap distribution $P(q)$ tends to a smoother, continuous function. Even if this effect is diluted for larger systems, it reveals a trend of the initially symmetric ensemble to break that symmetry and to organize internally in a spontaneous fashion.

The last question to ask is if the attractors of the dynamical system are hierarchically organized in an ultrametric way. As has been discussed, ultrametricity can be checked through the calculation of the distribution $H(\Delta q)$ of the differences between the two smallest overlaps out of a set of three. Though some finite size effects might produce a broad distribution for small systems, this function has to shrink as N grows. This is not observed in globally coupled logistic maps. Figure 9.9 shows a typical example of how $H(\Delta q)$ behaves for this system. Similarly to what occurs with the overlap distribution, this function seems to reach a fixed shape for each pair of parameters, and it remains invariant under increasing system size.

Ultrametricity is a very demanding condition which can be shown to hold in the simple system described by Eq. (9.3). However, it is very difficult to prove that exact ultrametricity holds in more realistic system models. The extensive numerical simulations carried out with globally coupled logistic maps indicate that the organization of the attractors is not ultrametric, even if some weaker form of hierarchical organization may be present. This fact is inferred from the shape of the function $H(\Delta q)$, with a maximum at $q = 0$ and a fast decaying tail for larger values of q. The question whether a dynamical system with all the properties of the simplest spin glass (including exact ultrametricity) exists, remains open.

The phenomenological behavior of globally coupled logistic maps parallels that characterizing spin glasses: symmetry breaking in the limit $N \to \infty$, continuous overlap functions, and a weak form of hierarchical organization in the space of attractors. Together with other related properties discussed in this chapter, these features make globally coupled logistic maps a paradigmatic example of a dynamical glass, with a role similar to that played by the model (9.3) for spin glasses.

PART 3
Selected Applications

Chapter 10

Chemical Systems

Nonequilibrium chemical systems are often chosen to test general predictions of nonlinear dynamics. Reactions are relatively slow and the characteristic times of chemical oscillations usually lie in the range of seconds, convenient for experimental observation. Diffusion of reactants provides natural local coupling between neighboring reaction volumes, each representing a chemical oscillator. Chemical reactions are easily controlled by varying the system composition and such parameters as temperature, illumination, etc. For some chemical systems, good theoretical models are available. But even in these cases, an experiment is never equivalent to running a computer simulation. Construction of any model involves reductions and simplifications. Compared with the models, real experimental systems possess not only noise, but also certain perturbations of their dynamics due to the processes and interactions neglected in the model. Thus, experiments allow to see how robust are the theoretical predictions. Generic theoretical results often retain qualitative validity even beyond the limit upon which they can be quantitatively justified.

Two inorganic chemical systems are selected in this chapter to illustrate the synchronization phenomena. Global coupling can easily be implemented and controlled for arrays of electrochemical oscillators, both when an individual oscillator is characterized by periodic or intrinsically chaotic dynamics. For such arrays, detailed experimental verification of general theoretical predictions on the onset of synchronization, clustering and dynamical order has recently been performed.

The second example refers to a catalytic surface reaction. Here, diffusive coupling between individual surface elements is present. Though local oscillations are periodic, diffusion leads in this system to destabilization of uniform oscillations and appearance of chemical turbulence. Introduction of artificial global delayed feedbacks via the gas phase allows to synchronize individual oscillations and thus suppress turbulence. Near the synchronization transition, clusters and other spatiotemporal patterns, such as oscillating cells and intermittent turbulence, have been observed. Theoretical results for a model of this system are in good agreement with the experimental data. Qualitatively, the development of turbulence and its control are described by the complex Ginzburg–Landau equation.

10.1 Arrays of Electrochemical Oscillators

If a piece of nickel is immersed into an aggressive water solution containing sulphuric acid, its dissolution begins. Since this process involves ions, it is accompanied by generation of currents and electrical fields which, in turn, affect the dissolution rate. The reaction proceeds in several stages and involves formation of intermediate products on the metal surface which block the surface sites and thus slow down the process. As a result of various nonlinearities, oscillations can develop in this system. Depending on the parameters, periodic or chaotic oscillations can take place. In the vicinity of a supercritical Andronov–Hopf bifurcation, such oscillations have small amplitudes and are approximately harmonical. Both periodic and chaotic oscillations are well reproduced by a mathematical model [Haim *et al.* (1992)].

To study collective dynamics in oscillator ensembles, arrays of nickel electrodes have been used [Kiss *et al.* (2002a)]. The experimental setup is schematically shown in Fig. 10.1. The electrodes are made from Ni wires of diameter 1 mm. They are embedded in epoxy, so that the reaction is taking place only at their ends. Typically, 64 electrodes in an 8×8 geometry were used. The electrodes are immersed into a sulphuric acid solution. All electrodes in the array are held at the same potential V with a potentiostat. The electrodes are connected to the potentiostat held at the potential V through one collective resistor (R_{coll}) and through individual resistors (R_i) connected to each electrode. By means of zero resistance ammeters inserted between the individual and collective resistors, it is possible to individually measure the currents passing through each of the electrodes.

The total resistance R_{tot} of the array is

$$R_{\text{tot}} = R_{\text{coll}} + \frac{1}{N}\sum_{j=1}^{N} R_j, \tag{10.1}$$

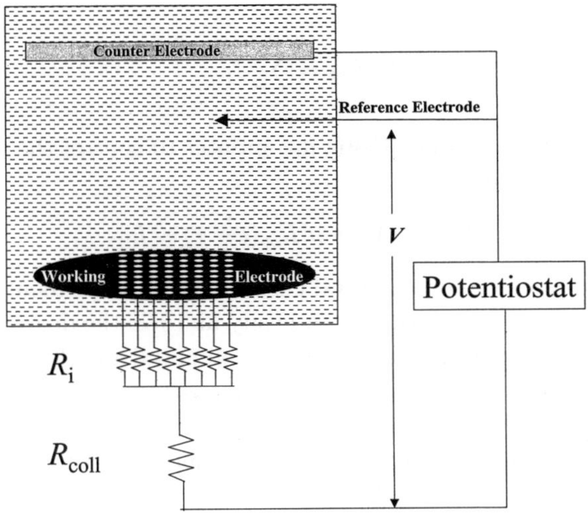

Fig. 10.1 The experimental setup.

where N is the number of electrodes. The collective potential V_{coll}, applied to the electrodes before the individual resistors, is therefore given by

$$V_{\text{coll}} = V - R_{\text{coll}} \sum_{j=1}^{N} I_j, \tag{10.2}$$

where I_j is the electrical current passing through the jth electrode. We see that this potential is determined by all the currents, which leads to global coupling between individual electrochemical oscillators. The coupling strength K can be defined as

$$K = \frac{R_{\text{coll}}}{R_{\text{tot}}}. \tag{10.3}$$

The global coupling is vanishing ($K = 0$) if the collective resistor is absent, $R_{\text{coll}} = 0$. It reaches its maximal strength $K = 1$ in absence of individual resistors (if $R_j = 0$).

Moreover, time-delayed feedbacks can also be introduced in the experiments. The potential V is maintained by a potentiostat and can be arbitrarily varied depending on some signal. As this signal, the total measured current

$$I_{\text{tot}} = \sum_{j=1}^{N} I_j \tag{10.4}$$

can, for instance, be chosen. Using a computer to control the potentiostat, a time-delayed dependence

$$V(t) = V_0 + \mu[I_{\text{tot}}(t-\tau) - I_0] \tag{10.5}$$

of the applied potential on the total current can be implemented. Here the coefficient μ specifies the feedback intensity and τ is the delay time; V_0 and I_0 are some constant reference levels of the potential and the current. When $\tau = 0$, the feedback represents just another way to introduce instantaneous global coupling, which can already be easily implemented by using the collective resistance. The advantage of the method is that it also permits to study the effects of controlled time delays on the collective behavior of the ensemble.

Electrochemical oscillations can be either periodic or chaotic, depending on the applied potential, the acid concentration and the external resistance. First, we present the results of experiments with periodic oscillators.

10.1.1 *Periodic oscillators*

The discussion of experimental evidence for synchronization behavior and clustering in this section is based on [Kiss *et al.* (2002b); Zhai *et al.* (2003)]. Figure 10.2a shows the current as a function of the applied potential for a single electrode with a relatively small external resistance. Oscillations begin with an Andronov–Hopf bifurcation (indicated as "H" in the figure). Inside the oscillatory region, the maxima and the minima of the current oscillations are displayed. The oscillations remain approximately harmonical even relatively far from the bifurcation point, as seen in Fig. 10.2b.

To facilitate the comparison with the general theory, presented in Part I, it is convenient to define oscillation phases. This can be done by employing the so called Hilbert transform method [Panter (1965)]. If $I(t)$ is the current and $\langle I \rangle$ is its temporal average, the Hilbert transform $H(t)$ is defined as

$$H(t) = \frac{1}{\pi} \int_{-\infty}^{\infty} \frac{I(t') - \langle I \rangle}{t - t'} dt'. \tag{10.6}$$

Now, the state of an oscillator at any moment t becomes specified by two variables, $I(t)$ and $H(t)$. As time goes on, this point moves along a certain

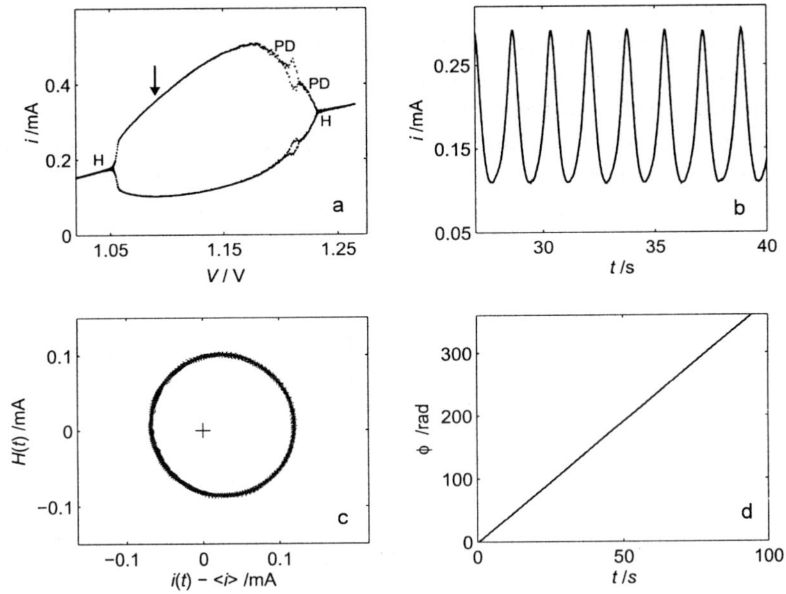

Fig. 10.2 Single periodic oscillator. (a) The experimental bifurcation diagram. Labels H and PD indicate supercritical Andronov–Hopf and period-doubling bifurcations. In oscillatory regimes, only maxima and minima of the current are displayed. (b) Time dependence of the current at the voltage denoted by the arrow in part (a). (c) Phase portrait of a single oscillator obtained with the Hilbert transform. (d) Phase of the oscillator as function of time. From [Zhai *et al.* (2003)].

trajectory in the plane (I, H). As an example, such a trajectory is shown for a single electrochemical oscillator in Fig. 10.2c. We see that, to a good approximation, it represents a circle, as should be expected for harmonical oscillators.

The phase $\phi(t)$ at time t is defined as

$$\phi(t) = \arctan \frac{H(t)}{I(t) - \langle I \rangle}, \tag{10.7}$$

that is, as the angle in Fig. 10.2c. In the experiments, it was an almost linear function of time (Fig. 10.2d). The frequency of an oscillator can be determined as the slope of this dependence.

In the experimental array with 64 oscillators, some natural dispersion of individual frequencies was observed, mostly due to the variations in the surface of electrodes and transport. This variation could be enhanced by using small random resistors R_j. Figure 10.3a shows the histogram of rescaled frequencies in the array in absence of global coupling ($K = 0$). When global coupling is introduced, the distribution of frequencies changes significantly. At a relatively low coupling strength ($K = 0.035$), a central synchronous group of oscillators is formed, but the rest of the population is not strongly affected (Fig. 10.3b). In contrast to this, stronger coupling ($K = 0.085$) induces full frequency synchronization (Fig. 10.3c).

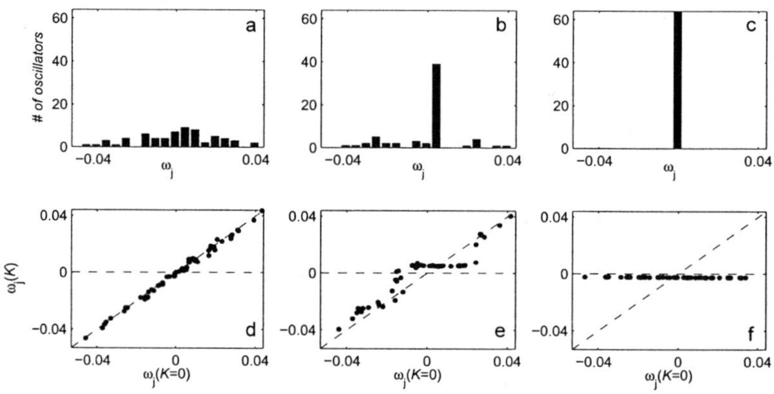

Fig. 10.3 Synchronization in the experimental array of 64 periodic electrochemical oscillators. Histograms of rescaled frequencies (a–c) and plots (d–e) of effective frequencies $\omega_j(K)$ as functions of natural frequencies $\omega_j(K = 0)$ for all oscillators at (a,d) $K = 0$ (without coupling), (b,e) $K = 0.035$ and (c,f) $K = 0.085$. The rescaled frequency ω_j of the jth oscillator is defined as $\omega_j = f_j/f_0 - 1$ where f_j is the dimensional frequency and f_0 is the mean frequency of all oscillators in the array in absence of coupling. From [Zhai *et al.* (2003)].

Synchronization can be further demonstrated by plotting the diagrams of effective vs. natural frequencies of individual oscillators. In absence of coupling, effective and natural frequencies are identical, and the diagram represents simply a diagonal line (Fig. 10.3d). When moderate coupling is introduced, the frequencies of the oscillators forming the central synchronous group undergo essential changes, but for other oscillators the distortions of their frequencies are only minor (Fig. 10.3e). In the fully synchronous state, frequencies of all oscillators become largely modified (Fig. 10.3f).

To further analyze the synchronization processes, phases $\phi_j(t)$ of all oscillators in the array over a long interval of time have been determined for various coupling strengths. Using them, the time-dependent global complex signal has been constructed as

$$Z(t) = \frac{1}{N} \sum_{j=1}^{N} \exp\left(i\phi_j(t)\right). \tag{10.8}$$

The mean modulus of the global complex signal, averaged over the entire observation interval, yields the synchronization order parameter $\sigma = \langle |Z(t)| \rangle$. The experimental dependence of this order parameter on the coupling strength is displayed in Fig. 10.4a.

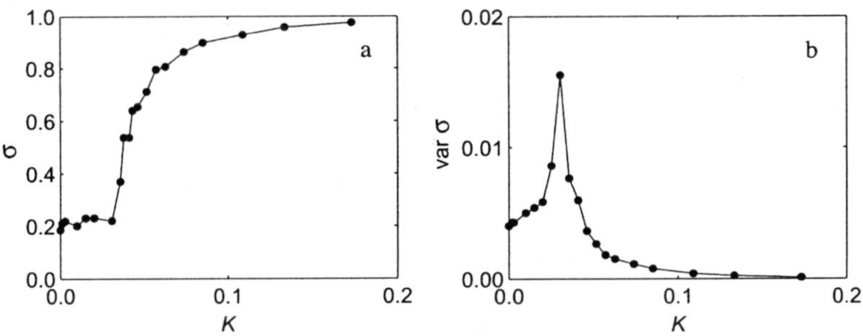

Fig. 10.4 Synchronization order parameter (a) and its statistical variation (b) as functions of the coupling strength. [Zhai *et al.* (2003)].

According to the theory of frequency synchronization for phase oscillators (Chapter 3), in the infinite ensemble ($N \to \infty$) the order parameter σ should remain zero until a critical coupling intensity K_c is reached. For $K > K_c$, in the vicinity of the transition point the dependence $\sigma \sim (K - K_c)^{1/2}$ should hold. Examination of Fig. 10.4a reveals that σ indeed remains relatively small until a certain critical coupling strength is reached. Above this critical strength, it rapidly grows and the growth law agrees with the dependence predicted by the theory.

Nonvanishing of the synchronization order parameter below the critical point is a finite-size effect. Suppose that coupling between the oscillators is absent and each oscillator evolves independently with its own frequency and the initial phase, $\phi_j(t) = \phi_j(0) + \omega_j t$. In this case, we have

$$\left\langle |Z(t)|^2 \right\rangle = \frac{1}{N^2} \sum_{j,k=1}^{N} \left\langle \exp\left[i\left((\omega_j - \omega_k)t + (\phi_j(0) - \phi_k(0))\right)\right]\right\rangle. \tag{10.9}$$

The terms in this sum with $j \neq k$ are oscillating and vanish after time averaging. The remaining terms with $j = k$ yield

$$\left\langle |Z(t)|^2 \right\rangle = \frac{1}{N}. \tag{10.10}$$

Therefore, in absence of coupling ($K = 0$) in a finite system of size N a synchronization order parameter with a magnitude $\sigma \sim N^{-1/2}$ is expected. This size scaling has been confirmed in the experiments.

Fluctuations of the amplitude $|Z(t)|$ of the global signal near the synchronization transition have been studied analytically [Daido (1989)]. The theory predicts that enhanced fluctuations should be observed at the transition point. Such behavior has indeed been found in the experiments. Fig. 10.4b shows the statistical dispersion of $|Z(t)|$ as a function of the coupling strength. Sharp increase of fluctuations near the synchronization transition is evident.

The above results correspond to the conditions where individual oscillations are approximately harmonical. Under a different choice of its parameters, the same system can also exhibit relaxational oscillations. The synchronization experiments with such relaxational oscillators yield qualitatively the same behavior as in the case of harmonical oscillators.

Experimental data has been compared in this section with the predictions of the general theory, formulated in Chapter 3. It should be, however, noted that realistic models of the considered electrochemical oscillators are also available [Haim *et al.* (1992)]. Simulations of the collective synchronization behavior, based on such models, have been carried out [Zhai *et al.* (2003)] and show similar results.

10.1.2 *Chaotic oscillators*

Synchronization experiments with arrays of chaotic electrochemical oscillators have been performed [Wang *et al.* (2000); Wang *et al.* (2001); Kiss *et al.* (2002c); Kiss *et al.* (2002a); Kiss and Hudson (2003)]. Chaos in an individual oscillator is experimentally reached via a period-doubling bifurcation sequence as the applied potential is changed. The information dimension of the respective attractor is 2.2 and, therefore, this attractor can be embedded into a three-dimensional space. In the experiments, only a single time series $I(t)$ of electric current was recorded. The two additional variables, needed to display the attractor, can be generated by taking the values of the current at two delayed moments, $I(t-\tau)$ and $I(t-2\tau)$. Figure

10.5a shows the chaotic attractor of a single oscillator, reconstructed using delayed coordinates.

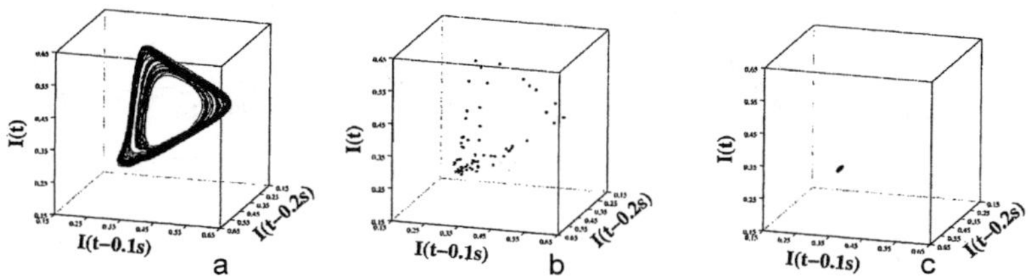

Fig. 10.5 Reconstructed chaotic attractor of a single electrochemical oscillator (a) and instantaneous states of all 64 oscillators (b) in absence of coupling and (c) under full synchronization ($K = 1.0$). From [Wang *et al.* (2000)].

Arrays of 64 chaotic electrochemical oscillators have been experimentally investigated. In the absence of global coupling, the states of the oscillators are randomly distributed over the attractor, as seen in Fig. 10.5b. Strong global coupling resulted in full chaotic synchronization, so that at any given time moment the states of all oscillators were almost identical (Fig. 10.5c). As the coupling strength was increased from zero to its maximum possible level, the following sequence of behaviors was generally found: weak phase synchronization → intermittent chaotic clusters → stable chaotic clusters → intermittent chaotic clusters → fully synchronized chaotic state.

Figure 10.6 shows a series of snapshots in a regime with two stable clusters (with sizes 23 and 41). Occasionally, the noise is strong enough to drive one element (see $t = 20$ s) from its cluster, but then this element quickly returns. The clusters change their positions relative to each other and sometimes come very close. However, mixing of the elements does not occur.

The cluster organization of a population can be analyzed by computing pair distances between the elements. The pair distance $d_{ij}(t)$ between elements i and j at time t is defined for this system as

$$d_{ij}(t) = \left[\, (I_i(t) - I_j(t))^2 + (I_i(t-\tau) - I_j(t-\tau))^2 + (I_i(t-2\tau) - I_j(t-2\tau))^2 \, \right]^{1/2}, \qquad (10.11)$$

where τ is a fixed delay time (typically 0.1 s). For an array of size $N = 64$, there are $N(N-1)/2 = 2016$ pair distances.

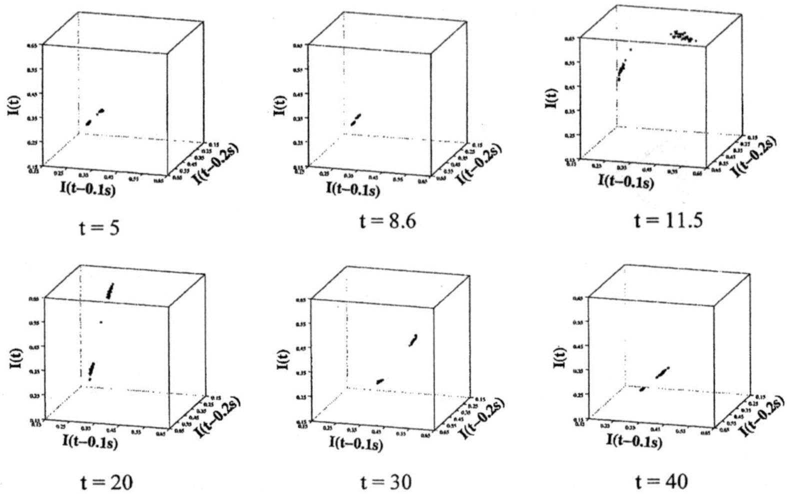

Fig. 10.6 Snapshots of instantaneous states of all 64 oscillators at subsequent time moments in the regime with two stable clusters ($K = 0.725$). From [Wang *et al.* (2000)].

Histograms of pair distance distributions at different coupling strengths are given in Fig. 10.7. Without coupling ($K = 0$), the pairs are uniformly distributed over the range of distances corresponding to the attractor diameter. Full synchronization (that is, a single cluster) is found at $K = 1$; the finite width of this distribution is due to the relatively strong noise and some intrinsic heterogeneity in the experimental system. Two clusters are clearly seen at $K = 0.725$. The distributions at $K = 0.67$ and $K = 0.78$ correspond to the regimes of intermittent clusters.

Since clusters move with respect to each other, the distributions of pair distances change with time. Two examples of their evolution are shown in Figs. 10.8 and 10.9. When clusters are stable, two groups are seen at all times in the pair distance distribution (Fig. 10.8). In the intermittent regime (Fig. 10.9), the distribution alternates between states with one or two clusters.

The time-dependent condensation order parameter $r(t)$ is defined (see also Chapter 8) as the ratio of the number of pairs whose distance at time t in the three-dimensional state space is less than some threshold δ (taken to be 0.06 mA here). It may vary from zero, in absence of any clusters, to unity in the regime of full synchronization. When several clusters are

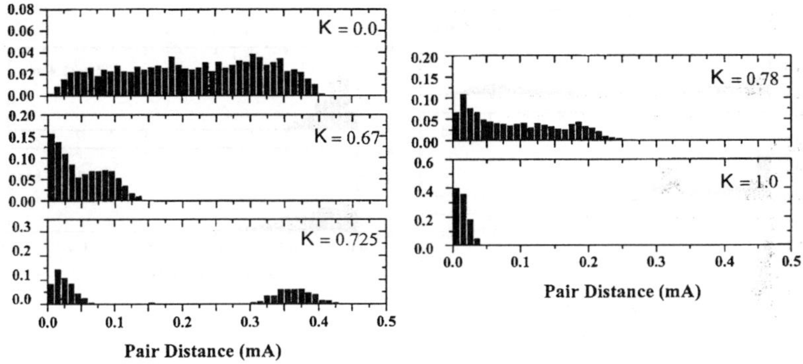

Fig. 10.7 Normalized instantaneous histograms of pair distances at different coupling strengths. From [Wang *et al.* (2000)].

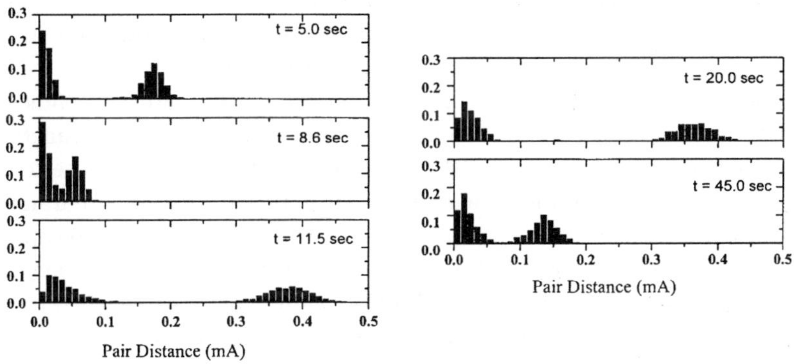

Fig. 10.8 Evolution of the distribution of pair distances with time in the regime with two stable clusters ($K = 0.725$). From [Wang *et al.* (2000)].

formed, this parameter is less than unity since the pair distances are (close to) zero only for pairs of elements belonging to the same cluster.

If a population of size N has two clusters with sizes M and $N - M$, the order parameter is

$$r = \frac{M(M-1) + (N-M)(N-M-1)}{N(N-1)}. \tag{10.12}$$

Therefore, the minimum value $r_{min} = 0.5$ of the order parameter for two clusters is reached when both of them have the same sizes ($M = N/2$).

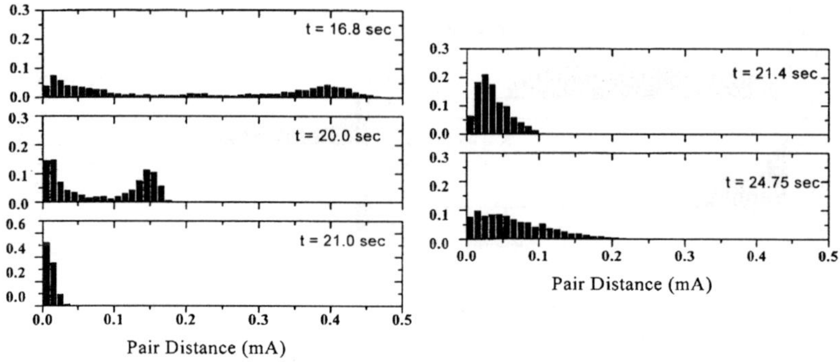

Fig. 10.9 Evolution of the distribution of pair distances with time in the intermittent regime ($K = 0.67$). From [Wang *et al.* (2000)].

Note that the order parameter can also be less than 0.5 if, in addition to two clusters, some non-entrained elements remain in the population. It can also be less than 0.5 if a larger number of clusters is formed.

Figure 10.10 displays the condensation order parameter as a function of time for different coupling strengths. For $K = 0$ the order parameter is close to zero at all times (Fig. 10.10a), whereas at $K = 1$ it is close to one (Fig. 10.10e). In the stable cluster regime (Fig. 10.10c), the order parameter is approximately constant. Occasional narrow peaks in Fig. 10.10e appear because the two clusters sometimes come at a distance less than δ one from another and, with the employed resolution, cannot be distinguished from a single-cluster state. In the intermittent regimes (Fig. 10.10b, d), the order parameter goes through large excursions above and below the value of 0.5 obtained for two equal clusters.

The average condensation order parameter $\langle r \rangle$ as a function of the coupling strength is shown in Fig. 10.11. It was obtained by determining temporal means $\langle d_{ij} \rangle$ of all pair distances and computing $\langle r \rangle$ from such mean distances according to Eq. (10.11), rather than by direct averaging of the time-dependent order parameter $r(t)$. The progression towards a fully synchronized state is clearly seen. Filled triangles correspond to the intermittent regimes and open symbols in the local maximum at $K = 0.725$ indicate the regimes with stable clusters.

The multiplicity of symbols corresponding to stable clusters is not due to an experimental error. Rather, it gives evidence that different stable cluster partitions can take place in the same system, depending on the

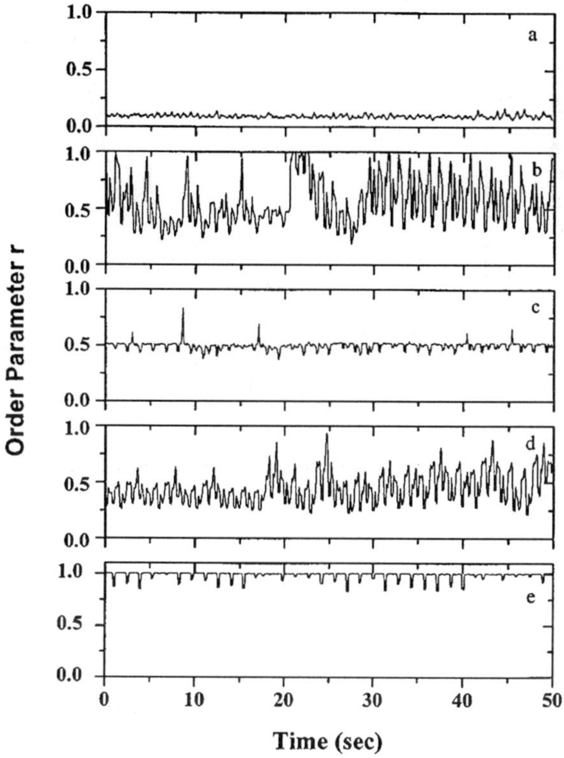

Fig. 10.10 Time dependences of the condensation order parameter r for different coupling strengths (a) $K = 0$, (b) $K = 0.67$, (c) $K = 0.725$, (d) $K = 0.78$, and (e) $K = 1.0$. From [Wang *et al.* (2000)].

initial conditions. In the experiments, only two stable clusters were always observed. However, their sizes varied considerably, from 18 to 46 out of a total number of 64. By applying sufficiently strong perturbations (that is, by breaking the electric circuit for short times), transitions between different stable configurations could be produced. All regimes with stable clusters had chaotic dynamics, which depended only on the cluster partition of the population.

Similar behavior was found [Wang *et al.* (2001)] in the experiments with feedback described by Eq. (10.5). When the feedback intensity was increased, dynamical clustering developed (Fig. 10.12). Starting from a certain feedback intensity, the clusters became stable, but the motion re-

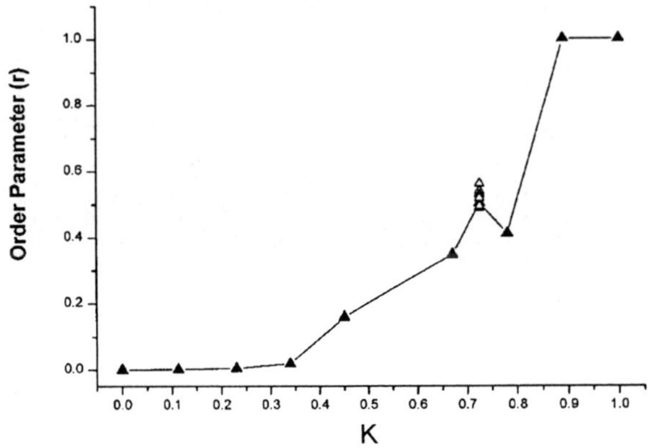

Fig. 10.11 Condensation order parameter based on mean distances as function of the coupling strength. Open triangles indicate regimes with stable clusters. From [Wang *et al.* (2000)].

mained chaotic. Further increase of the feedback resulted in a transition to periodic dynamics.

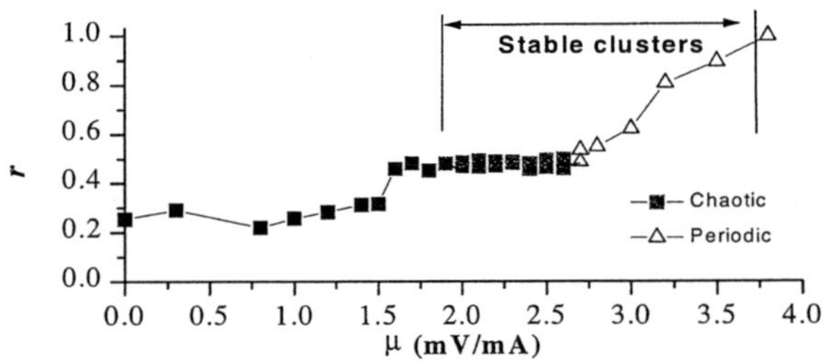

Fig. 10.12 Dependence of the average condensation parameter r on the feedback intensity μ. From [Wang *et al.* (2000)].

Statistical properties of clustering regimes have been systematically explored [Kiss and Hudson (2003)]. These studies were performed with a modification of the experimental setup, needed to record much longer time series without a significant parameter drift. Because of this modification, the level of noise was, however, increased.

The organization of the system was characterized by constructing the hierarchical cluster trees based on the experimental data. The pair distances between the states of all elements were computed each 2 seconds within a total observation time of 1000 s. A Matlab algorithm based on time-averaged pair distances was then used to construct cluster trees at different coupling strengths (see Fig. 10.13). This tree shows the number of clusters as a function of the clustering distance E (which is a pair distance below which two elements are classified as belonging to the same cluster). The number of points at a given E indicates the number of clusters within that spatial resolution.

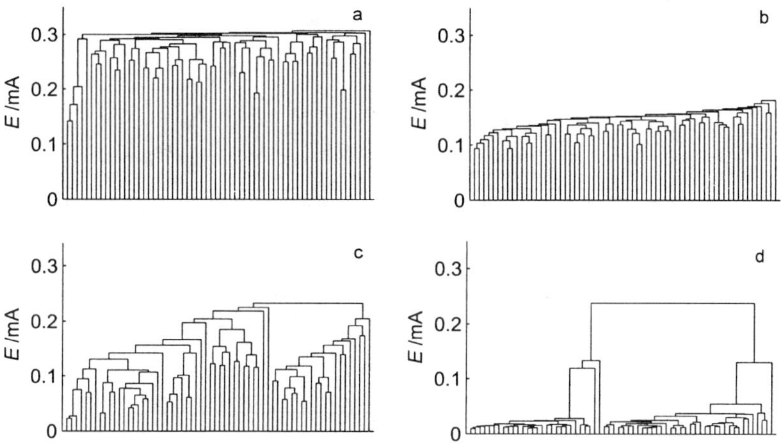

Fig. 10.13 Hierarchical cluster trees constructed from time series containing about 1500 chaotic oscillations at different coupling strengths: (a) $K = 0$, (b) $K = 0.6$, (c) $K = 0.7$, and (d) $K = 0.75$. From [Kiss and Hudson (2003)].

Without coupling (Fig. 10.13a), the cluster tree consists of individual elements which seem to cluster only at large distances, comparable to the attractor diameter. On the other hand, in the stable clustering regime at $K = 0.75$, two persisting clusters are clearly seen (Fig. 10.13d). They appear to split into an aggregation of smaller clusters only at fine resolution (this splitting is an effect of noise and some remaining heterogeneity in the array). The transition to such stable clusters under increase of the coupling strength proceeds through intermittent regimes characterized by labile cluster organization. The trees, based on averaging over a long time period, show a strong variation in the number of clusters (Fig. 10.13b) and gradual emergence of a relatively stable cluster organization (Fig. 10.13c).

The clusters could also be defined dynamically, as groups of elements separated by pair distances less than the threshold $E = 0.13$ mA for a time of at least ten seconds. At $K = 0.75$, the system stays in the two-cluster state, with only rare excursions to a single cluster due to an increased noise level. In contrast to this, the number of clusters in the system is strongly fluctuating with time in the intermittent regime. Figure 10.14 shows the mean number of clusters and statistical dispersion in this number as functions of the coupling strength.

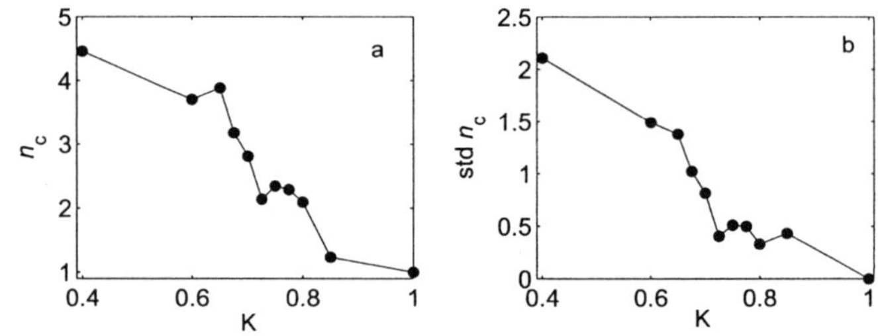

Fig. 10.14 Mean number n_c of clusters (a) and its statistical dispersion (b) as functions of the coupling strength. From [Kiss and Hudson (2003)].

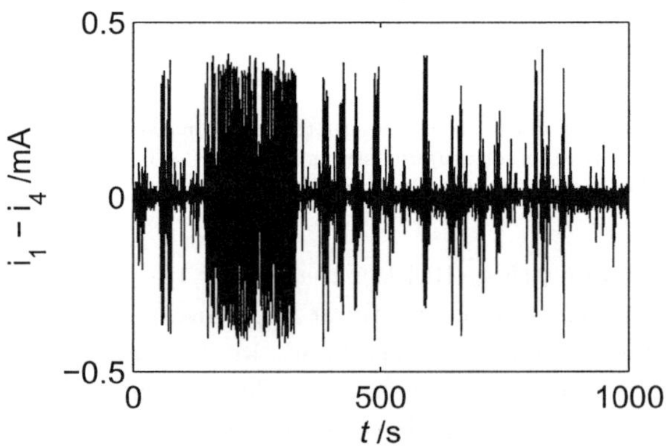

Fig. 10.15 Intermittent clustering. Time dependence of the current difference between elements $j = 1$ and $k = 4$ at $K = 0.4$.From [Kiss and Hudson (2003)].

The behavior of the population in the regime of labile clustering exhibited statistical properties characteristic for on–off intermittency. Figure

10.15 shows how the difference between the currents of two arbitrarily chosen elements evolved in time in the intermittent regime. The evolution can be described as alternating laminar and bursting states, where the difference is, respectively, small or large. In on–off intermittency, the power spectrum $S(f)$ of the difference signal should scale as $S(f) \sim f^{-1/2}$ with the frequency f [Fujisaka and Yamada (1986); Venkataramani *et al.* (1996); Sameshima *et al.* (2001)]. The statistical distribution $P(T)$ of lengths T of the laminar states should obey the power law $P(T) \sim T^{-3/2}$ [Heagy *et al.* (1994b)]. In the experiments at $K = 0.4$ and 0.6, the average power spectrum of 63 pairs of independent current differences followed a power law with an exponent of 0.5, whereas for $K = 0.7$ a slightly larger exponent of 0.6 has been found. The experimental exponents for the distribution of laminar lengths were 1.4 and 1.8, respectively.

The interval of coupling strengths immediately preceding the onset of clustering is also interesting. Here, the phenomenon of weak phase synchronization has been observed [Kiss *et al.* (2002b)]. Though the dynamics of the considered electrochemical oscillators is chaotic, it represents a kind of distorted periodic motion, as seen from the shape of the attractor of an individual oscillator in Fig. 10.5a. Therefore, it is possible to define an oscillation phase $\phi(t)$ for this dynamical system. The phase is determined from the time series for the current by Eq. (10.7) using the Hilbert transform (10.6). The oscillator frequency is then calculated as the mean phase velocity $(2\pi)^{-1} \langle d\phi/dt \rangle$. Moreover, the cyclic phase differences $\Delta\phi_{ij}(t) = \phi_i(t) - \phi_j(t) \pmod{2\pi}$ can be determined for any pair of oscillators at each time moment.

The histrogram of natural frequencies for all 64 elements in the array in absence of coupling is shown in Fig. 10.16a and the distribution of cyclic phase differences for all 2016 pairs of elements in the array is presented in Fig. 10.16b. The dispersion of natural frequencies is due to intrinsic heterogeneities in the experimental system. The distribution of phase differences is not completely flat, as should have been for truly independent oscillators. A shallow maximum seen in this distribution could be explained by very weak coupling through the electrolyte, which persisted even at $K = 0$.

When weak global coupling ($K = 0.1$) is introduced, the collective behavior of the oscillator population is dramatically changed, though the dynamics of individual elements is not significantly modified. The histogram of the effective frequencies shows synchronization (Fig. 10.16c); the frequencies of 63 oscillators are identical and only one oscillator has a frequency that is slightly higher. The distribution of phase differences has a pro-

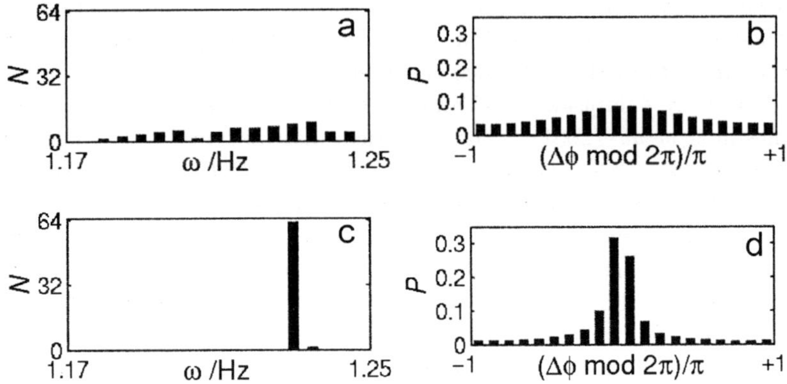

Fig. 10.16 Weak phase synchronization in a population of 64 chaotic oscillators. The histograms of frequencies (a,c) and the distributions of cyclic phase differences for all oscillators are shown (a,c) in absence of coupling and (b,d) at $K = 0.1$. From [Kiss *et al.* (2002a)].

nounced central maximum at $\Delta\phi = 0$. As the coupling strength increases from $K = 0$, this form of synchronization first appears at $K \approx 0.04$ and persists approximately up to $K = 0.15$. At larger strengths of global coupling, the simple periodicity of motion breaks up and more complex temporal variation, which cannot be characterized by cyclic phases, is observed. At still stronger coupling, dynamical clustering described above in this section takes place.

The behavior at weak global coupling looks similar to frequency synchronization of periodic oscillators (Chapter 3). Synchronization for periodic oscillators is also possible when external noise is applied, if the coupling strength exceeds a certain threshold. The above experiments with chaotic oscillators can be qualitatively interpreted by assuming that chaotic elements generate "intrinsic" noise, whose effects are however similar to those of external noise. Frequency synchronization persists while the chaotic component in the dynamics remains relatively small and it can still be described as distorted periodic motion. This kind of collective behavior in populations of chaotic elements is also known as phase synchronization (see Sec. 7.2.1).

10.2 Catalytic Surface Reactions

Some chemical reactions cannot occur in gas phase. If, however, the molecules are adsorbed on a surface of a metal catalyst, the energy barrier for the chemical transformation is greatly reduced and the reaction becomes possible. A good example is provided by the catalytic oxidation of carbon monoxide (CO). The reaction between molecules of CO and oxygen in the atmosphere is excluded. But in the presence of a platinum catalyst, the reaction proceeds at a high rate, converting carbon monoxide into carbon dioxide (CO_2). Though carbon monoxide is a dangerous poison, molecules of CO_2 represent a normal component of the atmosphere and are harmless. Therefore, catalytic conversion of CO into CO_2 is helpful in preventing environmental pollution. Indeed, this chemical process is employed in all car catalysts in order to eliminate CO from the exhaust. While being practically important, catalytic CO oxidation on Pt surfaces also yields a convenient experimental system where many phenomena of nonequilibrium pattern formation have been observed and investigated. In this section, our attention is focused on the use of this chemical reaction to study synchronization induced by delayed global feedbacks.

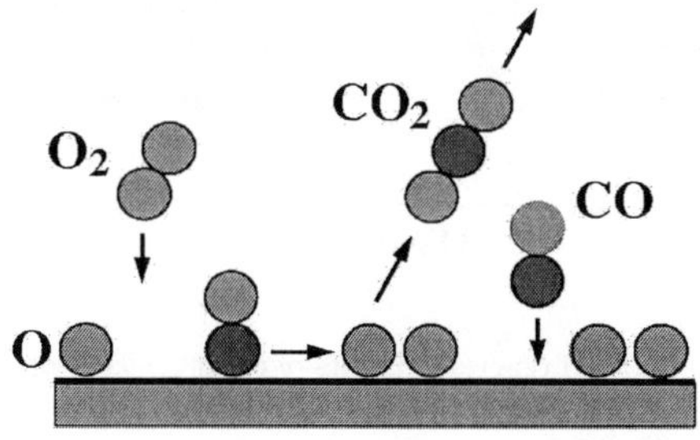

Fig. 10.17 Catalytic CO oxidation on platinum.

Figure 10.17 illustrates the mechanism of the considered chemical reaction. Molecules O_2 of oxygen dissociatively adsorb on the platinum surface, so that individual oxygen atoms are present on it. At the same time, molecules CO of carbon monoxide also become adsorbed. The adsorbed molecules of CO diffuse on the metal surface. When they meet adsorbed oxygen atoms, a reaction event takes place and a molecule of CO_2 is produced. The product immediately leaves the surface. The reaction is accompanied by the release of heat. To prevent strong thermal effects, experiments are typically performed at very low partial pressures of CO and oxygen, so that the overall reaction rate and, therefore, the total heating are small.

In the experiments, single crystals of platinum with perfect surfaces are used. The reaction process depends on the crystallographic orientation of the surface. Oscillations are only observed when the reaction takes place on the crystallographic plane Pt(110). A special property of this plane is that adsorption of CO induces a structural phase transition in the top layer of the platinum crystal. When surface concentration (or coverage) of the adsorbed CO molecules exceeds a certain threshold, the platinum atoms in the top layer rearrange themselves into a different 1×1 structure (see Fig. 10.18). If the CO coverage is decreased, the surface returns to its original 1×2 "missing row" structure.

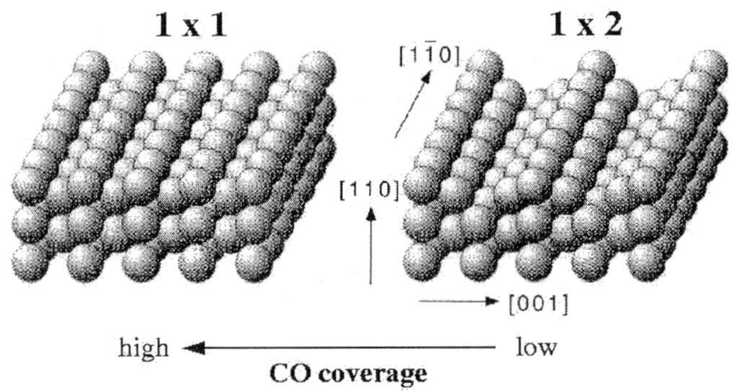

Fig. 10.18 Adsorbate-induced rearrangements of atoms in the top layer of the Pt crystal.

Oxygen adsorption and therefore the effective reaction rate are stronger for the 1×1 surface structure. When much carbon monoxide has adsorbed, this leads to the surface phase transition and to the acceleration of the reaction. But the reaction removes CO from the surface, decreasing its coverage and eventually causing the back phase transition to the original less reactive surface structure. Thus, persistent oscillations become possible. Note that the structural surface transition is a relatively slow process and therefore the reaction switching is inertial, which is important for the development of oscillations.

The reaction can be observed in real time with high spatial resolution by using photoemission electron microscopy (PEEM). This method allows to visualize local adsorbate coverages. In the PEEM images, the regions covered by oxygen look darker than the regions where CO molecules are locally in excess. Already the first PEEM observations of the CO oxidation reaction on Pt(110) have revealed complex spatiotemporal pattern formation on the reactive surface [Jakubith *et al.* (1990)]. Uniform oscillations, traveling or standing waves, pacemakers and rotating spiral waves were observed. Moreover, the reaction also showed regimes with spatiotemporal chaos, or chemical turbulence.

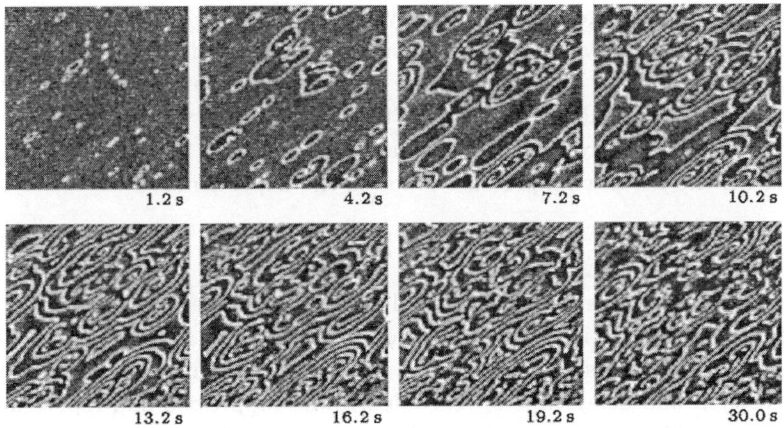

Fig. 10.19 Spontaneous development of chemical turbulence. Eight subsequent PEEM images of size 330 × 330 μm^2 are shown. From [Bertram *et al.* (2003)].

Figure 10.19 presents spontaneous development of turbulence from the initial uniform oxygen covered state. At $t = 0$, the reaction was started by opening the supply of CO molecules. Eight subsequent snapshots of PEEM images of size 330 × 330 μm^2 are displayed. A characteristic property of this process is the creation of multiple fragments of rotating spiral waves. The spiral waves repeatedly undergo breakups, leading to the formation of new spiral fragments at different locations. This type of turbulence is found in a wide range of temperatures for the appropriate choice of the partial pressures p_{CO} and p_{O_2} of the reactants. In the turbulent regime, each element of the surface is involved in oscillations, but their phases and amplitudes vary irregularly over the surface and with respect to time.

10.2.1 *Experiments with global delayed feedback*

To synchronize the oscillations (and thus to suppress the turbulence), global delayed feedbacks can be used. The reaction is highly sensitive to the concentration of CO in the gas phase, characterized by its partial pressures p_{CO}. This pressure can be varied by changing the rate of supply of CO into the reaction chamber. To implement global feedback, the supply of CO should be made instantaneously dependent on global properties of observed patterns.

In the experiments [Kim *et al.* (2001); Bertram *et al.* (2003)], the surface patterns were continuously monitored with PEEM. At each time moment, the average PEEM intensity $I(t)$ of the image was electronically determined and used for the generation of the control signal. The signal was further delayed by a certain time τ, inverted and amplified by a factor determining the feedback intensity. Finally, this signal was applied back to the system by controlling the automated inlet system for the CO gas. As a result, the CO partial pressure followed the dependence

$$p_{\text{CO}}(t) = p_0 - \mu\left[I(t-\tau) - I_{\text{ref}}\right], \tag{10.13}$$

where p_0 and I_{ref} are the CO partial pressure and the mean base level of the integral PEEM intensity in absence of feedback.

The experiments were conducted with different values of the feedback intensity μ and delay time τ. Turbulence was suppressed and replaced by uniform oscillations for any delay (delays up to $\tau = 10$ s have been probed) provided that the feedback intensity was sufficiently high. Typically, synchronization was reached already at feedback intensities corresponding to about 5% variation of p_{CO}, though in some cases it was increased up to about 20%. For lower feedback intensities, the feedback does not transform turbulence into stable uniform oscillations, but leads to the formation of new spatiotemporal patterns.

In the initial turbulent state in absence of feedback, the integral PEEM intensity is almost constant except for small random fluctuations. As the feedback intensity is increased starting from zero, global oscillations in $I(t)$ set in and a state of *intermittent turbulence* is first established. This state is characterized by turbulent cascades of localized objects on a uniformly oscillating background. Intermittent turbulence is found for any choice of the time delay. By further increasing the feedback intensity from the state of intermittent turbulence, additional spatiotemporal patterns—*clusters, oscillating cellular structures* and *standing waves*—are observed in the delay interval 0.5 s $< \tau <$ 1.0 s below the transition to uniform oscillations. The existence regions of such patterns sensitively depend on the choice of reaction parameters.

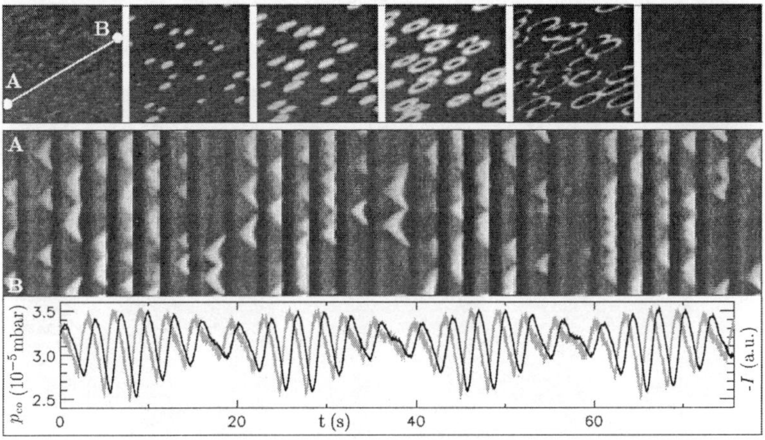

Fig. 10.20 Intermittent turbulence. Top: Six subsequent PEEM images of size 360×360 μm^2 during a single cycle of local oscillations. Time intervals between the images are 0.7 s. Middle: Temporal evolution of the pattern along the line AB indicated in the first image. Bottom: Corresponding temporal variations of CO partial pressure (black line) and inverted integral PEEM intensity (gray line). From [Bertram *et al.* (2003)].

An example of intermittent turbulence is displayed in Fig. 10.20. The PEEM images in the top row in Fig. 10.20 are snapshots taken within one cycle of the pattern evolution. Starting from a dark, uniform state, bright spots appear at different locations. When the growing spots reach a certain size, darker regions develop in the middle of these objects, transforming them into ring-shaped structures ("bubbles"). After some time, the whole pattern fades away and is replaced by the uniform dark state. Then the entire cycle repeats.

The temporal evolution of the pattern is further analyzed in the middle part in Fig. 10.20, showing the space-time diagram along the line AB indicated in the first image in the row above. Expanding bubbles are represented by triangular structures in the cross section. Examining the diagram, we see that the bubbles can die or reproduce. When the bubbles have reproduced until many of them are found, massive annihilation occurs and only a few of them survive. Thus, an irregular behavior of repeated

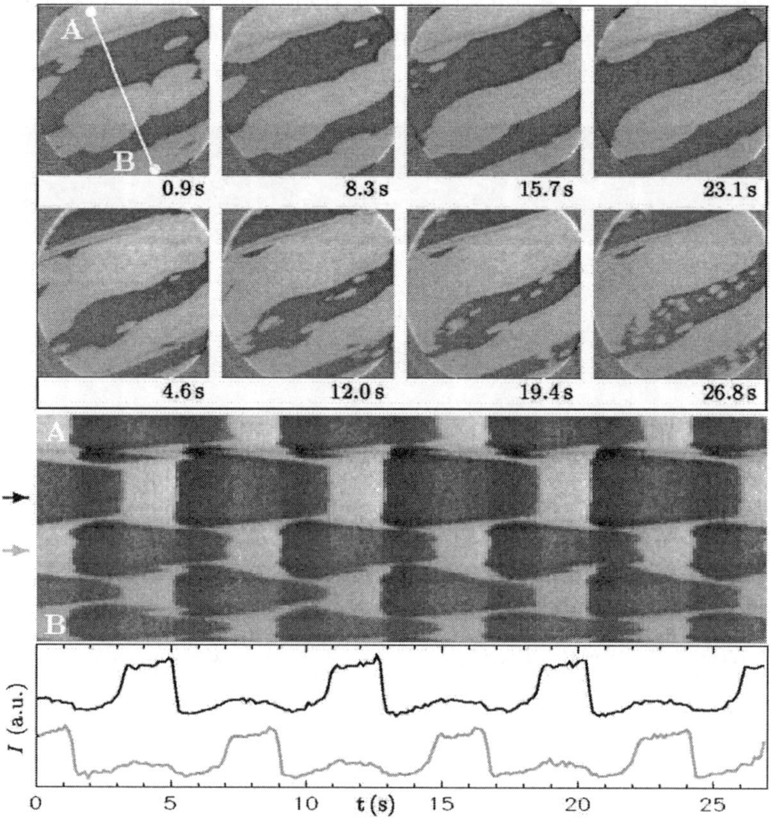

Fig. 10.21 Phase clusters. Top: Subsequent PEEM images of diameter 500 μm. Middle: Temporal evolution along the cross section AB in the first image. Bottom: Variation of the local PEEM intensity at two different points indicated by arrows in the space-time diagram. From [Bertram *et al.* (2003)].

annihilation cascades is observed. During intermittent turbulence, the variations of the CO partial pressure are aperiodic but rigidly correlated with the evolution cycles of the pattern (bottom row in Fig. 10.20). A similar intermittent behavior is observed with the spiral-wave fragments. They also reproduce until they occupy almost the entire monitored surface area, and then annihilate such that only a few of them survive.

When clusters develop, the catalytic surface splits into large regions belonging to either one of two different oscillatory states where frequencies are equal but phases are shifted by half a period. Usually, each of the two antiphase states occupies multiple spatial domains on the surface. An intrinsic spatial wavelength is missing in such a pattern. In the top and the second row in Fig. 10.21, snapshots of such a pattern are shown at time intervals of one oscillation period between subsequent frames. Snapshots lying one upon another are separated by half an oscillation period. The temporal evolution of the pattern along the cross section AB is shown in the third row of Fig. 10.21. The shape of the cluster domains undergoes small periodic variations, but on long time scales the average locations of the interfaces between the antiphase domains are almost stationary. Finally, the curves at the bottom of Fig. 10.21 display the temporal variations of the PEEM intensity at two sample points indicated by arrows on the left side of the space-time diagram. Each maximum of the PEEM intensity in the two curves is followed by a second, smaller peak which is characteristic of period-two oscillations.

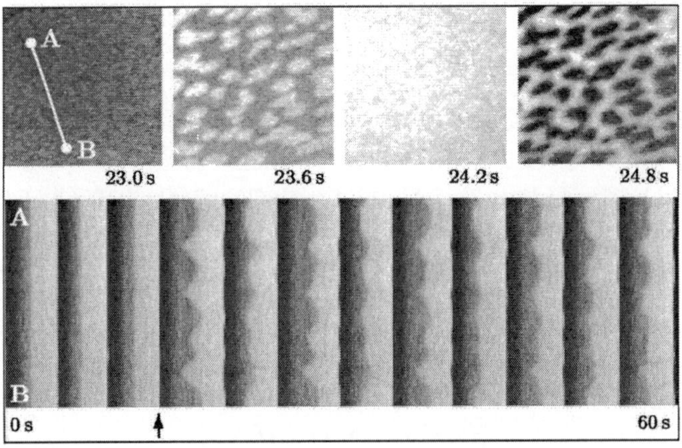

Fig. 10.22 Oscillating cellular structures. Top: Four subsequent images of size 270×270 μm^2 during a single oscillation period. Bottom: Development of cellular structures in the cross section indicated by the line AB in the first image. At the moment indicated by the arrow the feedback intensity was abruptly decreased. From [Bertram *et al.* (2003)].

A different type of patterns seen near the transition from turbulence to uniform oscillations is represented by oscillatory arrays of cells. Four snapshots of such a pattern, sampled within a single oscillation period, are displayed in the top row in Fig. 10.22. The cellular structure is visible only during short time intervals within each period. The space-time diagram in the lower part of Fig. 10.22 shows the development of the cellular structure after an abrupt decrease of the feedback intensity at the time moment indicated by the arrow below. Cellular structures usually occupy the entire imaged surface area and are irregular. The local oscillations in this pattern are in harmonic resonance with the almost periodic variations of the global control signal.

Oscillatory standing waves are characterized by the repeated development of bright stripes from the dark uniform state. They form a spatially periodic array and, depending on the chosen parameters, have a wavelength of 20 to 50 μm. The stripes are mainly oriented into the direction of fast CO diffusion (due to the surface anisotropy). The local properties of such patterns and their space-time diagram are similar to those of the cellular structures.

Further analysis of spatiotemporal patterns near the synchronization transition was performed using the Hilbert transform. For the time series $I(\mathbf{x}, t)$ of the local PEEM image intensity at an observation point $\mathbf{x}$, its Hilbert transform $\widetilde{I}(\mathbf{x}, t)$ was computed as

$$\widetilde{I}(\mathbf{x}, t) = \frac{1}{\pi} \int_{-\infty}^{\infty} \frac{I(\mathbf{x}, t')}{t - t'} dt' \tag{10.14}$$

(this could be easily done by determining the Fourier transform of I, shifting each complex Fourier coefficient by a phase of $\pi/2$, and performing the reverse Fourier transform). This was repeatedly done for all pixels $\mathbf{x}$ in an 100×100 array covering the respective pattern. Using $I(\mathbf{x}, t)$ and its Hilbert transform $\widetilde{I}(\mathbf{x}, t)$, a complex variable, known as the *analytic signal* [Panter (1965)],

$$\zeta(\mathbf{x}, t) = I(\mathbf{x}, t) + i\widetilde{I}(\mathbf{x}, t), \tag{10.15}$$

was defined.

Afterwards, the time-dependent spatial distributions of phase $\phi(\mathbf{x}, t)$ and amplitude $R(\mathbf{x}, t)$ were determined from the analytic signal in the following way: The phase was directly computed as $\phi = \arg \zeta$, thus representing the polar angle in the plane spanned by the variables I and $\widetilde{I}$. The amplitude was defined as $R = \rho/\rho_{\text{ref}}(\phi)$ where $\rho = |\zeta|$ is the standard definition of the amplitude modulus in the analytical signal approach. The normalization to $\rho_{\text{ref}}(\phi)$ was introduced to approximately compensate for deviations from harmonicity in the observed oscillations.

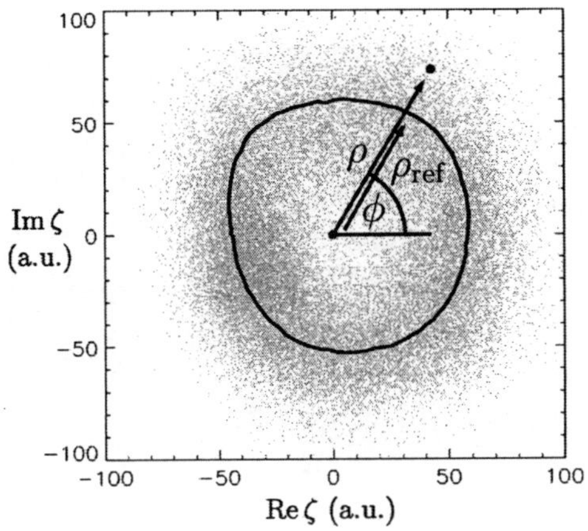

Fig. 10.23 Definition of phases and amplitudes.

To obtain $\rho_{\mathrm{ref}}(\phi)$, the statistical distribution of $\zeta(\mathbf{x}, t)$ for all 100×100 pixels and at all 250 time moments was plotted into the complex plane, as illustrated in Fig. 10.23 for a set of spatiotemporal data corresponding to turbulence. The reference amplitude $\rho_{\mathrm{ref}}(\phi)$ was then determined as the statistical average of $\rho = |\zeta|$ inside each of 200 equidistant narrow intervals of the polar angle ϕ. Note that the closed curve $\rho = \rho_{\mathrm{ref}}(\phi)$ can be viewed as representing a reference orbit deduced from the experimental data. In this way, a different reference orbit was constructed for each set of spatiotemporal data.

By applying this transformation separately to each of the different types of observed patterns, time-dependent spatial distributions of phase ϕ and amplitude R in each pattern were constructed [Bertram *et al.* (2003)]. Figure 10.24 shows snapshots of PEEM images (top row) for various patterns, and the corresponding snapshots of the phase (second row) and amplitude distributions (third row). Additionally, the bottom row of Fig. 10.24 shows a phase portrait of each pattern, obtained by displaying the amplitudes and phases for all resolving pixels in polar coordinates. The phase ϕ of a point is represented by the polar angle and the amplitude R is the distance to the coordinate origin.

In the regime of spiral-wave turbulence (Fig. 10.24a), fluctuations of amplitude and phase are strong, as revealed by the broad-band structure in the phase portrait. For intermittent turbulence (Fig. 10.24b), the amplitude

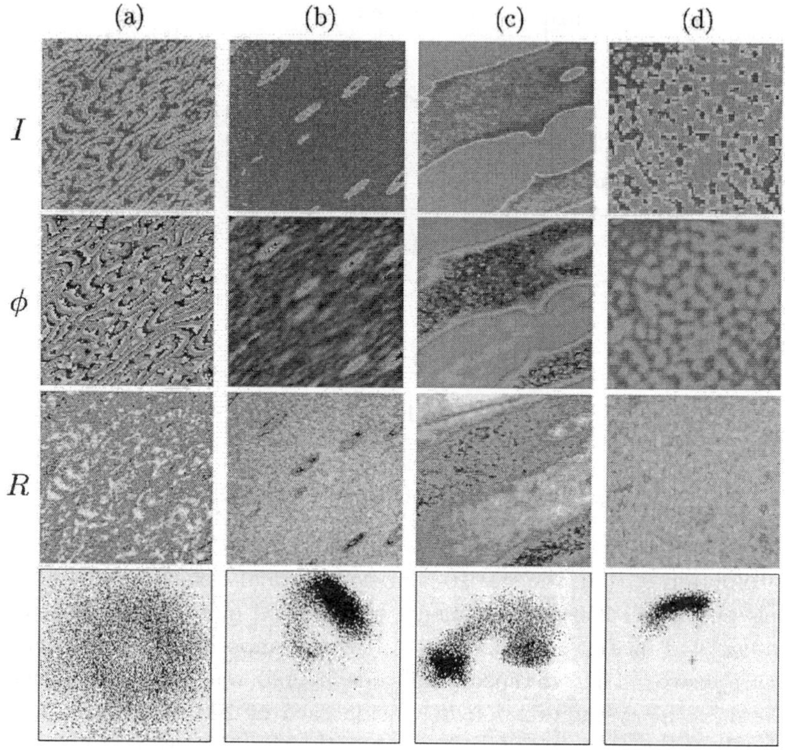

Fig. 10.24 Summary of the experimental patterns. (a) Turbulence in absence of feedback, (b) intermittent turbulence, (c) phase clusters, and (d) oscillating cellular structures. The bottom row displays phase portraits of the respective patterns. From [Bertram *et al.* (2003)].

and the phase are almost constant in the main part of the medium where uniform oscillations take place. The amplitude is significantly decreased in the bubble-shaped objects, which can therefore be viewed as extended amplitude defects. Across such objects, the oscillation phase varies rapidly in space. The phase portrait of intermittent turbulence shows a spot corresponding to the uniform state and a tail corresponding to the amplitude defects. When cluster patterns (Fig. 10.24c) develop, the medium breaks into two phase states seen as spots in the corresponding phase portrait. The amplitudes of the two oscillatory states differ because the local oscillations exhibit period-two behavior. The "bridge" in the phase portrait connecting the two spots corresponds to the interfaces between the cluster domains; note that the phase varies smoothly and the amplitude is not significantly reduced at the interface for the cluster patterns. In cellular structures (Fig. 10.24d), small phase modulations are observed, while the amplitude remains approximately constant.

Experiments with catalytic surface reactions have also been performed using a different global feedback scheme, known as time-delay autosynchronization [Beta *et al.* (2003)].

10.2.2 *Numerical simulations*

Mathematical modeling of the experiments has been done using a realistic model of catalytic CO oxidation on Pt(110) [Krischer *et al.* (1992)]. The model takes into account adsorption of CO and oxygen molecules, reaction, desorption of CO molecules, the structural phase transition, and surface diffusion of adsorbed CO molecules. Some minor processes are neglected, to simplify the description. The model equations are

$$\dot{u} = k_1 s_{\mathrm{CO}} p_{\mathrm{CO}} \left(1 - u^3\right) - k_2 u - k_3 uv + D\nabla^2 u \tag{10.16}$$

$$\dot{v} = k_4 p_{\mathrm{O}_2} \left[s_{\mathrm{O},1\times1} w + s_{\mathrm{O},1\times2} (1 - w)\right] (1 - u - v)^2 - k_2 uv \tag{10.17}$$

$$\dot{w} = k_5 \left(\frac{1}{1 + \exp\left(\frac{u_0 - u}{\delta}\right)} - w \right). \tag{10.18}$$

Here, the variables u and v represent the surface coverages of CO and oxygen, respectively. The variable w denotes the local fraction of surface area found in the 1×1 structure. The three variables vary in an interval from 0 to 1. All coefficients in the model are positive. Depending on its parameters, the model exhibits monostable, bistable, excitable, and oscillatory behaviors.

For the oscillatory regime, corresponding to the experiments, turbulence spontaneously develops through the modulational instability of uniform oscillations [Falcke and Engel (1994)]. The instability is caused by diffusion and oscillations of an isolated element are still stable under such conditions. Therefore, the system effectively represents a population of locally coupled periodic oscillators, where desynchronization (i.e.e, turbulence) is a result of interactions among them.

To model the global delayed feedback, it was assumed that the CO partial pressure p_{CO} in Eq. (10.16) is not constant but varies according to

the equation

$$p_{\mathrm{CO}}(t) = p_0 - \mu \left[\overline{u}(t-\tau) - u_{\mathrm{ref}} \right], \tag{10.19}$$

where

$$\overline{u}(t) = \frac{1}{S} \int_{(S)} u(\mathbf{x}, t) d\mathbf{x} \tag{10.20}$$

is the spatial average of the CO coverage at time t. The parameter μ specifies the feedback intensity, τ is the time delay, and p_0 is the base value of the partial CO pressure. The reference value u_{ref} is chosen as the CO coverage in the unstable uniform steady state in the absence of feedback. Note that not the CO coverage, but the local brightness I of the PEEM image was used in the experiments to construct the control signal. The PEEM image displays a combination of both CO and oxygen coverages. Its brightness increases with u and decreases with v, but the detailed functional form of these dependences is not known. Thus, the description (10.19) represents a simplification of the experimental setup.

Systematic numerical investigation of feedback effects on turbulence in the oscillatory state of this model has been carried out [Kim *et al.* (2001); Bertram and Mikhailov (2003)]. Spatiotemporal patterns yielded by the simulations were analyzed using a transformation from the local variables $u(\mathbf{x}, t)$ and $w(\mathbf{x}, t)$ to the local instantaneous phases and amplitudes of oscillations $\phi(\mathbf{x}, t)$ and $R(\mathbf{x}, t)$.

The transformation method is illustrated in Fig. 10.25. As a reference orbit in the projection plane (u, w), the limit cycle corresponding to stable periodic oscillations of an isolated element is chosen; the unstable fixed point of the model is taken as the origin O. The amplitude and the phase are determined for any state P. The radius vector of length $\rho = \overline{OP}$ and the point Q where this vector (or its extension) intersects with the reference orbit are constructed. The amplitude for the point P is then defined as $R = \rho/\rho_{\mathrm{ref}}$ where the length $\rho_{\mathrm{ref}} = \overline{OQ}$ is used for the normalization, similar to the analytical signal approach applied for the experimental data. The definition of the phase is, however, slightly different. Some "initial" point Q_0 is fixed on the reference orbit and the time $\widetilde{T}$ needed to reach the point Q along the reference cycle is computed. The phase is then defined as $\phi = 2\pi/T_{\mathrm{ref}}$ where T_{ref} is the period of the reference orbit. According to this definition, the amplitude is $R = 1$ and the phase ϕ increases at a constant velocity with time as long as the system stays on the reference orbit.

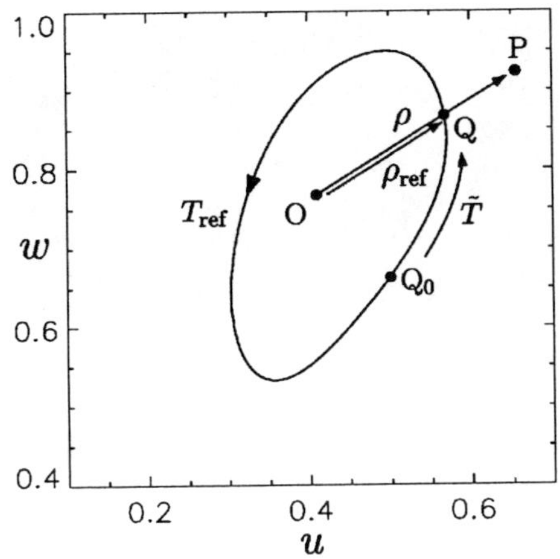

Fig. 10.25 Transformation to phase and amplitude variables, used in numerical simulations.

For the parameter values $k_1 = 3.14 \times 10^5$ s^{-1}mbar^{-1}, $k_2 = 10.21$ s^{-1}, $k_3 = 283.8$ s^{-1}, $s_{CO} = 1.0$, $s_{O,1\times1} = 0.6$, $s_{O,1\times2} = 0.4$, $u_0 = 0.35$, $\delta = 0.05$, $D = 40$ μm^{2}s^{-1}, and $p_0 = 4.81 \times 10^{-5}$ mbar, an isolated reaction element performs stable limit-cycle oscillations with period $T_0 = 2.73$ s. However, due to the destabilizing effect of diffusive coupling, uniform oscillations are unstable with respect to small spatial perturbations and chemical turbulence spontaneously develops. Figure 10.26 gives an example of such turbulence in a two-dimensional system. The oscillation amplitude R is strongly decreased inside narrow extended regions (strings) that represent extended amplitude defects. Across the strings, the phase undergoes a strong variation. The ends points of a string correspond to topological defects, such that the phase changes by 2π around them.

To study the effects of global feedback, the parameters were fixed at the values specified above and the feedback described by Eq. (10.19) was introduced. The feedback intensity μ and the delay time τ were varied, and the reference CO coverage was fixed at $u_{\text{ref}} = 0.3358$. The synchronization diagrams displayed in Fig. 10.27 summarize the results of many numerical simulations of the one-dimensional system. The simulations represented in Fig. 10.27a were started from the turbulent state in absence of feedback, and then the feedback intensity was gradually increased until synchronization occurred. In contrast to this, in the simulations of Fig. 10.27b the initial

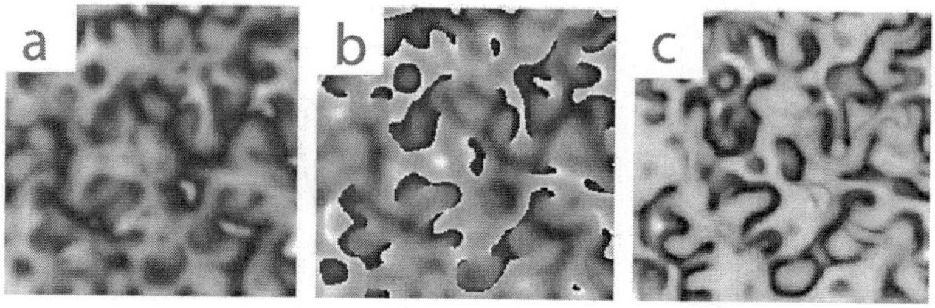

Fig. 10.26 Turbulence in absence of feedback. Instantaneous spatial distributions of (a) CO coverage u, (b) phase ϕ, and (c) amplitude R are displayed. From [Bertram and Mikhailov (2003)].

state represented uniform oscillations at a sufficiently high feedback intensity. Small random perturbations were added to this initial state and the feedback intensity was then gradually decreased, until desynchronization and transition to turbulence had taken place. We see that the synchronization and desynchronization boundaries do not coincide, i.e. hysteresis is observed.

If μ is sufficiently large, the feedback allows to suppress turbulence and synchronize oscillations in a wide range of delays, inside the light gray-shaded region in Fig. 10.27a. The synchronization threshold undergoes strong variation with the delay time. At very small delays, synchronization is not at all possible. At slightly higher values of μ, even weak feedbacks are, however, enough to synchronize the system. The next region of easy synchronization is reached when the delay time is close to the oscillation period (it should be noted that the feedback generally modifies the oscillation period, so that it is different from the period T_0 of free oscillations). In some narrow intervals of the feedback intensity (e.g., for $0.03 < \tau/T_0 < 0.1$), increasing μ first leads to synchronization, which is then followed by desynchronization at a higher intensity.

Even when global feedback is too weak to completely suppress turbulence, it can still substantially change the properties of the turbulent state. In a narrow region just below the synchronization boundary in Fig. 10.27a, *intermittent turbulence* is found. It is characterized by the occurrence of turbulent bursts on a laminar background of almost uniform oscillations. An example of such behavior in a one-dimensional system is displayed in Fig. 10.28. Repeated cascades of amplitude defects are visible. The defects reproduce until nearly the entire system is covered with turbulence. Then,

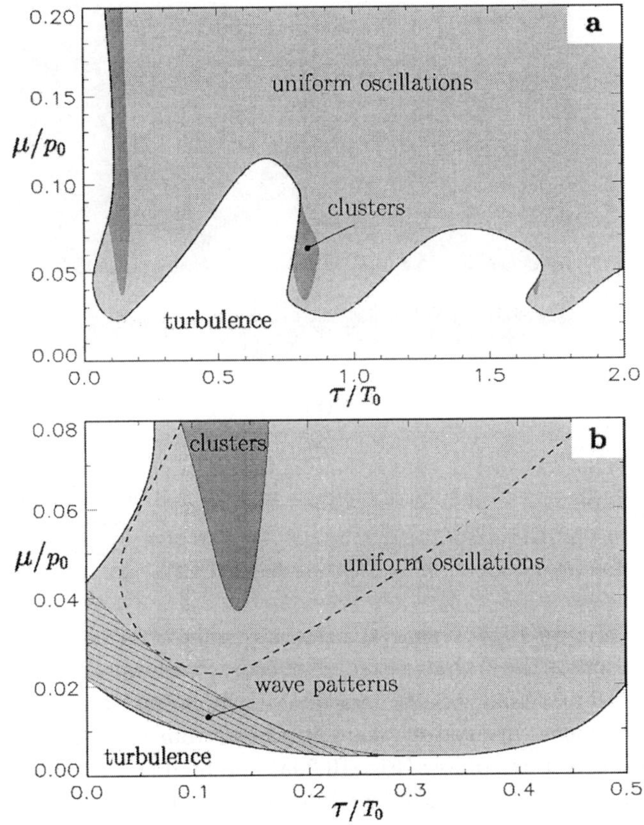

Fig. 10.27 Synchronization diagrams under gradual increase (a) or decrease (b) of the feedback intensity. The delay time is measured in multiples of the oscillation period in absence of diffusion and feedback, $T_0 = 2.73$ s. The feedback intensity is normalized to the base CO partial pressure p_0. For convenience, the synchronization boundary from part (a) is also shown as a dashed line in part (b). From [Bertram and Mikhailov (2003)].

most of them annihilate, but a few remaining ones give rise to the next reproduction cascades. The intermittent bursts are clearly seen in the temporal dependence of the partial CO pressure (i.e. of the control variable) at the bottom of Fig. 10.28.

In two space dimensions, intermittent turbulence exhibits irregular cascades of nearly circular structures (bubbles) on the background of uniform

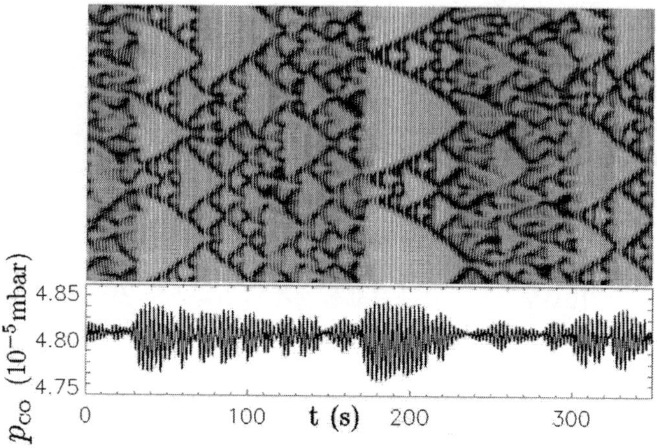

Fig. 10.28 Space-time diagram of intermittent turbulence. The amplitude R is displayed; dark color indicates low amplitude value. Below, the corresponding temporal variation of the CO partial pressure is presented. The feedback parameters are $\tau/T_0 = 0.293$ and $\mu/p_0 = 0.043$. From [Bertram and Mikhailov (2003)].

oscillations. Figure 10.29 displays three subsequent snapshots of the spatial distribution of the CO coverage u, phase ϕ, and amplitude R in such a pattern. Additionally, phase portraits of the system at the respective three time moments are shown below in the bottom row. The system alternates between states of low (Fig. 10.29a) and high (Fig. 10.29c) activity. In Fig. 10.29a, individual turbulent bubbles on a background of uniform oscillations are seen. The bubbles grow with time (Fig. 10.29b) and new bubble structures appear until the turbulent state covers almost all of the medium (Fig. 10.29c). The subsequent annihilation brings the system back to a state similar to that shown in Fig. 10.29a. On the borders of the bubbles, the oscillation amplitude is strongly decreased and the phase variation is strong. The borders seem to be formed by extended amplitude defects, or strings, which are already possible without the feedback.

At the feedback parameters corresponding to the dark gray regions in Fig. 10.27a *cluster patterns* are observed. Such patterns consist of large, homogeneously oscillating domains that are separated by narrow domain interfaces. No intrinsic spatial wavelength is characteristic for these patterns.

For most choices of the feedback parameters inside the dark-gray regions in Fig. 10.27a, two-phase clusters develop. Their space-time diagram is

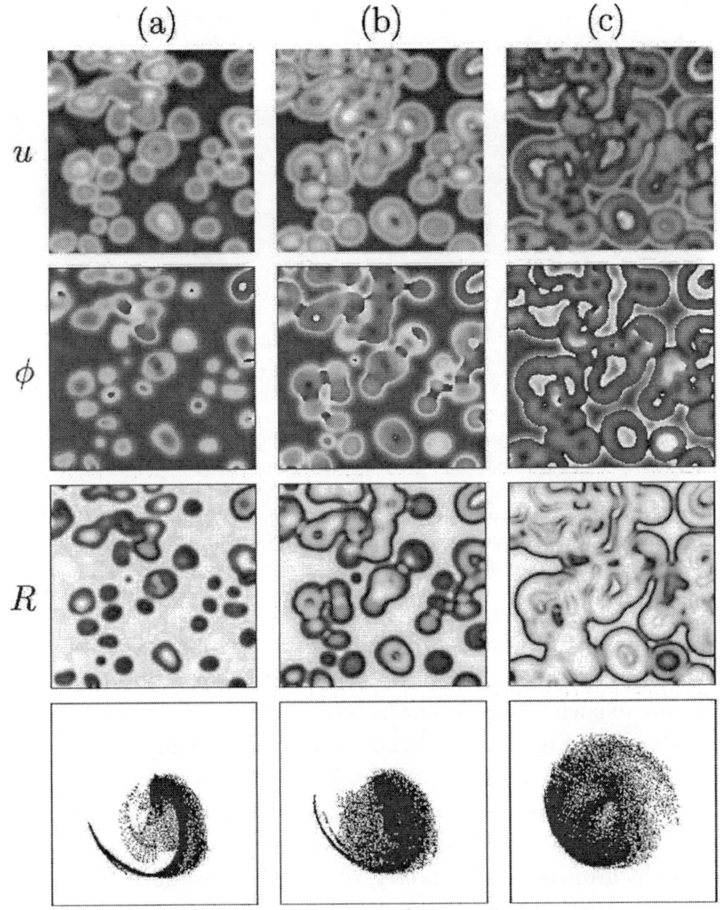

Fig. 10.29 Intermittent turbulence in the two-dimensional system. Subsequent snapshots (a, b and c) are separated by time intervals of 5.2 s. Phase portraits at the respective time moments are displayed at the bottom. The feedback parameters are $\tau/T_0 = 0.293$ and $\mu/p_0 = 0.056$, the system size is 600×600 μm^2. From [Bertram and Mikhailov (2003)].

shown in Fig. 10.30a. Inside the cluster regions, period-two oscillations are found. The interface between the clusters is characterized by period-one oscillations. The phases of oscillations in different domains are opposite. An important property of phase clusters is the phase balance: the total areas occupied by the domains with the opposite phases are equal. The average CO coverage $\overline{u}$ and, therefore, the control signal are characterized by period-one oscillations.

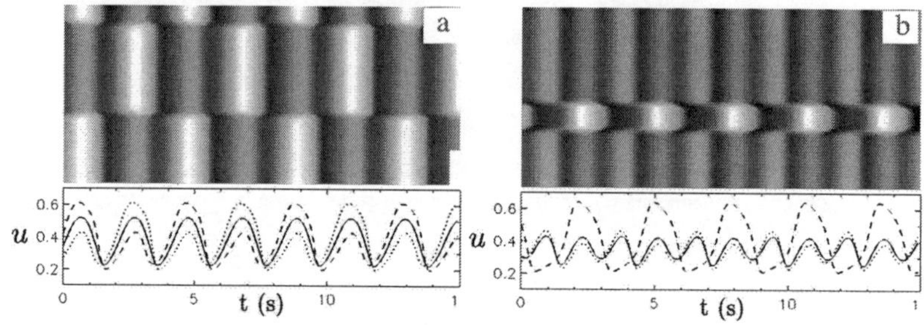

Fig. 10.30 Space-time diagrams of cluster patterns. (a) Phase clusters at $\tau/T_0 = 0.17$ and $\mu/p_0 = 0.083$. The dashed and dotted curves show temporal variation of u within different cluster domains. The solid curves presents variation of the spatial average $\overline{u}$. (b) Clusters with different limit cycles at $\tau/T_0 = 0.088$ and $\mu/p_0 = 0.2$. The dashed and dotted curves show variations of u within the small and the large cluster domains, the solid curve displays variation of $\overline{u}$. From [Bertram and Mikhailov (2003)].

Additionally, clusters of a different type are found inside the large left gray region in Fig. 10.27a (at $\tau/T_0 \simeq 0.15$ and $\mu/p_0 > 0.17$). The space-time diagram of such a cluster, characterized by the coexistence of two limit cycles, is shown in Fig. 10.30b. Inside the small domain, oscillations are of period one and have a large amplitude. In contrast to this, the surrounding region is occupied by period-two oscillations with a much smaller amplitude. The phase balance is absent in such a pattern.

The desynchronization boundary, obtained under decrease of the feedback intensity starting from the uniform state, lies significantly lower than the synchronization boundary. In the shaded region near the desynchronization boundary in Fig. 10.27b, *standing wave patterns* are observed. In two spatial dimensions, they represent *oscillatory cellular structures.* Such patterns (see Fig. 10.31) consist of periodic spatial modulations of both the phase and the amplitude, but the magnitude of variation of the amplitude is much less than that of the phase. The wavelength of these patterns is fixed and does not depend on the initial conditions or the size of the medium.

Three different types of cellular structures are encountered. Close to the border to uniform oscillations, the cell arrays are regular and show hexagonal symmetry (Fig. 10.31a). These stationary patterns are a result of nonlinear interactions between triplets of modes of wave vector $\mathbf{k}$ with

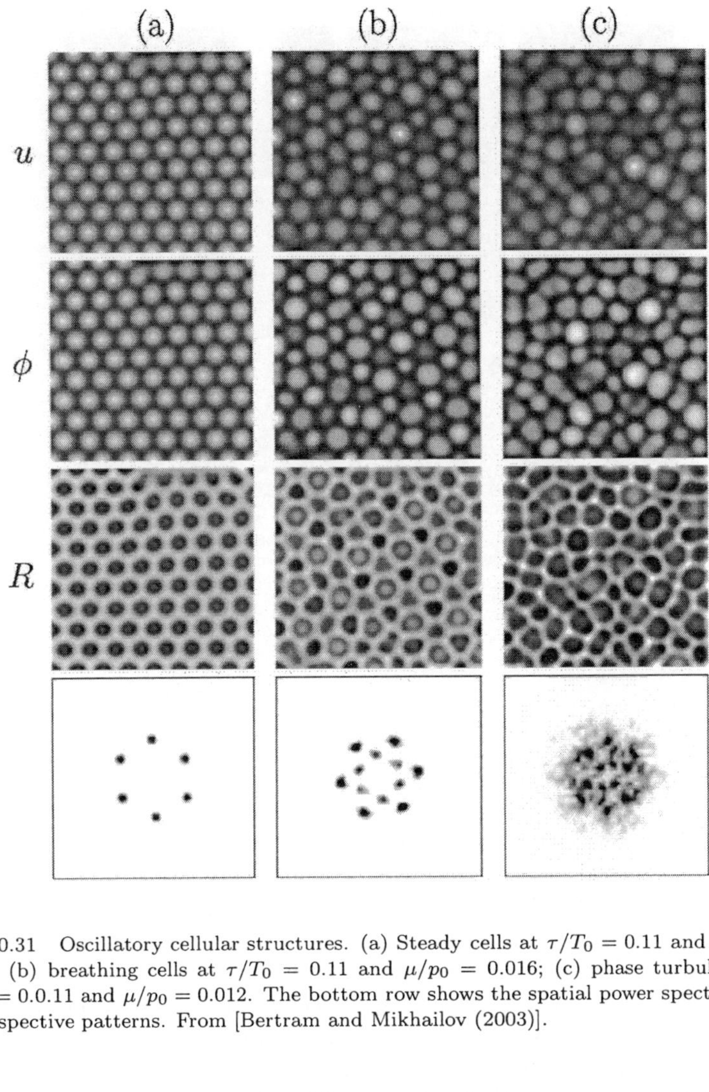

Fig. 10.31 Oscillatory cellular structures. (a) Steady cells at $\tau/T_0 = 0.11$ and $\mu/p_0 = 0.019$; (b) breathing cells at $\tau/T_0 = 0.11$ and $\mu/p_0 = 0.016$; (c) phase turbulence at $\tau/T_0 = 0.0.11$ and $\mu/p_0 = 0.012$. The bottom row shows the spatial power spectrum for the respective patterns. From [Bertram and Mikhailov (2003)].

the same wavenumber $k_0 = |\mathbf{k}|$. When the feedback is decreased, stationary regular cells become unstable at a delay-independent critical value of μ. Individual cells then periodically shrink and expand, so that an array of breathing cells is formed (Fig. 10.31b). In the spatial Fourier spectrum of such a pattern, two independent frequencies are present.

Phase turbulence (Fig. 10.31c) develops under further decrease of the

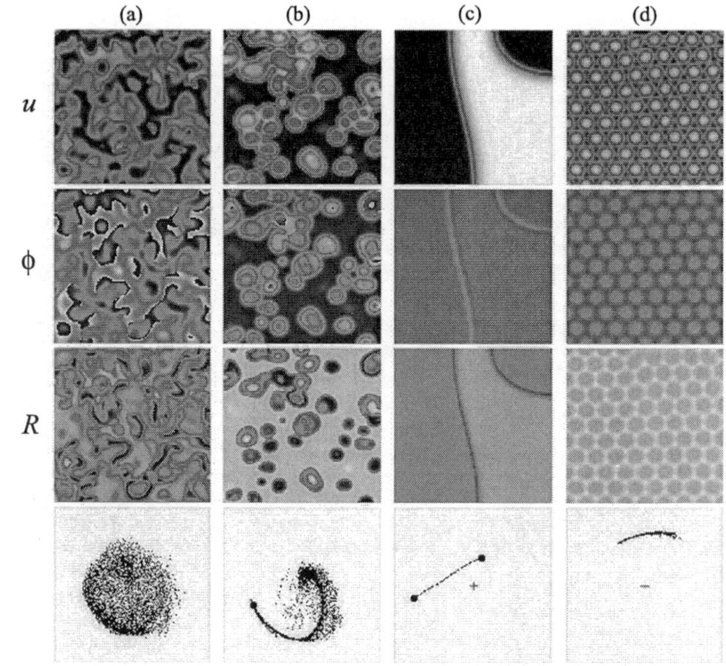

Fig. 10.32 Summary of numerical simulations. (a) Turbulence in absence of feedback; (b) intermittent turbulence; (c) phase clusters; and (d) cellular structures. From [Bertram and Mikhailov (2003)].

feedback intensity. The cells become mobile, the long-range order in the array is lost, and the shapes of the cells are less regular. Individual cells shrink or expand aperiodically while they slowly travel through the medium. Occasionally, some cells die out or, following an expansion, reproduce through cell splitting.

A graphic summary of various observed patterns is presented in Fig. 10.32. Here, the images in the top, second, and third rows display snapshots of spatial distributions of the CO coverage u, phase ϕ, and amplitude R. Additionally, the bottom row shows a phase portrait of each pattern. Comparing these results with the respective experimental data in Fig. 10.24, we see that the model yields all principal kinds of patterns seen in the experiments with global delayed feedbacks. Not only the qualitative aspects of the experimental patterns, but also their characteristic time and space scales are correctly reproduced.

Experiments and numerical simulations give an example of synchronization induced by global delayed feedbacks in a particular chemical system. Below we show, however, that this behavior is general and found in any array of weakly nonlinear limit-cycle oscillators with local coupling under global delayed feedback.

10.2.3 *Complex Ginzburg–Landau equation with global delayed feedback*

Oscillators near a supercritical Andronov–Hopf bifurcation have been considered in Chapter 5. In the continuous limit, an array of such oscillators with local coupling is described by the complex Ginzburg–Landau equation

$$\frac{\partial Z}{\partial t} = (1 - i\omega) Z - (1 + ib) |Z|^2 Z + (1 + i\epsilon) \nabla^2 Z, \tag{10.21}$$

where Z is the complex oscillation amplitude. The last term takes into account diffusive coupling between neighboring oscillators.

Comparing this equation with Eq. (5.5), one can notice that we have dropped here the coupling constant K. It can be eliminated by appropriate rescaling of spatial coordinates. Moreover, time is also rescaled in this equation in such a way that the growth rate of perturbations around the unstable fixed point $Z = 0$ becomes equal to unity. This increment should vanish when the Andronov–Hopf bifurcation takes place. Therefore, the rescaled time is very slow in the vicinity of this bifurcation where the complex Ginzburg–Landau equation is applicable. On the other hand, the oscillation frequency remains finite at the bifurcation. This means that, after a transition to the slow rescaled time, the frequency should become high, i.e. $\omega \gg 1$. This condition is usually not important, because the term proportional to ω in Eq. (10.21) can be eliminated by going to a rotating coordinate frame. It becomes, nonetheless, essential when effects of time delays are considered and this invariance breaks down.

The complex Ginzburg–Landau equation is a classical model in nonlinear science [Kuramoto (1984); Mikhailov and Loskutov (1996)]. An interesting property is that diffusive coupling can desynchronize uniform oscillations and give rise to a regime of spatiotemporal chaos, or turbulence. Uniform oscillations are unstable and turbulence spontaneously develops if the Benjamin–Fair condition $1 + b\epsilon < 0$ is satisfied. In a narrow parameter region near the onset of instability, phase turbulence characterized by irregular variations of oscillation phases and an almost constant oscillation amplitude is observed. Typically, the turbulent state exhibits large variations of both the phase and the amplitude of local oscillations. At some points, known as amplitude defects, the oscillation amplitude is greatly reduced or even vanishes.

We want to study effects of global delayed feedback on turbulence in the complex Ginzburg–Landau equation. Suppose that each oscillator in the medium additionally experiences the action of a certain time-dependent force $F(t)$, the same for all oscillators. Then, Eq. (10.21) takes the form [Mikhailov and Battogtokh (1996)]

$$\frac{\partial Z}{\partial t} = (1 - i\omega) Z - (1 + ib) |Z|^2 Z + (1 + i\epsilon) \nabla^2 Z + F(t). \tag{10.22}$$

Global feedback is realized if this force is collectively determined by the states of all oscillators in the medium at a delayed time moment, so that

$$F(t) = \mu \exp(i\chi_0) \frac{1}{S} \int_{(S)} Z(\mathbf{x}, t - \tau) d\mathbf{x}. \tag{10.23}$$

Here, μ is the feedback strength, χ_0 is the phase shift, and τ is the delay time. The integration is taken over the entire medium.

It is convenient to define a slowly varying complex amplitude $\eta(\mathbf{x}, t)$ as

$$\eta(\mathbf{x}, t) = Z(\mathbf{x}, t) \exp(i\omega t). \tag{10.24}$$

The new variable obeys the equation

$$\frac{\partial \eta}{\partial t} = \eta - (1 + ib) |\eta|^2 \eta + (1 + i\epsilon) \nabla^2 \eta + \mu \exp\left[i\left(\chi_0 + \omega\tau\right)\right] \overline{\eta}\left(t - \tau\right), \tag{10.25}$$

where

$$\overline{\eta}\left(t\right) = \frac{1}{S} \int_{(S)} \eta(\mathbf{x}, t) d\mathbf{x} \tag{10.26}$$

is the average slow oscillation amplitude.

Since rapid oscillations with frequency $\omega \gg 1$ are already eliminated from Eq. (10.25), the characteristic time scale for variation of η is of order unity (provided that the coefficients b and ϵ are also of this order). If the delay is short ($\tau \ll 1$), the slowly varying amplitude η does not significantly change within such a time interval and $\overline{\eta}\left(t - \tau\right)$ can be replaced by $\overline{\eta}\left(t\right)$ in Eq. (10.25).

Thus, dynamics of an array of locally coupled, weakly nonlinear oscillators with a delayed global feedback is effectively described by the equation

$$\frac{\partial \eta}{\partial t} = \eta - (1 + ib)\,|\eta|^2\,\eta + (1 + i\epsilon)\,\nabla^2\eta + \mu\exp(i\chi)\overline{\eta}, \tag{10.27}$$

where the delay contributes only to the renormalization of the phase shift, $\chi = \chi_0 + \omega\tau$. Note that, though the delay is short, it can lead to relatively large changes of the phase shift, because $\omega \gg 1$.

The model equation (10.27) was formulated for the analysis of global coupling through the gas phase in catalytic surface reactions [Veser *et al.* (1993)]. It gives a general description for the considered oscillator system, if the feedback is sufficiently weak. The complex Ginzburg–Landau equation is obtained by a decomposition in powers of the oscillation amplitude and includes terms up to third order in such amplitudes. When effects of global feedbacks in this equation are considered, it is generally necessary also to retain terms up to third order in the average complex oscillation amplitude $\overline{\eta}$. The weakness of the feedback allows to neglect all terms which are nonlinear in $\overline{\eta}$.

It should also be noted that Eq. (10.27) with global feedback is related to the equations describing external periodic forcing of oscillator arrays [Coullet and Emilsson (1992a); Coullet and Emilsson (1992b)]. The principal difference between the two systems is that the forcing collectively generated by all oscillators in Eq. (10.27) remains always resonant. In contrast to this, equations with external forcing include a detuning parameter, specifying the difference between the forcing frequency and the natural frequency of the oscillators. Even when detuning is zero, the system of oscillators can go away from the resonance with the external force by changing its collective oscillation frequency.

First, we derive the conditions under which global feedback stabilizes uniform oscillations, i.e. leads to synchronization in this system. The uniform oscillations $\eta(t) = \rho_0 \exp(-i\Omega t)$ are characterized by the frequency

$$\Omega = b - \mu\,(\sin\chi - b\cos\chi) \tag{10.28}$$

and the amplitude

$$\rho_0 = (1 + \mu\cos\chi)^{1/2}\,. \tag{10.29}$$

Stability of uniform oscillations is investigated by adding small perturbations to the local amplitudes and phases, $\eta(x,t) = (\rho_0 + \delta\rho)\exp[-i(\Omega t + \delta\phi)]$, substituting into Eq. (10.27), and linearizing with respect to the perturbations $\delta\rho(x,t)$ and $\delta\phi(x,t)$. The solution of the linear equations is sought in the form $\delta\rho(x,t) = \delta\rho_k \exp(\gamma_k t + ikx)$ and $\delta\phi(x,t) = \delta\phi_k \exp(\gamma_k t + ikx)$.

The growth γ_k of the mode with the wavenumber k satisfies the algebraic equation

$$\begin{aligned}&\left(\gamma_k + 2 + 3\mu\cos\chi + k^2\right)\left(\gamma_k + \mu\cos\chi + k^2\right)\\&+ \left(\epsilon k^2 + \mu\sin\chi\right)\left[2b\left(1+\mu\cos\chi\right) + \epsilon k^2 + \mu\sin\chi\right] = 0. \quad (10.30)\end{aligned}$$

Depending on the parameters, it can have either two real or two complex conjugated roots γ_k. At the instability onset, one of the perturbation modes should begin to grow. Hence, the instability boundary is determined by the conditions

$$\mathrm{Re}\,\gamma_k = 0, \quad \frac{\partial \mathrm{Re}\,\gamma_k}{\partial k} = 0, \quad \frac{\partial^2 \mathrm{Re}\,\gamma_k}{\partial k^2} < 0 \qquad (10.31)$$

for a mode with a certain wavenumber $k = k_0$.

Figure 10.33 shows the synchronization diagram of the complex Ginzburg–Landau equation, yielded by the linear stability analysis of uniform oscillations [Battogtokh *et al.* (1997)]. Uniform oscillations are stable above the boundary formed by the curve DABCE. Instabilities of different kinds are found when different parts of this boundary are crossed.

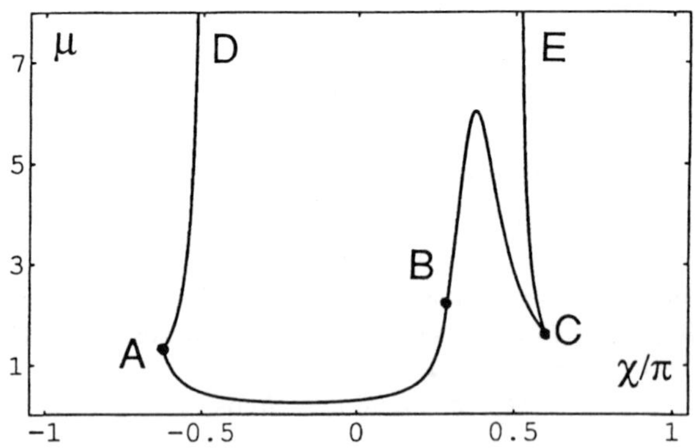

Fig. 10.33 Synchronization diagram of the complex Ginzburg–Landau equation with global feedback. Uniform oscillations are stable in the region above the curve DABCE; $\epsilon = 2$ and $b = -1.4$. From [Battogtokh *et al.* (1997)].

Along the curve AB, the first unstable mode is characterized by $\mathrm{Im}\,\gamma_k =$

0. Its wavenumber is approximately given by

$$k_0^2 = \frac{\nu}{1+\epsilon^2} - \frac{\epsilon\nu^2\left(\cos\chi + \epsilon\sin\chi\right)}{2\left(1+\epsilon^2\right)^2\left(\epsilon\cos\chi - \sin\chi\right)}, \tag{10.32}$$

and destabilization of uniform oscillations takes place approximately at the critical feedback intensity

$$\mu_c = \frac{\epsilon\nu^2}{2\left(1+\epsilon^2\right)^2\left(\epsilon\cos\chi - \sin\chi\right)}. \tag{10.33}$$

In these expressions, the notation $\nu = -1 - \epsilon b$ is used. They hold when ν is positive and relatively small.

Along the curves AD and CE, the instability corresponds to the growth of an oscillatory mode with $\operatorname{Im}\gamma_k \neq 0$ and a vanishingly small wavenumber $k_0 \to 0$. Its boundary is given by

$$\mu_c = -\frac{1}{\cos\chi}. \tag{10.34}$$

Along the curve BC, a static long-wavelength instability with $\operatorname{Im}\gamma_k = 0$ and $k_0 \to 0$ takes place. The instability boundary is given by

$$\mu_c = -\frac{2\left(b\sin\chi + \cos\chi\right)}{1 + 2\cos\chi\left(b\sin\chi + \cos\chi\right)}. \tag{10.35}$$

The modes with $k_0 \to 0$ correspond to nonuniform perturbations with the largest wavelength possible for a system, that is, with a wavelength equal to the system size L. Therefore, the respective instability gives rise to the formation of large-scale domain structures. In point B, the wavenumber given by Eq. (10.32) reaches zero.

Thus, the uniform stationary state of Eq. (10.25) looses its stability via a Turing bifurcation along the curve AB, via an Andronov–Hopf bifurcation along the curves AD and CE, and via a pitchfork bifurcation along the curve BC. Point A corresponds to a codimension-2 Turing–Hopf bifurcation, whereas points B and C correspond to the codimension-2 pitchfork–Turing bifurcations. It should be, however, remembered that this state is made stationary by going to the rotating coordinate frame, i.e. through a transformation to slow amplitudes. In terms of the original oscillation amplitudes Z, it represents a limit cycle. If this original formulation is used, the nomenclature of the respective bifurcations should be appropriately modified.

In the one-dimensional system, nonlinear dynamics of patterns in the vicinity of the curve AB can be approximately analyzed by keeping only three modes, i.e. the uniform mode and the unstable spatial modes with the wavenumbers $\pm k_0$ [Lima *et al.* (1998)]. In this case,

$$\eta(x,t) = \exp(-i\Omega t)[H + A_+ \exp(ik_0 x) + A_- \exp(-ik_0 x)], \tag{10.36}$$

where the complex amplitudes obey the following equations:

$$\begin{aligned}\dot{H} &= (1+i\Omega)H + \mu\exp(i\chi)H - (1+ib)|H|^2 H \\ &\quad -2(1+ib)(|A_+|^2 + |A_-|^2)H - (1+ib)A_+A_-H^*,\end{aligned} \tag{10.37}$$

$$\begin{aligned}\dot{A}_\pm &= (1+i\Omega)A_\pm - (1+i\epsilon)k_0^2 A_\pm - (1+ib)|A_\pm|^2 A_\pm \\ &\quad -2(1+ib)(|A_\mp|^2 + |H|^2)A_\pm - (1+ib)H^2 A_\mp^*.\end{aligned} \tag{10.38}$$

Inside the synchronization window, $|A_+| = |A_-| = 0$ and $H = \rho_0$. Below the curve AB (i.e. for $\mu < \mu_c$), this uniform solution becomes unstable and is replaced by a solution with $A_+ = A_- = T_s \exp(i\phi_s)$ and $H = \rho_s$. It describes *standing waves*,

$$\eta(x,t) = \exp(-i\Omega_s t)\left[\rho_s + 2T_s \exp(i\phi_s)\cos(k_0 x)\right], \tag{10.39}$$

with amplitude $T_s \sim \sqrt{\mu_c - \mu}$.

In two dimensions, the situation is more complicated, because all modes with $|\mathbf{k}| = k_0$ should grow at $\mu < \mu_c$. Interactions between these modes favor the development of *hexagonal cellular structures*

$$\begin{aligned}\eta(x,t) &= \exp(-i\Omega_H t)\rho_H + 2T_H \exp[i(\phi_H - \Omega_H t)] \\ &\quad \times \left[\cos(k_0 x) + 2\cos\left(\frac{k_0 x}{2}\right)\cos\left(\frac{\sqrt{3}k_0 y}{2}\right)\right],\end{aligned} \tag{10.40}$$

which represent superpositions of three modes with the wavenumbers satisfying the relationship $\mathbf{k}_1 + \mathbf{k}_2 + \mathbf{k}_3 = 0$. These structures exist even somewhat below $\mu < \mu_c$, so that hysteresis is observed [Lima *et al.* (1998)]. For $\chi/\pi < -0.5$, cellular structures are absent and standing waves develop even in the two-dimensional system. When the feedback intensity is further decreased, standing waves and cellular structures undergo subharmonic instabilities which lead to periodic breathing of such structures.

If the boundary curve BC in Fig. 10.33 is crossed, numerical simulations show the development of large-scale cluster patterns [Battogtokh *et al.* (1997)]. The oscillation amplitudes ρ are different inside different domains, and therefore such structures can be described as amplitude clusters. Numerical simulations also yield regimes of intermittent turbulence with cascades of amplitude defects [Mikhailov and Battogtokh (1996); Battogtokh *et al.* (1997)].

Thus, the phenomena observed in the experiments with catalytic oscillatory surface reactions and found in numerical simulations of a realistic model of such reactions are close to the theoretical predictions based on the complex Ginzburg–Landau equation valid for any reaction-diffusion system near a supercritical Andronov–Hopf bifurcation.

Chapter 11

Biological Cells

Cells are chemical reactors where a large number of reactions are simultaneously taking place. Both periodic and chaotic oscillations in biochemical reactions are known. An important example is provided by glycolysis, which represents the basic metabolic reaction network of any cell. Because some chemicals can penetrate cellular membranes and go into the extracellular medium, chemical communication between cells is possible. This leads to the experimentally observed synchronization of glycolytic oscillations in large populations of cells, which we present in the first section of this chapter.

Chemical cell-to-cell communication plays an important role in biological morphogenesis and differentiation of living cells. A change in the pattern of genetic expression, determining differentiation into a particular cellular type, can be triggered by variations in chemical compositions of the cells. How can such variations spontaneously develop in an initially uniform cellular population? One possible solution was offered by A. Turing who showed in the middle of the twentieth century that a sufficiently strong difference in diffusion constants of reactants can destabilize the uniform state of a system and lead to spontaneous development of static spatial concentration patterns. Another possible mechanism of symmetry breaking in cell populations is based on the effect of dynamical clustering in ensembles of globally coupled oscillators.

By investigating abstract artificial cells that contain a random network of catalytic reactions, one finds that, with a relatively high probability, the cells exhibit chaotic oscillations. When such cells are able to replicate and chemically communicate with each other, their globally coupled growing population becomes unstable with respect to spontaneous dynamical clustering. As a result, different stable types of cells appear and the growing population acquires a definite structure (Sect. 11.2).

Synchronization phenomena can also be essential inside individual biological cells. A characteristic feature of proteins is that these macromolecules can take various conformations, i.e. different shapes. Transitions between different conformations and processes of conformational relaxation are therefore accompanied in proteins by intramolecular mechanical motion. When such motions are functional, a protein operates as a molecular machine.

An enzyme is a protein representing a single-molecule catalyst. Its catalytic cycle can also involve functional conformational changes, so that the enzyme indeed acts like a machine. External synchronization of individual enzymic cycles by periodic optical forcing has been experimentally demonstrated. Inside a cell, allosteric enzymes communicate via small regulatory molecules. This intracellular communication may lead to synchronization of such molecular machines and formation of coherently acting enzymic groups, considered in Sect. 11.3.

11.1 Glycolytic Oscillations

A living cell is a chemical reactor where a great number of chemical reactions are simultaneously taking place. These reactions are responsible for various functions of a biological cell. Though all of them are to a certain extent coupled to each other, it is possible to distinguish relatively independent groups of reactions forming structural modules with definite biological functions [Harwell *et al.* (1999)]. One of such modules consists of a network of enzymic reactions that produce, starting from sugars, adenosine tri-phosphate (ATP) molecules which transfer chemical energy inside a cell. The network, known as the *glycolytic pathway*, is ubiquitous for living beings. Because energy is needed for operation of any molecular machine, glycolytic enzymes are found in large concentrations and represent the dominant component of cytosol. In yeast cells, their concentrations range between 10^{-5} and 10^{-4} M. Summing up all enzymes involved in glycolysis, one obtains a total concentration of 1.3 mM. This corresponds to an average distance of just about 50 Å between any two neighboring glycolytic enzymes inside a cell, smaller than the size of a single enzyme molecule [Hess (1973)].

The glycolytic structural module of biological cells exhibits oscillations [Ghosh and Chance (1964)]. They have been studied in single yeast cells, in suspensions of such cells and in cell-free yeast extracts (see the review [Hess (1997)]). Such oscillations are recorded by measuring fluorescence of reaction products and their characteristic time scale is about 30 s.

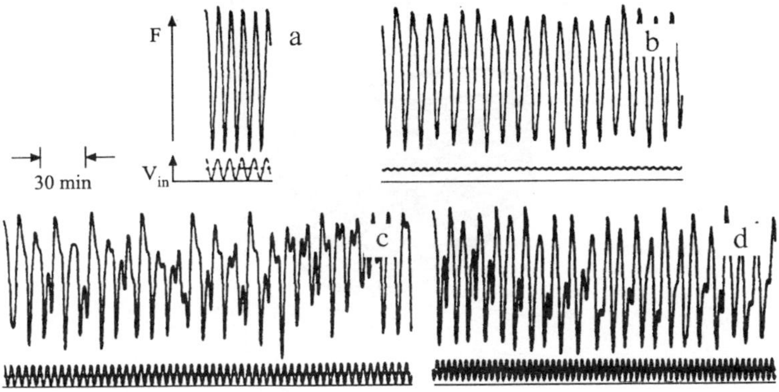

Fig. 11.1 Periodic external forcing of glycolytic oscillations in yeast extract. From [Hess (1997)].

Experiments with periodic forcing of glycolytic oscillations in yeast extracts have also been performed. Figure 11.1 shows some temporal patterns induced by periodic harmonic variation of the glucose input flux in glycolysing yeast extracts. Resonant 1:1 entrainment (Fig. 11.1a), quasi-periodic oscillations (Fig. 11.1b) and two different chaotic regimes (Fig. 11.1c,d) are displayed. The dynamical signal F represents NADH fluorescence of the extract; the temporal variation of the input flux V_{in} is shown at the bottom of each part. An extensive analysis of chaotic regimes has been undertaken [Markus *et al.* (1984); Markus *et al.* (1985)]. An example of a strange attractor, reconstructed from the experimental data, is given in Fig. 11.2. The information dimension of attractors was always smaller than three, indicating that only three effective variables are sufficient to describe the complex dynamics of glycolysis, despite the much larger number of involved metabolites. Besides uniform glycolytic oscillations, traveling waves and rotating spiral waves are also observed in yeast extracts [Mair and Müller (1996)].

In contrast to cell-free extracts, glycolytic reactions in cell suspensions are localized inside individual cells. The cells communicate with each other, because some intermediate products cross cellular membranes and go into the solution. Subsequently, such product molecules enter other cells in the

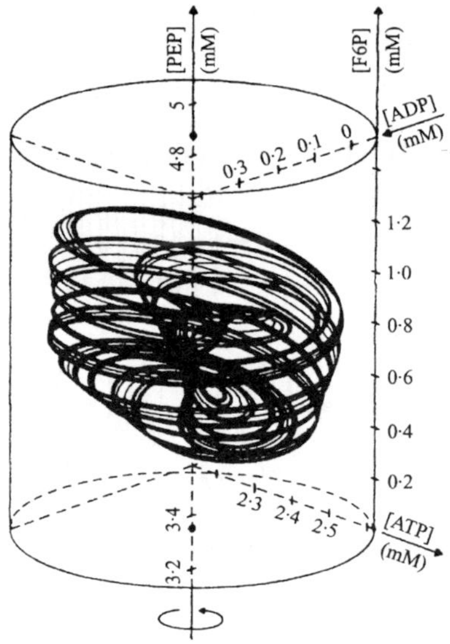

Fig. 11.2 Strange attractor of glycolytic oscillations. From [Hess and Markus (1985)].

suspension and influence reactions inside them. The communication leads to coupling of oscillations in different yeast cells. Experiments indicate that the principal coupling factor is acetaldehyde, whose extracellular concentration oscillates at the frequency of intracellular glycolytic oscillations [Richard *et al.* (1996)]. Observed bulk oscillations depend on the ability of individual cells to synchronize their oscillations [Ghosh *et al.* (1971); Richard *et al.* (1996)]. This was demonstrated by mixing two suspensions oscillating at opposite phases: the bulk oscillations disappeared immediately after the mixing, but reappeared after some time.

Most of the early experimental investigations of oscillations in yeast suspensions were performed by applying a pulse of glucose needed for the reaction. Therefore, only transient oscillation trains could be observed. A significant advancement in the experimental technique was the employment of a continuous-flow stirred reactor (CSTR) which allowed to generate sustained oscillations [Danø *et al.* (1999)]. This experimental setup is shown in Fig. 11.3. Fresh reactants (glucose and cyanide) and fresh starved yeast cells (cell suspension) are supplied at a controlled rate to the reactor. Stirring leads to efficient mixing, so that global coupling between individual cells is realized. The oscillations are monitored by measuring NADH fluorescence with a photomultiplier.

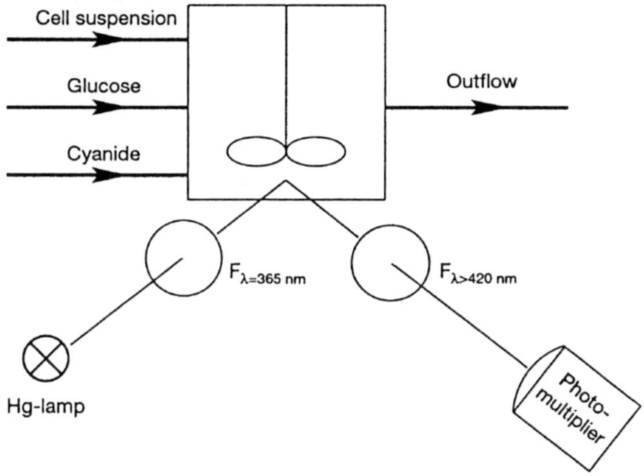

Fig. 11.3 The experimental setup. From [Danø *et al.* (1999)].

By changing the glucose flow rate, a transition from the stationary state to periodic oscillations was observed. Near the transition point, the square of the oscillation amplitude is proportional to the deviation from the critical value, as should be expected for a supercritical Andronov–Hopf bifurcation (Fig. 11.4). On the oscillatory side, the system has an almost elliptic stable limit cycle.

Limit-cycle oscillations can be perturbed by instantaneous addition of a chemical substance involved in the reaction. After a perturbation, the oscillator returns to the limit cycle after some transient behavior. Appropriately choosing the moment of the control pulse and its intensity, the oscillations can be quenched for a short time. When they reappear, the memory of the initial phase state becomes lost. This control method is known as *phase resetting* [Winfree (1972)]. By applying it to ensembles of limit-cycle oscillators, their degree of synchronization can be probed. Indeed, the response of an oscillator to a chemical pulse perturbation is highly sensitive to the phase in which it is found at the pulse moment. If only a fraction of the population is synchronized and the phase distribution is broad, only a weak response of the population to the resetting pulse would take place. On the

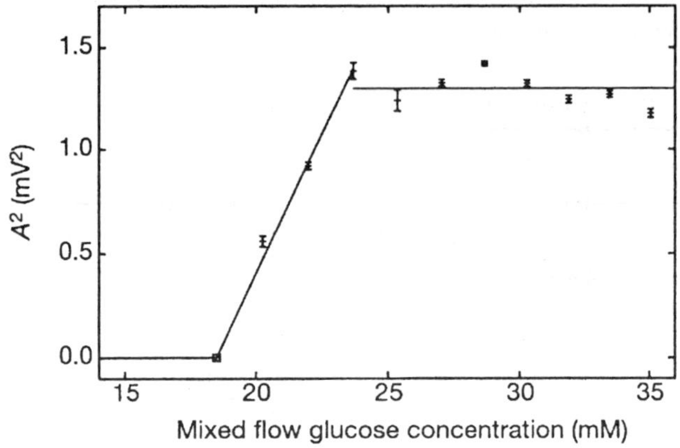

Fig. 11.4 Dependence of the square of the oscillation amplitude on the glucose flow rate. From [Danø *et al.* (1999)].

other hand, fully synchronized ensembles should respond exactly as a single oscillator. Moreover, only the response of a fully synchronized ensemble should be strongly sensitive to the perturbation phase.

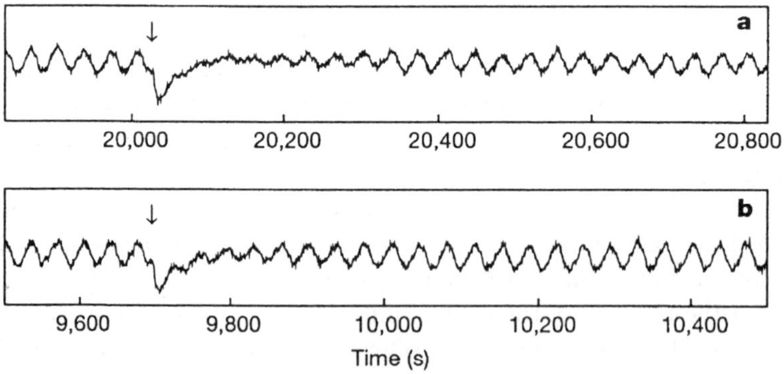

Fig. 11.5 Phase resetting of glycolytic oscillations. Equal amounts of acetaldehyde were added at the moments indicated by arrows and corresponding to different oscillation phases. From [Danø *et al.* (1999)].

Figure 11.5 shows phase resetting of bulk oscillations in yeast cell suspensions by instantaneous addition of acetaldehyde. When the pulse of acetaldehyde is applied at an oscillation phase of 172°, complete quenching of the oscillations is achieved (Fig. 11.5a). If the perturbation is instead applied at the phase of 180°, the oscillations diminish in their amplitude, but do not vanish (Fig. 11.5b). This provides experimental evidence of strong phase synchronization of oscillations in individual cells inside the suspension. Quenching by phase resetting remains possible even near the transition corresponding to the disappearance of bulk oscillations. Therefore, this transition is related to the disappearance of oscillations through an Andronov–Hopf bifurcation in each single cell, rather than to the desynchronization of individual oscillators.

Though glycolytic oscillations were most extensively studied in yeast, they were also observed in many other cell types [Hess (1997)]. Such intrinsic oscillations of energy metabolism were found in the heart cells (cardiomyocytes), where they lead to oscillations of the electrical membrane potential [O'Rourke *et al.* (1994)]. Another important example is provided by the pancreatic β-cells where glucose-induced oscillations, accompanied by periodic variation of the membrane potential, were observed [Matthews and O'Connor (1979)]. The bursts of membrane depolarization can control the pattern of insulin secretion by such cells [Tornheim (1997)].

11.2 Dynamical Clustering and Cell Differentiation

If a cell behaves like a single oscillator, populations of globally coupled cells should show not only synchronization, but also the regimes of dynamical clustering which were discussed in Chapters 2 and 8. In such regimes, a uniform cell population spontaneously breaks into several coherent groups, each characterized by a well-defined phase. Direct experimental proof of clustering would require observation of phase states of individual cells, as it has been done, for example, for electrochemical oscillators (Sect. 10.1). Such experiments are not yet available. Nonetheless, there are theoretical studies which suggest that dynamical clustering is indeed taking place and may play an important role in cell differentiation.

Though all cells in a macroorganism possess the same genetic information, they must have different properties in order to build various body parts. The diversity of genetic expression originates from variations in the extracellular environment and in the internal chemical composition of the cells. In turn, such composition and environment are strongly dependent on the pattern of genetic expression in a given kind of cells. This means that various cell types should essentially correspond to different *attractors* of a complex dynamical process.

Cell differentiation normally occurs during the development of a macroorganism from a primary egg cell. Initially, the egg cell undergoes multiple divisions and a uniform cell population is thus produced. A general question is how the symmetry between the cells becomes broken, so that their differentiation may begin. One mechanism, proposed a long time ago [Turing (1952)], makes use of an instability of uniform stationary states in reaction-diffusion systems. The Turing instability takes place when chemical components of a system are characterized by a strong difference in their diffusion rates. It leads to the development of a static spatial pattern of chemical concentrations. The spatial variation of some chemicals can then trigger different kinds of genetic expression, and give way to the differentiation of cells.

In mature organisms, differentiation of stem cells is observed. These cells are used for continuous renewal of such tissues as blood or epidermis, which are composed of cells with a finite lifespan. In primitive animals and plants, proliferation of stem cells allows regeneration of parts which were lost or damaged through an injury. When a tissue is damaged, active stem cells appear through the activation of quiescent stem cells or the de-differentiation of neighboring cells. If a fraction of some cell types (e.g. red blood cells) has decreased due to an external influence, their production through appropriate differentiation of the stem cells becomes enhanced so that the original population distribution is recovered [Alberts *et al.* (1994)]. Experiments with cell colonies originating from a single stem cell have shown spontaneous differentiation of cells in absence of any externally applied heterogeneity or growth signals.

It has been suggested that spontaneous dynamical clustering can explain differentiation in homogeneous cellular populations [Kaneko and Yomo (1994); Kaneko and Yomo (1997); Kaneko and Yomo (1999); Furusawa and Kaneko (2001)]. This has been demonstrated by considering abstract models of cellular populations. Suppose that we have a population of N cells, each containing a copy of the same cross-catalytic network of biochemical reactions with M different molecular species. The network topology is characterized by a matrix $\mathbf{J}$, whose elements J_{ijk} are equal to unity if chemical species j catalyzes a reaction converting species k to species i, and are zero otherwise. The state of a cell l at time t is specified by a set of chemical concentrations $c_l^{(i)}(t)$ with $i = 1, 2, \ldots, M$. Taking only the reactions in a given cell, evolution of chemical concentrations inside it would be described by the equations

$$\frac{dc_l^{(i)}}{dt} = \sum_{j,k=1}^{M} \nu J_{ijk} \left(c_l^{(j)}\right)^{\alpha} c_l^{(k)} - \sum_{j,k=1}^{M} \nu J_{kji} \left(c_l^{(j)}\right)^{\alpha} c_l^{(i)}. \tag{11.1}$$

Here α is the degree of catalysis, equal to $\alpha = 2$ for the considered example of quadratic catalysis (models with other catalytic laws have also been investigated). For simplicity, the rate constants of all reactions in the network are assumed to be the same and are given by the coefficient ν. The first term takes into account increase of the concentration of chemical i inside cell l due to its catalytic production, whereas the second term corresponds to the consumption of this chemical in all catalytic reactions inside this cell.

Some of the reactants can penetrate the membrane separating the cell from the extracellular medium. In this way, the cell is supplied with fresh reactants. On the other hand, certain cell products can go into the extracellular medium and subsequently enter other cells, establishing chemical communication between them. In the model, the transmembrane transport obeys the diffusion law. This means that the rate of transfer of some mobile reactant i is proportional to the difference $C^{(i)} - c_l^{(i)}$ of its concentrations in the extracellular medium $C^{(i)}$ and inside the considered cell l. Ideal mixing is assumed to take place in the extracellular medium. The reactants able to penetrate cellular membranes represent a subset of M' out of M species participating in the reactions. For simplicity, diffusion constants of all M' penetrating reactants are taken to be the same and are given by D. To indicate whether a given chemical species i is able to cross the membrane, it is convenient to introduce coefficients ζ_i, such that $\zeta_i = 1$ for penetrating species and $\zeta_i = 0$ otherwise. Note that $\sum_i \zeta_i = M'$. When exchange of reactants through the extracellular medium is incorporated into the model, its kinetic equations take the form

$$\frac{dc_l^{(i)}}{dt} = \sum_{j,k=1}^{M} \nu J_{ijk} \left(c_l^{(j)}\right)^{\alpha} c_l^{(k)} - \sum_{j,k=1}^{M} \nu J_{kji} \left(c_l^{(j)}\right)^{\alpha} c_l^{(i)} + \zeta_i D \left(C^{(i)} - c_l^{(i)}\right). \tag{11.2}$$

These kinetic equations hold if the volume of each chemical reactor (that is, of each cell) is fixed. Actually, this volume changes in time because of cell growth. Increase of volume leads to dilution of reactants that should be taken into account in the kinetic equations.

Suppose that a volume V contains K molecules, so that their concentration is $c = K/V$. If both the number of molecules and the volume vary with time, the rate of change of concentration is

$$\frac{dc}{dt} = \frac{1}{V}\frac{dK}{dt} - \frac{K}{V^2}\frac{dV}{dt}. \tag{11.3}$$

In the first term on the right-hand side of this equation, the derivative dK/dt describes the rate of change of the number of molecules in the entire volume due to the reaction. Therefore, $dK/dt = wV$ where w is the reaction rate. The second term here corresponds to the dilution effect.

A biological cell is densely packed with biochemical molecules and it would be natural to assume that its volume is simply proportional to the total number of molecules, i.e. $V = uK$. In this case,

$$dV/dt = vdK/dt = uwV. \tag{11.4}$$

Substituting this into Eq. (11.3), we obtain

$$\frac{dc}{dt} = w - uwc. \tag{11.5}$$

This equation should hold for each cell. However, a cell actually contains many different chemicals i and therefore its volume V should be rather determined by the total number of molecules of all species, i.e. $K = \sum_i K^{(i)}$. Therefore, the concentration $c^{(i)}$ of a particular species i inside the cell should obey the equation

$$\frac{dc^{(i)}}{dt} = w^{(i)} - uc^{(i)} \sum_{i=1}^{M} w^{(i)}. \tag{11.6}$$

Here u is the mean volume element occupied by a single molecule inside the cell. It is convenient to choose it as a unit volume, so that $u = 1$.

Hence, when effects of cell growth are taken into account, the kinetic equations (11.2) should be modified to

$$\frac{dc_l^{(i)}}{dt} = w_l^{(i)} - c_l^{(i)} \sum_{i=1}^{M} w_l^{(i)}, \tag{11.7}$$

where

$$w_l^{(i)} = \sum_{j,k=1}^{M} \nu J_{ijk} \left(c_l^{(j)}\right)^{\alpha} c_l^{(k)} - \sum_{j,k=1}^{M} \nu J_{kji} \left(c_l^{(j)}\right)^{\alpha} c_l^{(i)} + \zeta_i D \left(C^{(i)} - c_l^{(i)}\right) \tag{11.8}$$

are the reaction rates for species i inside cell l. Note that according to these equations, the sum of all concentrations in any cell remains constant with time, $\sum_i c_l^{(i)} = 1$. This is because the cells adjust their volumes proportionally to the total number of molecules inside them.

The kinetic equations (11.7) and (11.8) should be complemented by the equations describing evolution of chemical concentrations $C^{(i)}$ in the extracellular medium. We assume that the extracellular medium represents a flow reactor: fresh reactants are continuously supplied and the solution is also pumped away. Therefore, we have

$$\frac{dC^{(i)}}{dt} = f\zeta_i\left(\overline{C^{(i)}} - C^{(i)}\right) - \gamma\sum_{l=1}^{N}\zeta_i D\left(C^{(i)} - c_l^{(i)}\right). \tag{11.9}$$

Here, the first term takes into account supply of fresh reactants and their removal by pumping at rate f. In the absence of cells, stationary concentrations $\overline{C^{(i)}}$ are established in the medium. The last term describes release or consumption of chemicals by the cells.

Additionally, the model allows division and death of the cells. Each cell receives mobile chemicals ("nutrients") from the medium and the reaction network inside the cell transforms them into other chemicals which cannot cross the cellular membrane. Because of this, the total number of molecules inside the cell increases and the cell grows. According to Eq. (11.4), the rate of volume growth (if we put $u = 1$) is given by

$$\frac{dV_l}{dt} = V_l\sum_{i=1}^{N} w_l^{(i)}. \tag{11.10}$$

Substituting Eq. (11.8) and taking into account that reactions inside the cell only transform chemicals one into another, we find that

$$\frac{dV_l}{dt} = DV_l\sum_{i=1}^{N}\zeta_i\left(C^{(i)} - c_l^{(i)}\right). \tag{11.11}$$

The volume V_l of a cell l cannot grow indefinitely. When it reaches a certain threshold V_c, the cell divides into two new cells, each with the volume $V_c/2$. During the division, all chemicals are almost equally divided, with a relative random imbalance of order 10^{-6}.

If the flow of chemicals out of a cell exceeds their supply from the extracellular medium, a cell can also decrease its volume. A cell dies if its volume becomes smaller than a certain minimum value $V_{\min}$.

Thus, the total number N of cells in the population changes with time because of divisions and deaths. Each cell $l = 1, 2, \ldots, N(t)$ is characterized by its volume $V_l(t)$ and a set of chemical concentrations $c_l^{(i)}(t)$ inside it ($i = 1, 2, \ldots, M$). The state of the common extracellular medium is characterized by a set of M' chemical concentrations $C^{(i)}(t)$ of the substances that are able to penetrate cell boundaries. Evolution of the variables $N(t), V_l(t), c_l^{(i)}(t)$ and $C^{(i)}(t)$ is determined by Eqs. (11.7), (11.9) and (11.11) complemented by the rules specifying division and death of the cells.

The cross-catalytic reaction network occupying each cell is characterized by the reaction matrix $\mathbf{J}$. Depending on its structure, different dynamical regimes are possible inside a cell [Furusawa and Kaneko (2001)]. Random networks of size $M = 32$ with a fixed number of connections per single node were considered. When the connectivity of the network was low, the cellular dynamics usually fell into a steady state without oscillations, where a small number of chemicals were dominant and most of other chemicals vanished. If the network was highly connected, any chemical could be generated by almost any other chemical. In this case, a steady reaction state was again typically established where, however, all chemicals were present in nearly equal concentrations. At intermediate connectivities, nontrivial oscillatory regimes became possible. Another important factor was the number of autocatalytic species (that catalyze their own production) in the network: the probability to observe oscillations increased when such species were more numerous. Nonetheless, even when such conditions were fulfilled, the fraction of reaction networks exhibiting oscillatory dynamics remained relatively low. Out of thousands of randomly generated networks, only about 10% showed oscillations, either periodic or chaotic. For the simulations of collective population dynamics, a random network leading to chaotic oscillations was always chosen.

Extensive numerical investigations of the model (11.7), (11.9), (11.11) and its modifications were performed. The modifications consisted in taking the Michaelis–Menten law instead of the second-order catalysis in Eq. (11.8) and additionally including active transport of some chemicals through the cellular membranes [Kaneko and Yomo (1997); Kaneko and Yomo (1999)]. All variants of the model exhibited similar behavior, as described below.

The simulations start with a single cell and random initial conditions for the concentrations of chemicals inside it. The cell grows and undergoes division. The daughter cells also repeatedly divide and a cellular population is rapidly established. In the initial stage of this process (up to eight divisions), all cells are identical in their chemical composition and concentration oscillations inside them are synchronous (while being chaotic). Hence, this initial stage can be described as *full chaotic synchronization.* Note that, because of such synchronization, all cells divide at the same time in this stage.

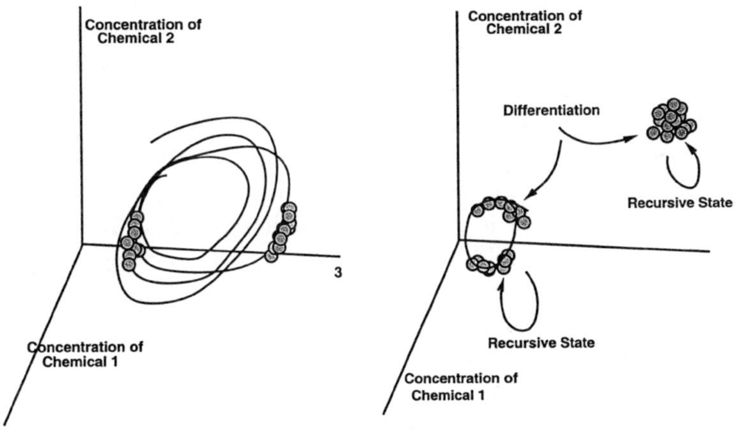

Fig. 11.6 (a) Dynamical clustering and (b) irreversible differentiation in a cell population. Courtesy of K. Kaneko.

As the population increases, full synchronization is replaced by a stage of *dynamical clustering.* Though the cells are still identical, they form several coherent groups characterized by different oscillation phases (Fig. 11.6a). While the instantaneous states of the cells are different at this stage, time averages of concentrations remain almost identical in all of them. This means that dynamical clustering itself does not yet represent cell diversification. Note that oscillations are still chaotic in the clustered states.

Further growth of the population brings however a qualitative change: chemical compositions of the cells in different clusters become different and irreversible *differentiation* takes place (Fig. 11.6b). Several mechanisms contribute to the differentiation process.

As clustering progresses, not only the phases but also the amplitudes of oscillations and their profiles in different clusters start to differ. This means that, depending on the cluster to which it belongs, a cell would experience slightly different chemical environments. Moreover, division of cells would now occur at different phases and therefore under different conditions for each of the clusters.

When a cell divides, the chemicals are not exactly equally distributed between the two daughter cells. Each replication event produces therefore a weak heterogeneous perturbation of chemical concentrations. In the early stages of synchronization and dynamical clustering, such perturbations introduced by cell divisions are damped and the population remains uniform. At the differentiation onset, some perturbations cannot be any longer damped and lead to the divergence of kinetic regimes inside the cells.

Figure 11.7 shows an example of differentiation in a population of cells whose chemical networks consist each of $M = 32$ chemicals with 9 connections for each chemical. For clarity, only the time dependence of the concentrations of 6 arbitrarily chosen chemicals inside a cell is given here. Chaotic concentration oscillations in the initial nondifferentiated cells (Type 0) are displayed in Fig. 11.7a. Spontaneous transitions to the regimes of types 1, 2 and 3 are seen in Fig. 11.7b, c, d. The model parameters are $\nu = 1$, $D = 0.001$, $f = 0.02$, $\gamma = 0.1$, and $\overline{C}^{(i)} = 0.2$ for all 10 mobile species.

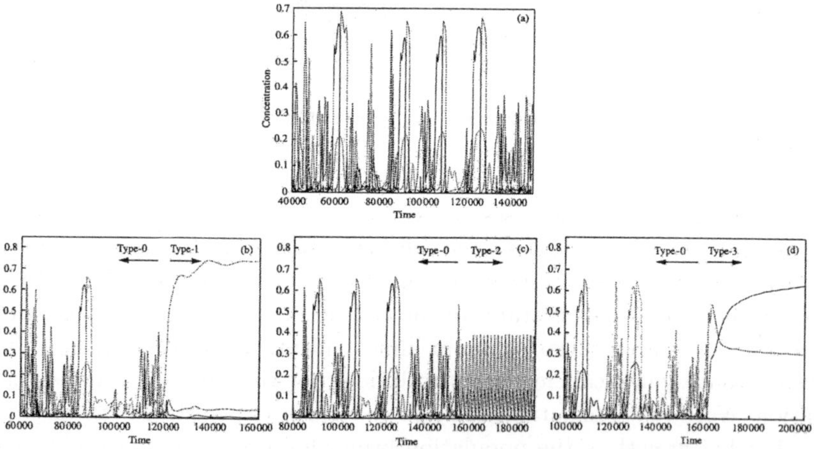

Fig. 11.7 Chaotic chemical oscillations in the initial "stem" cell of type 0 (a) and its spontaneous differentiation to cells of types 1,2, and 3 (b-d). Time dependences of concentrations for 6 different internal chemicals are displayed. From [Furusawa and Kaneko [2001)].

The transition to a new kinetic regime inside a cell is related to the intracellular extinction of some chemicals which are not able to cross the cellular membrane. This loss leads to a modification in the chemical composition of a cell and to a change (i.e. reduction) of its reaction network. Because of this, the transition is irreversible and a new kind of cell is effectively produced by it. Typically, the differentiated cells are characterized by much more simple dynamics, which corresponds either to a steady state or to simple periodic oscillations. They also have a lower chemical diversity, as compared with the initial "stem" cells of type 0.

Figure 11.8 presents the cell lineage diagram corresponding to such cell differentiation events. The four different types of cells are shown here by different shades of gray. The spontaneous development of a structured cellular population from the single initial cell is clearly seen.

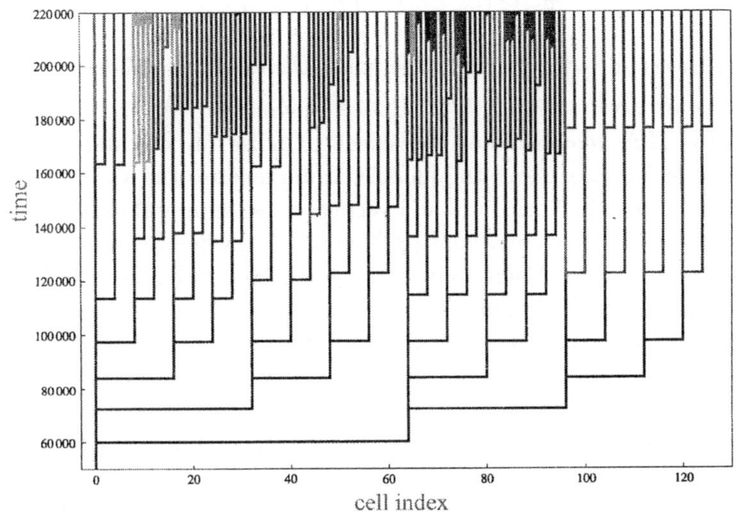

Fig. 11.8 Cell lineage diagram. Different shades of gray color indicate different cell types. From [Furusawa and Kaneko [2001)].

Thus, even a simple abstract model of interacting and reproducing cells is able to yield differentiation of cells into several distinct types. Importantly, the differentiation takes place under stirred conditions and spatial heterogeneities in the medium are not needed for it. At the stage immediately preceding irreversible differentiation, clustering of cells in terms of their internal concentration variables, i.e. with respect to the internal states of a cell, develops. This can play a role similar to spatial isolation, creating different "environments" for different cell clusters and therefore opening different channels for cellular evolution.

The differentiation proceeds through enhancement of small concentration imbalances in cell division events and, hence, it is important to check whether its outcome is robust with respect to initial conditions, macroscopic perturbations, and noise [Furusawa and Kaneko (2001)]. To include the effects of molecular fluctuations, weak multiplicative noise was added to the reaction rates, so that they become

$$w_l^{(i)} = \sum_{j,k=1}^{M} \nu J_{ijk} \left(c_l^{(j)}\right)^{\alpha} c_l^{(k)} - \sum_{j,k=1}^{M} \nu J_{kji} \left(c_l^{(j)}\right)^{\alpha} c_l^{(i)} + \zeta_i D \left(C^{(i)} - c_l^{(i)}\right) + \sqrt{c_l^{(i)}}\eta_l^{(i)}(t). \tag{11.12}$$

Here $\eta_l^{(i)}(t)$ is an independent white noise of intensity σ, whose correlation functions are given by $\left\langle \eta_l^{(i)}(t)\eta_{l'}^{(i')}(t') \right\rangle = \delta_{ll'}\delta_{ii'}\delta(t-t')$.

Figure 11.9 demonstrates the effect of noise on cell differentiation. The chemical network and the model parameters are the same here as in Fig. 11.7. Along the horizontal and vertical axes, temporal averages of two arbitrarily chosen concentrations ($c^{(19)}$ and $c^{(28)}$), when the number of cells is 200, are displayed. Each point corresponds to a different cell. Without noise, the cells split into four cell types shown in Fig. 11.7. The intensities of noise are $\sigma = 3 \cdot 10^{-4}$ (Fig. 11.9a), 10^{-3} (Fig. 11.9b), $3 \cdot 10^{-3}$ (Fig. 11.9c), and 10^{-2} (Fig. 11.9d). The four distinct groups, corresponding to different cell types, are preserved as long as the noise intensity remains smaller than 0.01. For stronger noise, differentiation into well-defined groups does not take place and all cells fall into the Type-0 dynamics.

Thus, both the cell types and the frequencies of these types in the emerging cellular population are determined by the parameters of the cells and the outcome of the differentiation process is definite, if noise is not too strong. The dominant role in specifying the course of differentiation is played by the network of catalytic chemical reactions inside the initial cell.

Similar results are found when ideal mixing of reactants in the extracellular medium is absent and they are allowed to form spatial concentration patterns [Furusawa and Kaneko (2000)]. As the population grows and cell differentiation takes place, various types of cells develop under such conditions in different parts of the medium. This can be described as the formation of a multicellular organism. A heterogeneous ensemble of cells with a variety of dynamics and stable states (cell types) has usually a larger growth speed than a uniform population of simple cells. Apparently, such

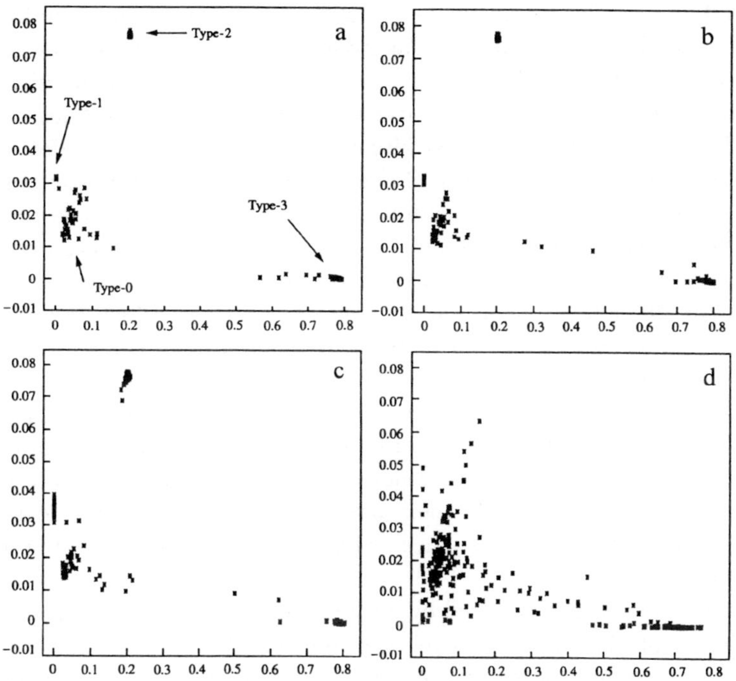

Fig. 11.9 Effect of molecular noise on cell differentiation. Temporal averages of concentrations $c_l^{(19)}(t)$ and $c_l^{(28)}(t)$ over a time of 200000 time steps for each cell in a population of size 200 are plotted along the horizontal and the vertical axes. Every point corresponds to a particular cell (some points are overlapped). The noise intensities are $\sigma = 3 \cdot 10^{-4}$ (a), 10^{-3} (b), $3 \cdot 10^{-3}$ (c), and 10^{-2} (d). From [Furusawa and Kaneko [2001)].

differentiated cellular ensembles can better utilize the nutrients than populations of identical cells. This may have provided a decisive advantage giving way to the emergence of macroorganisms in the process of biological evolution.

11.3 Synchronization of Molecular Machines

As we have already seen, whole cells can behave as periodic or chaotic oscillators, and their synchronization is functionally important. Now, we turn our attention to the processes that go on inside a single biological cell. The operation of a cell is based on the highly coordinated action of a large population of molecular machines. Such machines, representing individual proteins or their complexes, are far from equilibrium because they receive energy in the chemical form. This allows them to act autonomously, overcoming restrictions set by thermodynamics for equilibrium systems. Active protein machines are immersed into water solution that provides a passive medium needed for supply of energy and for communication between the machines. The communication is realized through diffusion of small molecules released by a machine and able to affect the operation of other machines. Small molecules are also employed to transfer energy.

In this section, we shall mainly consider enzymes, which are proteins acting as single-molecule catalysts. Their function is to convert substrate S into product molecules P in a reaction $S + E \to E + P$, that cannot proceed in absence of an enzyme (E). Hence, enzymes are analogous to inorganic metal catalysts (such as Pt catalyzing oxidation of CO into CO_2, see Sect. 10.2). This similarity is not purely formal: many enzymes would indeed possess a metal ion in their *active center*, where the chemical event of catalytic conversion takes place. This active center is integrated into a protein macromolecule.

An enzyme is characterized by its turnover rate, which is defined as the number of product molecules released per unit time by a single enzyme molecule, provided the substrate is present in abundance. The inverse of the turnover rate is the turnover time, needed on the average by an enzyme to convert a single substrate molecule. The turnover times can be as short as a microsecond, but typically they range from tens of milliseconds to a few seconds.

According to the classical Michaelis–Menten view, the operation principles of enzymes are basically the same as those of inorganic catalysts. A substrate molecule arrives at the active center and is converted there into a product which immediately dissociates. Hence, the reaction proceeds in two stages. In the first stage $S + E \to ES$, a substrate binds to the enzyme to form an enzyme-substrate complex (ES); this reaction is reversible and, with some probability, the enzyme-substrate complex can dissociate. In the second stage $ES \to E + P$, the enzyme-substrate complex is transformed into a free enzyme and a product molecule. Each stage is stochastic and characterized by the respective rate constant, determining the characteristic waiting time.

The Michaelis–Menten concept was formulated a long time ago, when very little was known about the properties of individual macromolecules. Today, when we know much more about them and can already observe processes in single proteins, it raises serious questions. If the rest of an enzyme macromolecule only provides support for the active center where catalysis takes place, why is the function of an enzyme so sensitive to the choice of such external support and to its physical shape? Why are often the enzymic reactions much slower than the processes of inorganic catalysis?

A protein can exhibit many conformations representing different shapes of this macromolecule. Most of them are metastable and, as time goes on, the protein would move towards its equilibrium conformation corresponding to the native state. The process of conformational relaxation is however extremely slow. Protein folding, which is relaxation from a distant unfolded state, can take minutes for a single molecule! Typical time scales of conformational relaxation for a folded molecule are of the order of tens or hundreds of milliseconds.

It is natural to expect that turnover cycles in some enzymes include conformational changes. Because conformations are different shape states of a molecule, this would mean that an enzymic cycle is accompanied by mechanical motions inside the protein molecule. The functional roles of such motions may be different, from bringing a substrate to the active center and putting it in an optimal position for a catalytic event to exporting a product out of the enzyme. When such conformational motions are involved, a single enzyme molecule already acts as a *machine* [Blumenfeld and Tikhonov (1994)].

Figure 11.10 gives a schematic illustration of one possible operation mechanism of an enzymic machine. The displayed enzyme is a protein with an active center lying in its center (Fig. 11.10a). A substrate molecule binds at a different location on the enzyme surface (Fig. 11.10b). Binding of a substrate initiates a sequence of conformational changes inside the enzyme-substrate complex, and the molecule changes its shape in such a way that the substrate is gradually transported towards the active center (Fig. 11.10c-e). When this center is reached, catalytic conversion takes place (Fig. 11.10f) and the product is expelled (Fig. 11.10g). Subsequently, the free enzyme molecule returns to its original conformational state (Fig. 11.10h,i). Note that all functional conformational motions are relaxation processes. Energy can be brought with the substrate or released when a substrate is converted into the product inside the molecule. It can also come from thermal fluctuations.

Thus, a distinguishing feature of enzymes, operating as protein machines, should be that their turnover cycles include many intermediate states which differ not by their chemical composition, but by the phys-

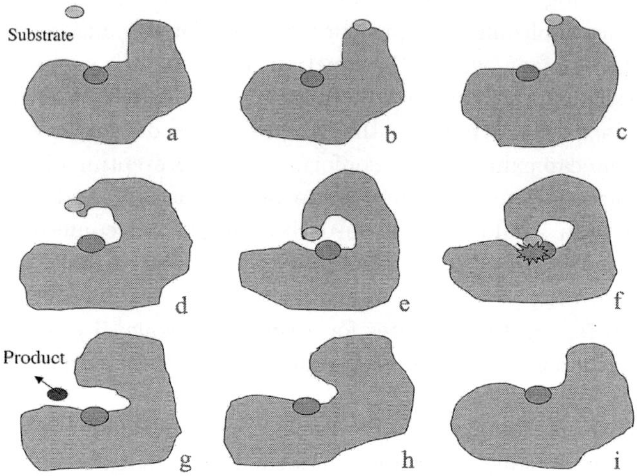

Fig. 11.10 Enzyme as a molecular machine.

ical configuration corresponding to different conformations of the same molecule. The transitions between individual functional states occur in an ordered way, as a relaxation process. A cycle is completed only when all states in the sequence are passed.

Because enzymic machines act in a cyclic manner, like an oscillator, *synchronization* of molecular cycles in enzymic populations should be possible. This has indeed been demonstrated in the experiments with a complex enzyme—the cytochrome P-450 monooxygenase system [Häberle *et al.* (1990); Gruler and Müller-Enoch (1991); Schienbein and Gruler (1997)]. The family of various P-450 enzymes plays an important role in all living organisms (and particularly in the liver cells) because it is responsible for the removal by oxidation ("burning") of various waste products of biochemical reactions. These enzymes are very slow, with a characteristic turnover times of the order of seconds.

The enzyme employed in the investigations [Häberle *et al.* (1990); Gruler and Müller-Enoch (1991)] was photosensitive and its catalytic activity could be enhanced by illumination with light of a certain wavelength. Moreover, its product was fluorescent and therefore its concentration could be optically recorded. In the experiments, the product was not removed from the reactor and thus gradually accumulated in the reacting solution.

To synchronize the enzyme molecules, a sequence of 10 intensive light flashes (of duration 0.1 s) with the required wavelength was applied. The repetition interval $T = 1.32$ s of the flashes was a little shorter than the turnover time $\tau = 1.54$ s of the enzyme. After the illumination was stopped, the catalytic activity of free running enzymes was determined by real-time measurement of product concentration in the medium. A typical result of such experiments is displayed in Fig. 11.11.

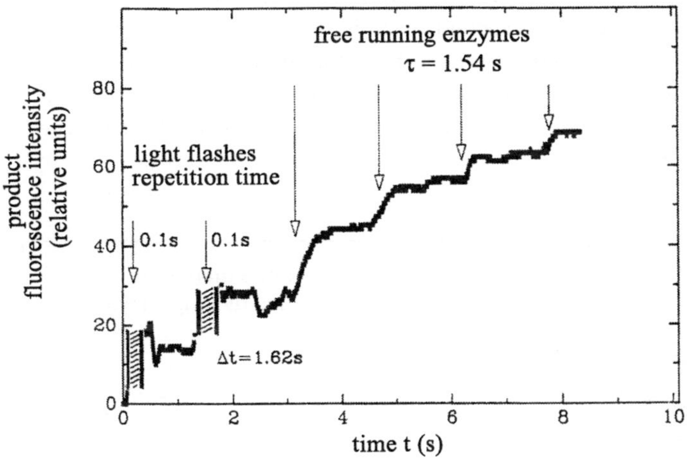

Fig. 11.11 Optical synchronization of enzymic turnover cycles. From [Gruler and Müller-Enoch (1991)].

Instead of a linear increase of the product concentration, expected for steady asynchronous operation of individual enzymes, a sequence of steps in the product concentration is observed. Such steps are formed because a large fraction of the enzymes is simultaneously releasing the product. Between the steps, the product concentration remains approximately constant, because the enzymes are inside their cycles preparing for the new firing of the product molecules. Remarkably, the interval between subsequent steps is close to the turnover time $\tau = 1.54$ s of the employed enzyme.

As time goes on, the steps become less pronounced and finally fade away. At the molecular level, all motions are accompanied by fluctuations. As a result, the enzymes cannot operate as precise clocks and the duration of their cycles is fluctuating. Hence, even if all enzymes in a population were initially synchronized, their cycles would slowly desynchronize in absence of external forcing. Using experimental data, a statistical dispersion of turnover times of 20% was deduced [Schienbein and Gruler (1997)].

The response of the enzymic population to external optical forcing was resonant. Figure 11.12 shows the fraction of coherently operating enzymes as a function of the repetition time of light flashes. This fraction was estimated by the height of the first step observed after a series of 10 light flashes. A narrow peak at the repetition time close to the turnover time of free enzymes is seen. Another maximum is found at roughly the double of that time, when each second cycle is optically stimulated. The difference between the optimal repetition time for resonant forcing and the turnover time of free enzymes can be explained by taking into account that the light may shorten the enzymic cycle by, for instance, facilitating the release of the product.

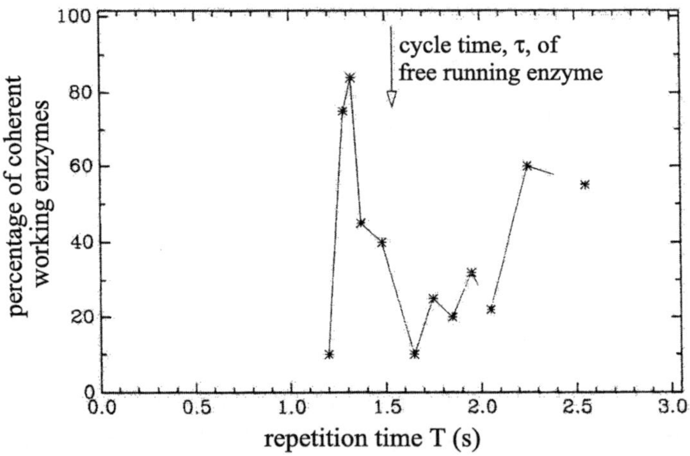

Fig. 11.12 Resonant response of enzymes to periodic optical stimulation. From [Gruler and Müller-Enoch (1991)].

In the above experiments, light was used to control the enzymic activity. Chemical regulation of enzymes is however also possible. Almost all of them are *allosteric*, so that the catalytic conversion rate is influenced (increased or decreased) by binding of small regulatory molecules. Several mechanisms of allosteric regulation are known. Sometimes, a regulatory molecule binds to the active center and blocks it for binding of the substrate, thus directly inhibiting the reaction. In many other enzymes, binding of a regulatory molecule occurs at a location which is different from the substrate binding site. Then, the regulatory molecule induces a transition to a different conformational state where binding of a substrate becomes more likely (allosteric activation) or is more difficult (allosteric inhibition).

Functioning of a cell is based on a large complex network of enzymatic reactions. The product of a particular enzyme usually not only serves as a substrate for further reactions, but also acts as a regulatory molecule affecting the activity of other enzymes in the network. Thus, different reaction pathways become integrated. Moreover, it is often found that an enzyme is allosterically regulated by its own product.

Each species in a biochemical reaction network of the cell is represented by a population of enzyme molecules. Small regulatory molecules are produced by enzymes, diffuse through the medium, bind to other enzymes and allosterically influence their operation. Thus, chemical communication and effective interactions between different molecular machines are established. Since the machines are cyclic, it should be possible that, under certain conditions, full or partial synchronization and clustering in this system take place.

Though experimental evidence of intracellular synchronization of enzymic activity is not yet available, this problem was theoretically analyzed in a series of publications [Hess and Mikhailov (1994); Hess and Mikhailov (1995); Hess and Mikhailov (1996); Mikhailov and Hess (1996); Stange *et al.* (1998a); Stange *et al.* (1998b); Stange *et al.* (1999); Stange *et al.* (2000); Lerch *et al.* (2002)]. The impetus for such studies was provided by the observation [Hess and Mikhailov (1994)] that chemical communication between different molecules in a volume of a micrometer size, characteristic for a biological cell, is extremely fast: any two molecules within such a volume would meet due to their diffusion *every second*! Thus, if a small regulatory molecule has to find by diffusion one of 1000 identical targets randomly distributed inside a micrometer volume, it can do this within one millisecond. This is much shorter than the characteristic time scale of individual molecular machines (i.e. the turnover time in case of enzymes). Therefore, communication through diffusing regulatory molecules can easily lead to global instantaneous coupling between molecular machines inside a biological cell [Hess and Mikhailov (1995)].

To investigate the synchronization phenomena, a simple model can be considered [Stange *et al.* (1999)]. We have a population of N identical enzyme molecules E, participating in the reaction

$$S + E \to E + P, \quad P \to 0. \tag{11.13}$$

The enzyme is allosteric and the product molecules P represent at the same time regulatory molecules that inhibit binding of substrate S. The substrate concentration is maintained constant and the product is gradually removed by some decay process. The reaction takes place in a sufficiently small volume, so that the conditions of global coupling are fullfilled. This means that any product molecule can with equal probability bind to any enzyme molecule in the volume and the time needed for diffusive transport to the target is negligibly small.

A single enzyme molecule can be modelled as a variant of a phase oscillator (Fig. 11.13). The phase corresponds to the conformational coordinate, specifying the configuration of this molecular machine. The dynamics of the enzyme inside its catalytic turnover cycle represents diffusive drift along this coordinate.

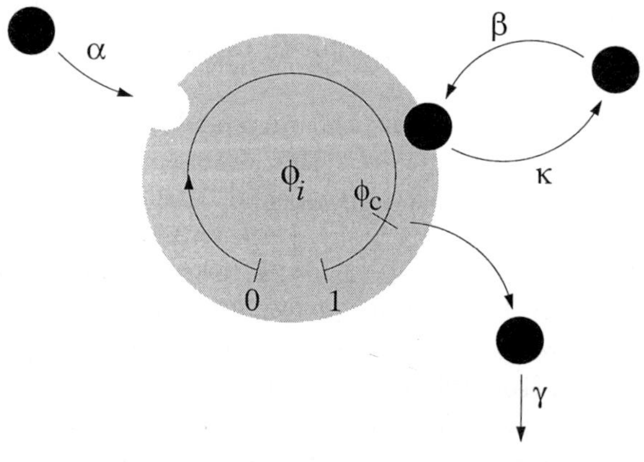

Fig. 11.13 Schematic representation of an enzymic turnover cycle. From [Stange *et al.* (1999)].

It is convenient to define for each enzyme i a binary variable s_i, such that $s_i = 0$ if the enzyme is in its free state ready to bind a substrate molecule. The formation of a substrate-enzyme complex is then described as a transition into the state with $s_i = 1$. This transition initiates the turnover cycle, which consists of the catalytic conversion of the substrate into the product and the subsequent return of the enzyme to its free state.

This process is modelled as diffusive drift through an energy landscape along the conformational coordinate ϕ_i. The coordinate $\phi_i = 0$ corresponds to the beginning of the cycle. The cycle ends when $\phi_i = 1$ and the enzyme returns to its free state with $s_i = 0$. The release of the product molecule takes place in the state $\phi_i = \phi_c$ inside the cycle. Thus the point ϕ_c on the reaction coordinate separates two different processes. In the coordinate interval $0 < \phi_i < \phi_c$, the substrate-enzyme complex exists, whereas later in the interval $\phi_c < \phi_i < 1$ the enzyme returns back to its free state. There, it can again bind a substrate molecule to start a new cycle.

Introducing the probability distribution $p(\phi_i, t)$ over the coordinate ϕ_i, we assume that this distribution satisfies the diffusion equation

$$\frac{\partial p}{\partial t} = -v\frac{\partial p}{\partial \phi_i} + \sigma\frac{\partial^2 p}{\partial \phi_i^2}. \tag{11.14}$$

The first term in this equation describes the drift and the second term takes into account thermal fluctuations inside the cycle. The diffusion equation is equivalent to the stochastic Langevin equation

$$\frac{d\phi_i}{dt} = v + \eta_i(t), \tag{11.15}$$

where v is the drift velocity, $\eta_i(t)$ is a white Gaussian noise with correlation function

$$\langle \eta_i(t)\eta_j(t')\rangle = 2\sigma\delta_{ij}\delta(t-t'), \tag{11.16}$$

and the parameter σ determines the noise intensity. For simplicity, it is assumed in this model that the energy landscape has a constant negative slope, so that the drift velocity v is constant.

The enzyme has two characteristic times $\tau_1 = \phi_c/v$ and $\tau_0 = 1/v$. These times are required, on the average, to release the product and to complete the cycle. Because of the intramolecular thermal fluctuations, the actual cycle duration (that is, the time needed to reach $\phi_i = 1$) is fluctuating from one realization to another. The fluctuations can be conveniently characterized by the relative mean statistical dispersion ξ of turnover times, defined as $\xi = \Delta\tau/\langle\tau\rangle$ where $\Delta\tau = \sqrt{\langle\tau^2\rangle - \langle\tau\rangle^2}$. For small noise intensities, $\langle\tau\rangle \approx \tau_0$ and $\xi \approx \sqrt{2\sigma/v}$.

In the considered enzyme, binding of the substrate is allosterically inhibited by product molecules. We assume that, in addition to the binding site for the substrate, the enzyme has another site where a regulatory molecule can bind. If the regulatory molecule sits there, the probability of binding a substrate molecule is reduced. To describe this process, we introduce for each enzyme i the second binary state variable r_i which is equal to 1, if the regulatory product molecule is bound to the regulatory site, and equal to zero otherwise. The probability α per unit time for an enzyme to bind a substrate molecule depends on r_i, i.e. $\alpha = \alpha_1$ if $r_i = 1$ and $\alpha = \alpha_0$ if $r_i = 0$ (where $\alpha_0 > \alpha_1$). Both rates are proportional to the substrate concentration which is maintained constant. Dissociation of substrate molecules from an enzyme is neglected.

Binding of a regulatory molecule to the enzyme occurs with probability β per unit time, if one regulatory molecule is present in the volume. When m such molecules are present, this probability rate raises to βm. Dissociation of regulatory molecules from enzymes occurs at the probability rate κ. Generally, both rates depend on the state of the enzyme molecule, i.e. on the variables s_i and ϕ_i. We assume here that binding of regulatory molecules occurs only in the free enzyme, not within its cycle. This means that the binding rate is zero when $s_i = 1$. A dissociation rate κ is assumed to be independent of the state of the enzyme. Variants of the model with other assumptions concerning binding and dissociation of regulatory molecules have also been considered [Stange *et al.* (1999)].

The number m of free product molecules in the reaction volume is influenced by several processes. Whenever an enzyme i reaches the phase $\phi_i = \phi_c$, a product molecule is released. Moreover, each binding or dissociation event increases (decreases) this number m by one. Product molecules also decay at a constant rate γ. The mean life time of product molecules with respect to their decay is shorter than the average cycle duration, $\gamma\tau < 1$.

Stochastic numerical simulations of this model have been performed [Stange *et al.* (1999)]. The enzymic population consisted of $N = 400$ molecules; it was always assumed that $v = 1$ so that the mean cycle duration is unity ($\tau_0 = 1$). The inhibition effect of regulatory molecules was very strong, $\alpha_1 = 10^{-4}\alpha_0$, so that binding of substrate was practically impossible in the inhibited state.

Numerical simulations revealed the existence of two qualitatively different regimes. Below a certain threshold value of the parameter β determining the probability rate for binding an inhibitory product molecule, the enzymes operate independently of each other. Figure 11.14a displays the distribution of enzymes over their phases in this case. To obtain the distribution, the phases of all enzymes at a certain time moment are determined. The interval $0 \leq \phi \leq 1$ is divided into 100 equal parts and the number of enzymes with phases inside each of them is counted. We see that the distribution is flat. This means that all phases are equally probable and there are no correlations between internal states of different enzymes. The corresponding time dependence of the number of free product molecules is shown in Fig. 11.14b.

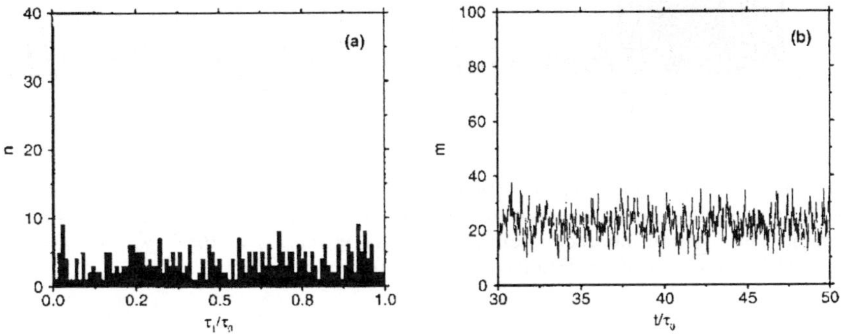

Fig. 11.14 Distribution over cycle phases (a) and time dependence of the number of product molecules (b) for the asynchronous reaction regime in a population of 400 enzymes. The reaction parameters are $\beta = 0.03$, $\alpha_0 = 10$, $\alpha_1 = 10^{-3}$, $\gamma = 15$, $\kappa = 20$, $\tau_1 = 0.55$, and $\sigma = 0$. From [Stange *et al.* (1999)].

The behavior of the system changes drastically when the parameter β is increased. Figure 11.15 a shows a typical distribution of phases in the resulting coherent regime. This distribution has a maximum, indicating synchronization of cycle phases of different enzymes. The synchronous enzymic activity is manifested in rapid spiking in the number of free product molecules (Fig. 11.15b).

The synchronization process is seen in Fig. 11.16. At the initial time moment, the enzymes are randomly distributed over their phases. After a transient, ranging from a few to hundreds of turnover cycles, the states of enzymes become synchronized and spiking in the number of product molecules develops.

To statistically characterize synchronization, we define the distribution

$$P(\Delta\phi) = \left\langle \left[\sum_{i,j=1,i\neq j}^{N} s_i s_j \right]^{-1} \sum_{i,j=1,i\neq j}^{N} s_i s_j \delta\left(\phi_i - \phi_j - \Delta\phi\right) \right\rangle, \quad (11.17)$$

that specifies the probability to find a phase difference $\Delta\phi$ between any

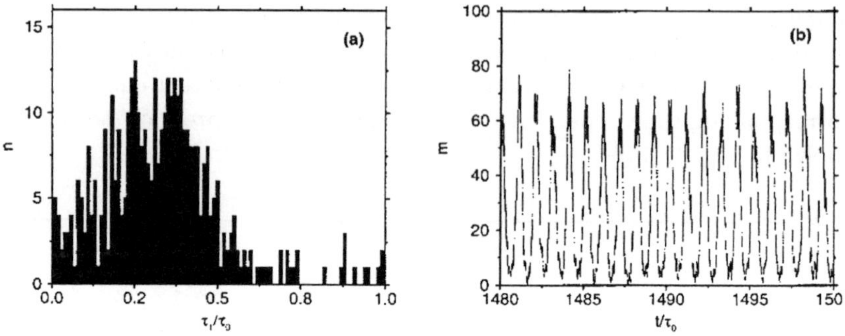

Fig. 11.15 Distribution over cycle phases (a) and time dependence of the number of product molecules (b) for the synchronous reaction regime in a population of 400 enzymes. The reaction parameters are $\beta = 0.1$, $\alpha_0 = 100$, $\alpha_1 = 10^{-2}$, $\gamma = 15$, $\kappa = 20$, $\tau_1 = 0.55$, and $\sigma = 0$. From [Stange *et al.* (1999)].

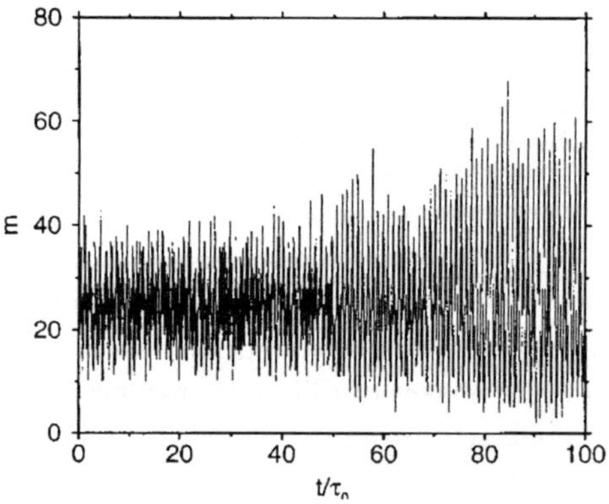

Fig. 11.16 Development of spiking in an enzymic population. From [Stange *et al.* (1999)].

two enzymes. Since $s_i = 0$ for enzymes in their free states, the summation is performed here only over enzymes inside their turnover cycles ($s_i = 1$). Angular brackets denote time averaging.

When the phase states of different enzymes are not correlated, all phase differences are equally probable. Then the distribution $P(\Delta\phi)$ is flat (see Fig. 11.17a). If, however, synchronization of enzymes takes place, this probability distribution displays a maximum at $\Delta\phi = 0$ (Fig. 11.17b).

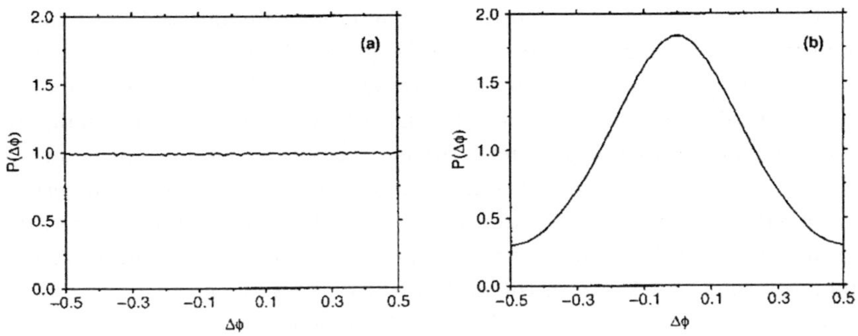

Fig. 11.17 Distributions over phase differences in the asynchronous (a) and synchronous (b) regimes. The same parameters as in Fig. 11.15. From [Stange *et al.* (1999)].

The synchronization order parameter θ can be defined as

$$\theta = \int_{-1/2}^{1/2} \left(P(\Delta\phi) - 1\right)^2 d\Delta\phi. \tag{11.18}$$

If correlations between the phases of different enzymes are absent, we have $\theta = 0$. Nonvanishing values of θ indicate presence of synchronization in the considered system.

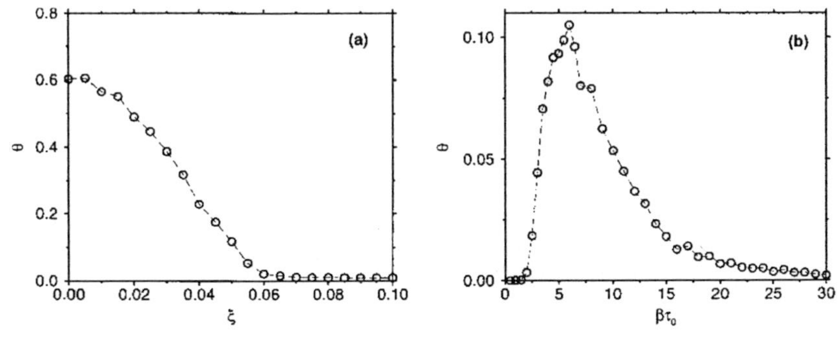

Fig. 11.18 The order parameter θ as functions of (a) relative statistical dispersion ξ of turnover times and (b) binding rate constant β for the regulatory molecules. The reaction parameters are (a) $\beta = 5$ and (b) $\sigma = 0.00125$ ($\xi = 0.05$). Other parameters are the same as in Fig. 11.15. From [Stange *et al.* (1999)].

Figure 11.18a illustrates the influence of intramolecular fluctuations on the synchronization phenomena. As the relative statistical dispersion ξ of turnover times increases, the order parameter gets smaller and, for ξ larger than 0.1, synchronization does not take place. In Fig. 11.18b, the noise intensity is kept constant and the parameter β, determining the binding rate of regulatory product molecules, is instead varied. If β is small, the inhibitory action of the product is weak and the individual molecular cycles are not correlated. Synchronization sets on when β exceeds a certain threshold. Remarkably, it again disappears when inhibition becomes too strong. This can be explained by the fact that very strong inhibition also implies strong sensitivity of the system with respect to noise.

A similar study of synchronization phenomena has been performed for enzymes with allosteric product activation [Hess and Mikhailov (1996); Mikhailov and Hess (1996); Stange *et al.* (1998a)]. Though complete synchronization in the case of allosteric activation is possible, the population typically divides into several synchronous clusters. Non-allosteric enzymes can also show mutual synchronization. For instance, it was found for enzymic reactions where a fraction of product molecules is converted back into the substrate [Stange *et al.* (2000)]. Many enzyme molecules consist of several identical functional subunits, each catalytically active. The turnover cycles in such subunits influence each other, and synchronization phenomena in populations of such enzymes are complex [Lerch *et al.* (2002)].

Chapter 12

Neural Networks

The human brain is the ultimate challenge for the theory of complex systems. Its level of organization exceeds by far anything that can be found in the inanimate Universe. Billions of neural cells are wired together in a huge ensemble of interconnected neural networks. Collectively, they are responsible for processing of information that arrives from the outside world and working out of the decisions, for motor responses and control of the human body. On top of that, the higher functions of consciousness, rational reasoning and emotional discourse are coming. Most of the brain functions cannot be reproduced even by the best modern computers—despite the fact that the operation frequency of these computers is more than 10^7 times greater than the spiking rate of a single neural cell.

The neurons building up the brain are essentially oscillators. Therefore, it is natural to expect that the concepts of dynamical order related to synchronization and dynamical clustering should play an important role in understanding neural networks. In this Chapter, we discuss some aspects of synchronization phenomena in such systems.

An individual neuron is as complicated as any other biological cell. It is however believed that, insofar as communication between such cells is involved, they behave as relatively simple dynamical units. From the viewpoint of nonlinear dynamics, many of them are found in states near a special bifurcation which is known as saddle-node bifurcation on the limit cycle. As we show in the next section, the canonical form of a dynamical system near this bifurcation corresponds to the phenomenological model of an integrate-and-fire neuron. Moreover, interactions in a network formed by such units are based on generation, propagation and reception of short pulses (spikes).

The experimental data indicating the presence of synchronization and clustering in brain activity is briefly reviewed in Sec. 12.2. The experiments with microelectrodes inserted into the visual cortex of animals have shown that synchronization of neuronal activity in this brain region leads to the integration of individual perceived features into a coherent visual scene. On the other hand, statistical analysis of electroencephalography (EEG) recordings indicates that synchronization also links together processes in distant parts of the brain. According to a popular hypothesis, development of transient synchronous clusters in neural networks spanning the whole brain is responsible for the appearance of distinct mental states which make up the flow of human consciousness.

When large-scale synchronization of neuronal processes is discussed, one should avoid the mistake of assuming that it merely results from synchronization of states of individual neurons. If this were the case, the whole brain or its large parts would have behaved just like a single neuron. Apparently, such synchronization rather involves the emergence of some temporal correlations in the activity patterns of different neural networks, responsible for particular mental functions. At the end of the chapter, a simple model of an ensemble of cross-coupled neural oscillatory networks is considered. We show that interactions between the networks can lead to mutual synchronization of their activity patterns and to spontaneous separation of the ensemble into coherent network clusters.

12.1 Neurons

Brain is the animal organ specialized on information processing. Like all other organs, it consists of biological cells and the ability of information processing is based on communication between them. The main difference is that communication between neurons takes place in the form of electrical signals and electrical activity of such cells is essential. A neuron has many protrusions that are like electrical cables and can spread out to significant distances from the cell body. One of them is always the axon, used to send signals. A neuron receives electrical signals through a large number of dendrites, making up the rest of protrusions. Though the detailed internal organization of neurons is as complicated as that of any other biological cell, they operate as relatively simple electrical devices. When the sum of the signals received through all dendrites over a certain interval time exceeds a threshold, an excitable neuron generates an electrical pulse (a spike) that is sent out through its axon. Oscillatory neurons periodically generate spikes even in absence of any input. However, the moment of the next spike firing can then be retarded or advanced depending on the signals received.

There are no direct electrical contacts between neurons. Instead, transmission of electrical signals from one cell to another occurs within synapses. In a synapse, a dendrite of one neuron reaches very closely an axon of another neural cell: they become separated only by a synaptic gap with a width of about 20 nanometers. When an electrical signal arrives through the axon, molecules of a special chemical substance (neuromediator) are released into the gap. They rapidly diffuse inside it and reach the dendrite. The dendrite responds by sending an electrical pulse to its central body. The polarity of generated signals depends on the kind of synaptic connection; it is positive for activatory and negative for inhibitory synapses.

From an evolutionary perspective, synaptic transmission has developed from chemical cell-to-cell communication discussed in the previous chapter. Direct communication between the cells became possible by bringing together some parts of the two cells very close to each other within a synapse. A chemical released in the synapse can affect only that other cell which is in the synaptic contact. Actually, neurons in the brain can also communicate in the "standard" chemical way, like other biological cells. They may release neuromediators into the common extracellular medium, which diffuse and affect the activity of other neural cells. Such form of communication is however slow and non-directional; it is employed in the neural system only for some special purposes.

Mathematical modeling of neural cells falls into two different classes. Some of the models are very detailed and attempt to incorporate many known processes that take place inside a single cell. They are more suited for the analysis of behavior of individual cells or their small groups. Alternatively, simple phenomenological models of neurons can be used. These models try to capture only the principal aspects of such cells, which are relevant for information processing in neural networks.

An individual neuron represents a nonlinear dynamical system. Persistent periodic oscillations should correspond to a stable limit cycle of a neuron. On the other hand, excitable neurons should have a fixed point stable with respect to sufficiently weak (subthreshold) perturbations. In response to a stronger superthreshold perturbation, a neuron performs a large excursion from the fixed point, but eventually returns to it.

Many neurons (belonging to the so-called "Class I") show a gradual transition from the oscillatory to the excitable behavior [Hodgkin (1948)]. As some control parameter is varied starting from the oscillatory state, the interval between subsequent generated spikes increases and becomes infinite at the bifurcation point. On the other side of this point, oscillations are absent and the neuron is excitable. Hence, this transition should correspond to a bifurcation where a stable limit cycle disappears and gives rise to a stable fixed point. This bifurcation must furthermore be characterized by vanishing of the oscillation frequency (i.e., divergence of the oscillation period) at the critical point.

In Chapter 5, we have considered the Andronov–Hopf bifurcation corresponding to the disappearance of a limit cycle. In this case, the limit cycle shrinks into a point. It means that the oscillation amplitude decreases and vanishes at the bifurcation point. However, the oscillation frequency remains finite near the Andronov–Hopf bifurcation. Thus, it cannot reproduce the behavior characteristic for neurons of Class I.

There is another kind of instability of limit cycles which instead takes place for such neurons. It is related to the *saddle-node bifurcation on a limit cycle*, illustrated in Fig. 12.1. Before the bifurcation, the system has a stable limit cycle (Fig. 12.1a). As the bifurcation is approached. motion along this cycle becomes increasingly slow inside a certain part of it. At the bifurcation, a fixed point appears on the cycle and oscillations are terminated (Fig. 12.1b). Immediately after the saddle–node bifurcation, there are two fixed points (one stable and the other unstable) that are both lying on the former limit cycle (Fig. 12.1c).

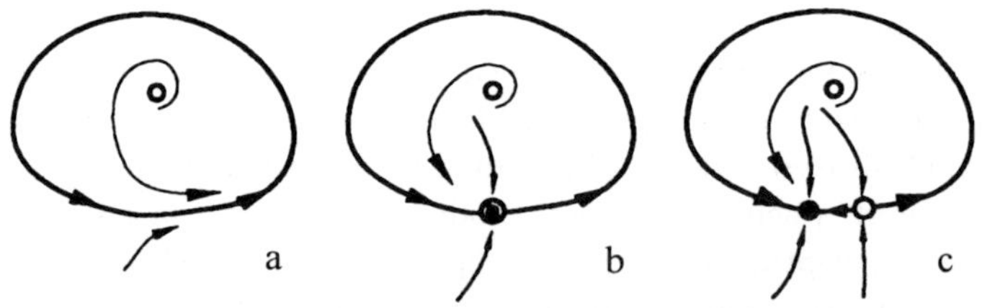

Fig. 12.1 Saddle–node bifurcation on a limit cycle. From [Izhikevich (2000)].

Suppose, for instance, that a dynamical element is described by two equations $\dot{x} = f(x, y)$ and $\dot{y} = \mu g(x, y)$ where $\mu \ll 1$. In this case, the motion is well characterized by the nullclines $f(x, y) = 0$ and $g(x, y) = 0$, that correspond to the lines on the plane (x, y) where $\dot{x} = 0$ or $\dot{y} = 0$. The intersections of these nullclines give the fixed point of the system. The limit cycle consists of the intervals of slow motion along the nullcline $\dot{x} = 0$

and rapid changes of the fast variable x during which the slow variable y remains approximately constant. Before the saddle–node bifurcation, the nullcline $\dot{y} = 0$ comes close to a certain part of the nullcline $\dot{x} = 0$, so that the dynamics gets very slow in this region (Fig. 12.2a). At the bifurcation, the nullcline $\dot{y} = 0$ touches the nullcline $\dot{x} = 0$ (Fig. 12.2b). Above the bifurcation, the two curves cross so that two fixed points are formed (Fig. 12.2c).

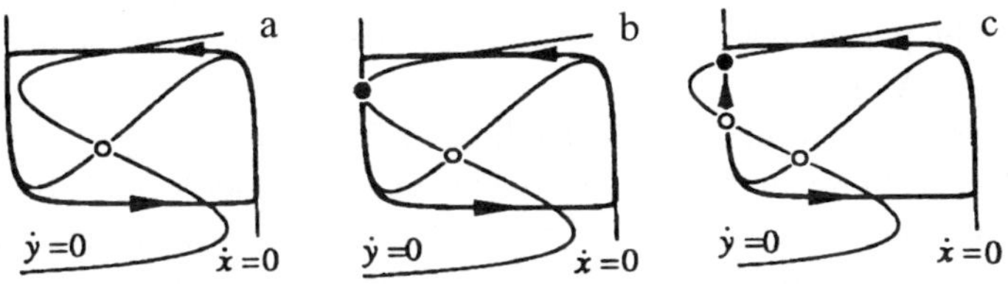

Fig. 12.2 Saddle–node bifurcation on a limit cycle in a model with relaxational oscillations. From [Izhikevich (2000)].

The saddle–node bifurcation on the limit cycle describes a transition from oscillatory to excitable dynamics. If λ is the control parameter, the oscillation period T diverges near the bifurcation point $\lambda = \lambda_0$ as $T \sim (\lambda_0 - \lambda)^{-1/2}$ and the oscillation frequency ω vanishes as $\omega \sim \sqrt{\lambda_0 - \lambda}$ there. For $\lambda > \lambda_0$, the oscillations are absent and the element has two closely lying fixed points. In the vicinity of such points, the motion is slow and has a characteristic timescale of $(\lambda - \lambda_0)^{-1/2}$. Large enough perturbations, moving the element from its stable (white) to the unstable (black) fixed points, are followed by a long excursion along the remaining outer part of the cycle. At the end of the excursion, the element returns to the stable fixed point. In the subsequent discussion, we put $\lambda_0 = 0$.

Populations of weakly coupled elements in the vicinity of the saddle–node bifurcation on the limit cycle have universal properties and allow a unified description [Hoppensteadt and Izhikevich (1997)]. Below in this section we follow the analysis given in the review article [Izhikevich (2000)].

Any dynamical element close to saddle–node bifurcation on the limit cycle is approximately described by the canonical model

$$\dot{\varphi} = (1 - \cos\varphi) + (1 + \cos\varphi)\, r, \tag{12.1}$$

where φ is an appropriate phase variable and r is a parameter. The reduction to this canonical form is based on the *Ermentrout–Kopell theorem*.

Suppose that a dynamical system

$$\dot{X} = Q(X, \lambda) \tag{12.2}$$

where X is a vector with m components has a saddle–node bifurcation on the limit cycle at $\lambda = 0$. Then, there is a mapping $\varphi = h(X)$ that projects all solutions of (12.2) in the neighborhood of the limit cycle to those of the canonical model (12.1). The time t in the corresponding canonical model is slow, that is $t = \sqrt{|\lambda|}t'$ where t' is the time variable in the original system (12.2). The parameter r in the canonical model depends on the form of the function $Q(X, \lambda)$.

The transformation h maps the limit cycle into a circle $\varphi \in [-\pi, \pi]$ (see Fig. 12.3). It blows up a small neighborhood of the saddle–node bifurcation point and compresses the entire limit cycle to a narrow interval near the point $\varphi = \pi$. Therefore, when X makes a rotation around the limit cycle (generates a spike), the phase variable φ crosses only a tiny interval at point π.

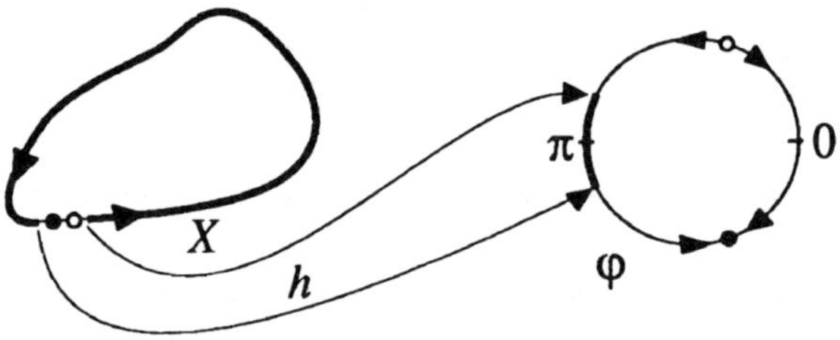

Fig. 12.3 Transformation to the phase variable. From [Izhikevich (2000)].

When $r > 0$, a neuron described by the canonical model (12.1) oscillates with the period $T = \pi/\sqrt{r}$. Since the points π and $-\pi$ are equivalent on the circle, we should reset φ to $-\pi$ every time when it crosses $\varphi = \pi$. If we plot now $\varphi(t)$, the graph shows a periodic sequence of discontinuities that look like spikes (Fig. 12.4a). If $r < 0$, it has a rest state (a stable fixed point) $\varphi = \varphi^-$ and a threshold state (an unstable fixed point) $\varphi = \varphi^+$, where

$$\varphi^{\pm} = \pm \arccos\left(\frac{1+r}{1-r}\right). \tag{12.3}$$

If a perturbation is so small that it leaves the element near the rest state (a subthreshold stimulus), the element immediately returns to the rest state. However, if the perturbation is so large that the threshold state becomes crossed (a suprathreshold stimulus), the element makes a rotation (fires a spike) and only then returns to the initial state of rest (Fig. 12.4b). Hence, the element behaves as an excitable neuron.

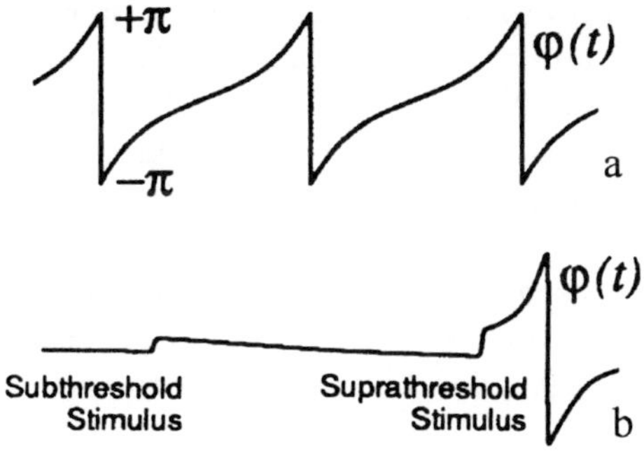

Fig. 12.4 Spiking activity in the neuron described by the canonical model (12.1). (a) Periodic spiking in the oscillatory neuron, (b) Response of the excitable neuron to subthreshold and suprathreshold stimuli. Adapted from [Izhikevich (1998)].

Let us consider a network of N such neurons with weak pair interactions which is described by the equations

$$\dot{X}_i = Q(X_i, \lambda) + \varepsilon \sum_{j=1}^{N} G_{ij}(X_i, X_j) \tag{12.4}$$

where $\varepsilon \ll 1$ is a small parameter. The functions $G_{ij}(X_i, X_j)$ in these equations can be arbitrary. We only require that $G_{ij}(X_i, X_j) = 0$ when the argument X_j is in some small neighborhood of the rest state. This means that a neuron does not act on other neurons in the population if it is currently found near the rest state. Hence, to exercise action on other elements, a neuron should make a rotation, i.e. generate a spike.

It can be shown [Hoppensteadt and Izhikevich (1997)] that the solutions of the general system (12.4) are well approximated by the following *pulse-coupled* canonical model:

$$\dot{\varphi}_i = (1 - \cos\varphi_i) + (1 + \cos\varphi_i)\, r + \sum_{j=1}^{N} w_{ij}(\varphi_i)\delta\left(\varphi_j - \pi\right), \tag{12.5}$$

where the functions $w_{ij}(\varphi_i)$ are

$$w_{ij}(\varphi_i) = 2\arctan\left(\tan\frac{\varphi_i}{2} + s_{ij}\right) - \varphi_i, \tag{12.6}$$

$\delta(x)$ is the Dirac delta-function and s_{ij} are constants determined by the interactions G_{ij}.

When a neuron j fires a spike (that is, the phase φ_j crosses π), the phase φ_i of another neuron i is changed by an amount $w_{ij}(\varphi_i)$. This change is positive for activatory ($s_{ij} > 0$) and negative for inhibitory ($s_{ij} < 0$) synaptic connections. A neuron is connected to many other neurons in the network and individual phase changes, caused by spiking of different neurons, are summed up or integrated. If the phase of a neuron crosses the threshold π, it fires itself a spike. Therefore, the pulse-coupled models of this type are also known as *integrate-and-fire models.*

If interactions are so weak that $\varepsilon \ll \sqrt{|\lambda|}$, the phase shifts s_{ij} are small and the functions $w_{ij}(\varphi_i)$ can be linearized with respect to them, so that we get

$$w_{ij}(\varphi_i) \approx s_{ij}\left(1 + \cos\varphi_i\right). \tag{12.7}$$

This yields a simpler model [Hoppensteadt and Izhikevich (1997)]

$$\dot{\varphi}_i = (1 - \cos\varphi_i) + (1 + \cos\varphi_i)\left(r + \sum_{j=1}^{N} s_{ij}\delta\left(\varphi_j - \pi\right)\right). \tag{12.8}$$

When neurons are oscillatory ($r > 0$), we can introduce new phase variables ϕ_i as

$$\phi_i = 2\arctan\left(\frac{1}{\sqrt{r}}\tan\frac{\varphi_i}{2}\right). \tag{12.9}$$

In terms of these new variables, the *standard form* of an integrate-and-fire model is obtained,

$$\dot{\phi}_i = \omega + (1 + \cos\phi_i)\sum_{j=1}^{N} c_{ij}\delta\left(\phi_j - \pi\right), \tag{12.10}$$

where $\omega = 2\sqrt{r}$ is the oscillation frequency and the coefficients $c_{ij} = \sqrt{r}s_{ij}$ are the rescaled phase shifts. Note that the factor $(1 + \cos\phi_i)$ describes the *refractory* effect: after a neuron has fired a spike (i.e., the phase ϕ_i has crossed π), this term is small and the neuron is temporarily not sensitive to the signals coming from other neurons.

Finally, if the interactions are so weak that the condition $\varepsilon \ll |\lambda|$ holds, the system (12.8) becomes equivalent to the model of coupled phase oscillators [Hoppensteadt and Izhikevich (1997)]

$$\dot{\phi}_i = \omega + \sum_{j=1}^{N} c_{ij} F(\phi_j - \phi_i), \tag{12.11}$$

where the function $F(x)$ is given by $F(x) = 1 - \cos x$.

Thus, we see that weakly coupled networks made of Class I neurons are described, close to their saddle–node bifurcation, by the general integrate-and-fire model (12.5) and its simplifications (12.8), (12.10) and (12.11). Remarkably, the detailed knowledge of the processes inside neurons and of the mechanisms of their interactions was not needed in the derivation of this canonical model.

Originally, models of integrate-and-fire neurons have also been formulated in a more phenomenological way, by constructing simple equations that would capture essential aspects of interactions between neurons [Knight (1972); Peskin (1975); Glass and Mackey (1979); Keener *et al.* (1981); Belair (1986)]. In addition to the considered neurons of Class I, the brain also includes neurons of Class II [Hodgkin (1948)]. Their distinguishing property is that frequency remains finite at the transition point where oscillations disappear. Such neurons are characterized by an Andronov–Hopf bifurcation. If this bifurcation is supercritical, the respective canonical model (i.e. the normal form of the bifurcation) is given by equation (5.1). Large locally-coupled arrays of such oscillators are described by the complex Ginzburg–Landau equation (5.5).

Because neurons behave like oscillators, their populations can exhibit synchronization and dynamical clustering. In the past, much of the theoretical research on synchronization phenomena has actually been motivated by studies of neural networks . As we have noted, networks of pulse-coupled neurons are described in the limit of very weak coupling by phase models. Therefore, the analysis in Chapters 2, 3 and 4 is also relevant for neural networks. An extensive study of the phase models from the viewpoint of neurophysiological and medical applications is given in the monograph [Tass (1999)]. At intermediate coupling strengths, models of interacting integrate-and-fire neurons should be explicitly investigated. Synchronization and clustering phenomena in such models have some special properties. Theoretical studies of such phenomena under various conditions have been performed in [Mirollo and Strogatz (1990b); Kuramoto (1991); Tsodyks *et al.* (1993); Gerstner (1995); Ernst *et al.* (1995); Gerstner (2000)].

12.2 Synchronization in the brain

The human brain consists of several billions of neurons. Its parts, specializing on particular functions of information processing and process control, are still by many orders of magnitude larger than the sizes of neural networks which can be currently modelled and analytically investigated. Therefore, even most advanced mathematical models can describe only some elementary aspects of neural activity. To get an insight into the principles of brain operation, experimental evidence should be analyzed.

The biological "hardware" on which the brain is based is extremely slow. A typical interval between the spikes of an individual neuron is about 50 ms and the time needed to propagate a signal from one neuron to another is not much shorter than such an interval. This corresponds to a characteristic frequency of merely 100 Hz. Recalling that modern digital computers should operate at a frequency of 10^9 Hz and yet are not able to reproduce its main functions, we are lead to conclude that the brain should work in a way fundamentally different from digital information processing.

Simple estimates indicate that spiking in populations of neurons must be synchronized in order to yield the known brain operations. "Humans can recognize and classify complex (visual) scenes within 400–500 ms. In a simple reaction time experiment, responses are given by pressing or releasing a button. Since movement of the finger alone takes about 200–300 ms, this leaves less than 200 ms to make the decision and classify the visual scene" [Gerstner (2001)]. This means that, within the time during which the decision has been made, a single neuron could have fired only 4 or 5 times! The perception of a visual scene involves a concerted action of a population of neurons. We see that exchange of information between them should take place within such a short time that only a few spikes are generated by each neuron. Therefore, information cannot be encoded only in the rates of firing and the phases (that is, the precise moments of firing) are important. In other words, phase relationships in the spikes of individual neurons in a population are essential and the firing moments of neurons should be correlated.

Synchronization phenomena in the visual recognition system have already been investigated. In the experiments by W. Singer and his coworkers [Gray *et al.* (1989)], such phenomena were studied by inserting electrodes in the visual cortex of cats. The neural cells in this part of the brain, which performs primary visual recognition, are specialized in detecting various graphical elements of a picture. Particularly, there are neurons which are sensitive only to bars (short line segments) with a certain spatial orientation. When such a bar is present, they generate an oscillatory response with frequencies ranging from 40 to 60 Hz. The neurons in the visual cortex are organized into an array, so that cells near a particular location in the array respond only to what happens in the respective small area in the viewed picture.

In the experiments [Gray *et al.* (1989)], two electrodes (1 and 2) were identified which were both sensitive to a moving light bar with a certain orientation. The electrodes were separated by a relatively large distance of 7 mm, so that the receptive fields of these two cortex sites did not overlap. When an appropriately oriented bar was presented at a location, corresponding to the first electrode site, an oscillatory signal was recorded by this electrode. The signal was noisy, and the presence of a periodic component with a mean frequency of 50±6 Hz was revealed by an autocorrelation analysis. If the same bar was presented at the location corresponding to the electrode 2, it showed a similar oscillatory response. Note that an electrode collectively probes the electrical states of an entire group of adjacent neurons. If an oscillatory signal is recorded by an electrode, this implies that spiking of neurons in this group is temporally correlated.

Then the setup was changed, and two moving bars were simultaneously presented (Fig. 12.5). In the experiment I, the two bars moved in opposite directions. Then, both electrodes showed oscillatory responses. They are seen in the temporal autocorrelation functions "1-1" and "2-2" in the first column in Fig. 12.5. However, cross-correlation between the signals recorded by the two electrodes was absent. Indeed, the cross-correlation function "1-2" displayed in the same column in Fig. 12.5 is almost flat. A similar behavior was observed when the directions of motion of both bars were interchanged (the respective correlation functions are displayed by the black-filled plots in Fig. 12.5).

When both bars moved in the same direction (experiment II), similar oscillatory responses were produced in each of the electrodes. However, oscillations in two electrodes were *correlated* in this case (as revealed by the cross-correlation function "1-2" in the second column in Fig. 12.5). The synchronization of signals recorded by two electrodes was seen for both possible directions of motion (left and right). Finally, a single long bar spanning both receptive areas was presented in the experiment III. This

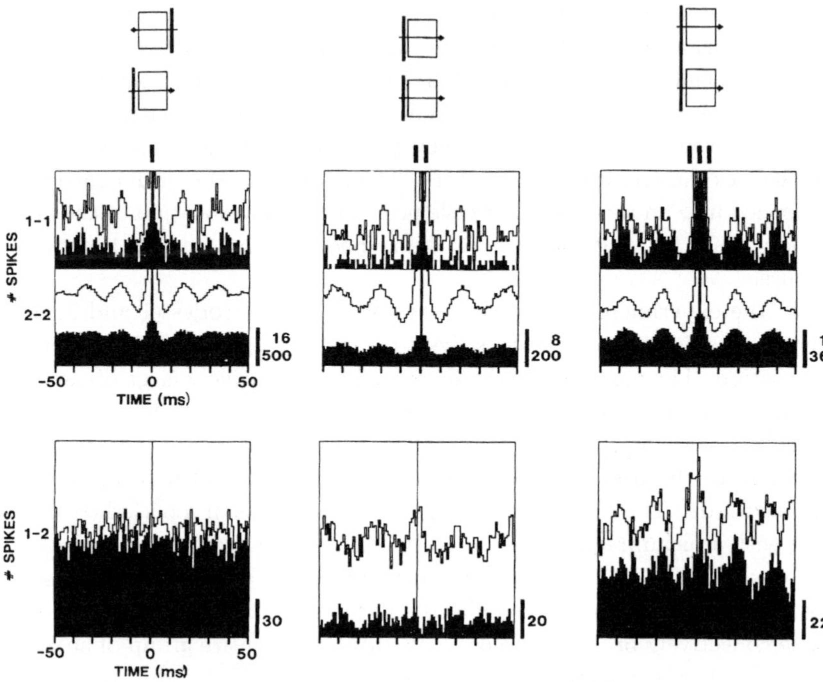

Fig. 12.5 Autocorrelation (1-1 and 2-2) and cross-correlation (1-2) functions of the oscillatory neuronal responses recorded by two electrodes under presentation of different stimuli (I, II and III). Adapted from [Gray *et al.* (1989)].

stimulus induced even a stronger synchronization of the recorded signals, shown in the third column in the figure.

Thus, synchronization apparently depends on global features of the stimuli such as coherent motion and continuity which are not reflected by the local responses alone. This has lead to the suggestion [von der Malsburg and Singer (1988)], [Gray *et al.* (1989)] that synchronization of oscillatory responses in spatially separated regions of the cortex may be used to establish a transient relationship between common but spatially distributed properties of a pattern. It can therefore serve as a mechanism for the extraction and representation of global and coherent features in a visual scene (see also [von der Malsburg and Schneider (1986)]).

While the experiments, which we have just described, yield evidence of synchronization within a single functional region of the brain (i.e., the visual cortex), there is also a large volume of experimental data indicating that synchronization of neural activity between distant and functionally different parts of the brain is also taking place. Such distant correlations have been detected by using microelectrode arrays [Roelfsema *et al.* (1997)] and by employing magnetoencephalography [Tononi *et al.* (1998)].

Interesting results have been obtained by analyzing the data yielded by electroencephalography (EEG). In this method, electrodes are attached on the skin, without penetration into the brain tissue. Therefore, the recorded electrical signals can represent only averages over rather large brain areas. The advantage of this method, on the other hand, is that it is not invasive and can be easily used with humans. The EEG signals are complex and have a broad frequency spectrum. Traditionally, they are viewed as consisting of several components. The fastest of them is the gamma band with the frequencies in the range from 30 to 80 Hz.

Taking into account that an EEG signal represents an average over the activities of millions of neurons, the very fact that this signal shows some temporal variation already indicates that correlations between firing of many neurons are present. As we shall see below, temporal correlations can also be detected in the EEG signals which are recorded by distant electrodes, attached to the left and the right brain hemispheres. Synchronization of brain activity in two hemispheres should result from communication between neurons in these two parts of the brain. There are numerous connections between the hemispheres and the conduction velocities are of the order of 10 m/s. Therefore, one cycle of spike exchanges between these two regions should take about 40 ms. This corresponds to the frequency of 25 Hz, which is near the gamma band. The studies of the EEG synchrony are usually focused on this gamma rhythm [Varela (1995)].

In the experiments by F. Varela and his coworkers [Rodriguez *et al.* (1999)], the so-called "Mooney faces" were shown to ten subjects (Fig. 12.6a, b). These are images easily recognized as faces when presented in upright orientation, but normally seen as meaningless when presented upside-down. Subjects were asked to report as quickly as possible whether they had seen a face or not by pressing one of the two keys. The EEG was recorded through 30 electrodes and the frequency analysis was carried up to 100 Hz.

A time-frequency transform of the EEG signals has been computed in each trial and then summed over all trials, subjects and electrodes. Thus, the temporal diagrams of spectral power, shown in Fig. 12.6c, d, were constructed. In these diagrams, intensities of various frequency components

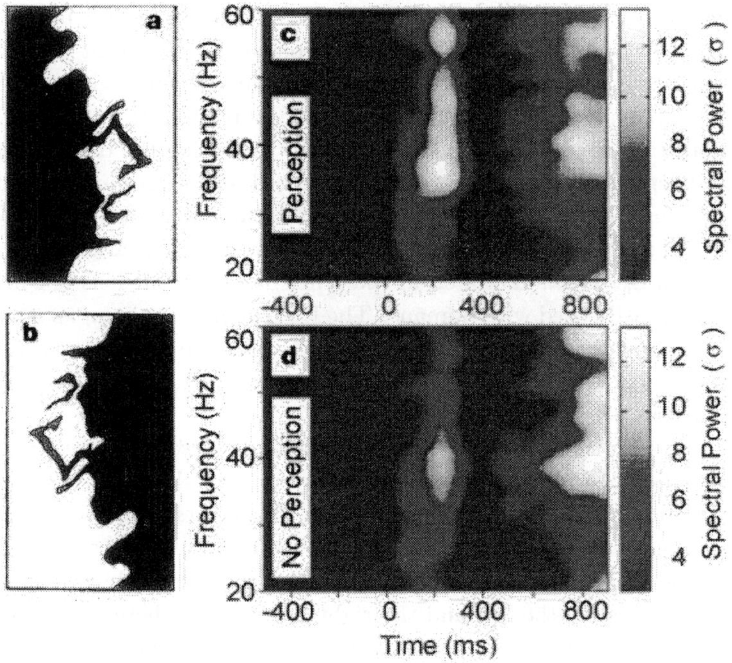

Fig. 12.6 Two Mooney faces (a,b) and the neuronal responses under perception (c) and non-perception (d) conditions. Black–and–white reproduction of the original color figure. From [Rodriguez *et al.* (1999)].

are shown as functions of time, by using a gray-color code with the white color corresponding to the highest intensity. The stimulus was presented at time zero. Fig. 12.6c displays the response to the upright image, perceived as a human face. In Fig. 12.6d, the response to the same image in the upside-down orientation, not perceived as a face, is shown. Both diagrams exhibit two periods of increased gamma activity. The first peak was located at approximately 230 ms after the image presentation; it was significantly stronger when an image was recognized as meaningful (Fig. 12.6a). The second peak was located at approximately 800 ms after the stimulus and corresponds to the ensuing motor reaction.

Thus, the responses of an individual electrode did not strongly differ when the image was recognized or not recognized as a face. However, as we shall see below, recognition of an image resulted in a special pattern of synchrony, which could be revealed only by considering phase differences in oscillations recorded by different electrodes.

Suppose that $x_{i,k}(t)$ is the original signal recorded in trial k by the electrode i in a given subject, whose gamma activity has a maximum at a certain frequency f_0. First, this signal is passed through a narrow-band filter ($f_0 \pm 3$Hz), so that a complex signal $z_{i,k}(t)$ retaining only the Fourier components in this frequency range is produced. The phase $\varphi_i(t)$ is then determined as $\varphi_{i,k}(t) = \arg z_{i,k}(t)$. The "phase-locking value" $\Theta_{ij}(t)$ between signals, recorded by electrodes i and j at time t (measured starting from the moment of stimulus application) and averaged over all N trials in a given set, is defined as

$$\Theta_{ij}(t) = \frac{1}{N}\left|\sum_{k=1}^{N} \exp\left[i\left(\varphi_{i,k}(t) - \varphi_{j,k}(t)\right)\right]\right|. \tag{12.12}$$

Since we are interested in detecting the phase locking induced by the stimulus and there may have been some phase locking between the considered two signals even before its application, a normalization procedure is further applied. The values of $\Theta_{ij}(t)$ within a baseline interval of 500 ms preceding the stimulus are determined for each trial. Using this data, the time average μ_{ij} of $\Theta_{ij}(t)$ over the baseline interval and the statistical dispersion σ_{ij} of $\Theta_{ij}(t)$ over this baseline time and over all trials are computed. The normalized phase synchrony parameter $Q_{ij}(t)$ is determined as

$$Q_{ij}(t) = \frac{1}{\sigma_{ij}}\left(\Theta_{ij}(t) - \mu_{ij}\right). \tag{12.13}$$

To characterize the overall phase synchrony in a particular set of experiments, the values of $Q_{ij}(t)$ were further averaged over all electrode pairs and all subjects.

Figure 12.7 shows this averaged phase synchrony parameter as a function of time for the two sets of trials which corresponded to the perception (thick line) and non-perception (thin line) conditions. For comparison, the dashed line shows the same parameter which was computed using shuffled (randomized) data. Under the perception conditions, the phase synchrony was significantly increased at approximately 230 ms after stimulus presentation, when the image was recognized as a human face. This increase was followed at 500 ms by an interval of *active desynchronization*, when the normalized phase synchrony parameter was negative and the degree of synchronization was lower than the base level. Finally, the synchronization again increased at about the reaction time, when the subjects were pressing the key. Under non-perception conditions (i.e., when the image was turn upside-down), the initial synchronization and subsequent desynchronization episodes were absent, and only the final synchronization increase, accompanying the motor reaction, was observed.

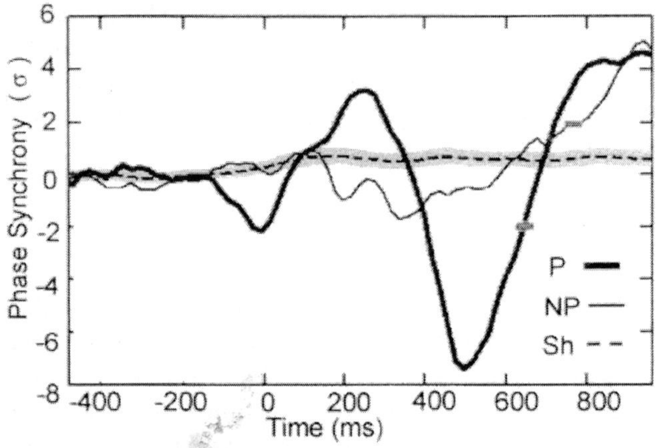

Fig. 12.7 Time courses of phase synchrony under perception (thick line) and non-perception (thin line) conditions. The dashed line shows the same property computed with the randomized (shuffled) experimental data. Adapted from [Rodriguez *et al.* (1999)].

Detailed spatiotemporal information is provided by the regional distribution of gamma activity and phase synchrony over the scalp (Fig. 12.8). The gamma activity, which is displayed by gray-color coding, is obtained by computing the total spectral density of the local signal in a frequency range from 34 to 40 Hz. Synchrony between pairs of electrodes is indicated by lines, which are displayed only if the phase synchrony is significantly changed with respect to those yielded by the randomized data.

To draw the lines, indicating phase synchrony or desynchronization in Fig. 12.8, the following statistical procedure was employed. Time was divided into several equal windows with a width of 180 ms. For each window T_l, the mean phase synchrony $Q_{ij}(T_l)$ between electrodes i and j was computed by averaging over the entire time window. To enhance synchronization changes between the windows, this mean phase synchrony was compared with its value $Q_{ij}(T_{l-1})$ for the preceding window T_{l-1} and the differences $\Delta Q_{ij}(T_l) = Q_{ij}(T_l) - Q_{ij}(T_{l-1})$ were computed for all pairs of electrodes. To estimate the statistical significance of such differences, they were also computed by using randomized (shuffled) data. Only if the measured value of $\Delta Q_{ij}(T_l)$ was significantly different from that predicted by the randomization, a line connecting electrodes i and j was drawn in the snapshot in Fig. 12.8 corresponding to the respective time window. The drawn line is black, if the phase synchrony is increased, and gray, if it is decreased within the considered time interval.

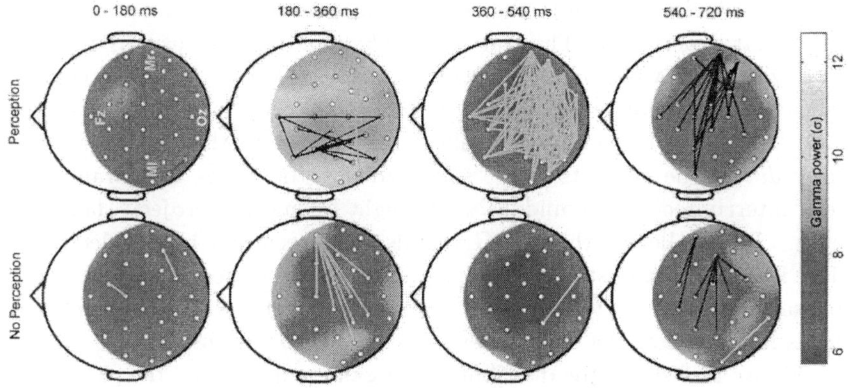

Fig. 12.8 Scalp distributions of gamma power and phase synchrony for four subsequent time windows following the stimulus presentation. Black–and–white reproduction of the original color figure. From [Rodriguez *et al.* (1999)].

The pattern of gamma activity was spatially homogeneous and was similar between perception and non-perception conditions, differing only in its amplitude. In contrast to this, strong differences were found in the patterns of phase synchrony corresponding to perception and non-perception conditions. When the image was not recognized as a face (non-perception), only little synchronization was observed. However, perception of a face was accompanied (180–360 ms) by the development of synchronization between different parts of the brain. This was followed by a massive desynchronization episode (360–540 ms). The second synchrony increase (540–720 ms) was linked with the motor response and was predominant between the right and the central regions. Only during this last stage, the synchrony patterns under perception and non-perception conditions were similar, because the subject had to press one of the keys in any case. Thus, different brain processes are clearly distinguishable in terms of the respective synchrony patterns.

Subsequent investigations have shown [Lutz *et al.* (2002)] that there are also strong correlations between the mental states, consciously experienced and described by the subjects, and the patterns of synchronization seen in the EEG recordings. In these experiments, the subjects were presented with a certain three-dimensional optical illusion and asked to press a button as soon as they have noticed it. At the same time, they had to report verbally their mental state. The verbal reports, produced by different subjects, were later classified according to the degree of preparation felt by a subject in a particular experiment. The state of *steady readiness* was usually reported by saying that a person feels "ready," "present," "here," "well-prepared" when the image appeared on the screen; the subjects responded in this case "immediately" and "decidedly." In contrast to this, in the state of *unreadiness* the subjects were "surprised" by the image appearance and felt "interrupted" in the middle of a thought (memories, projects, fantasies, etc.). In parallel to this, EEG signals were recorded by 62 electrodes. The EEG data was statistically processed, as described above, to detect phase synchrony patterns. Four different subjects, with the number of trials ranging from 200 to 350 per subject, were investigated.

Figure 12.9 shows the distributions of gamma power and the patterns of phase synchrony recorded in these experiments. When a subject reported that he was prepared and that he immediately saw the illusion (upper row in Fig. 12.9), a frontal pattern of synchrony gradually emerged several seconds before the stimulus. This activity was still present during the perception and motor response. In contrast, when the subject was unprepared and surprised by the arrival of the stimulus (bottom row in Fig. 12.9), there was no stable pattern in the gamma band before the stimulation. Only after the stimulation and in discontinuity with the prestimulation activity, patterns of synchrony emerged. The effect of surprise was associated with a special pattern in the neural response, which combined phase scattering (white lines) and an increase in synchrony (black lines). The final motor response was accompanied by patterns of synchrony that were similar to those seen during preparation, but delayed by 300 ms.

Above, we have presented several examples of synchronization phenomena in the brain. The concept of synchronization plays an important role in modern theoretical constructions, intended to explain the principles of brain operation. It has been suggested that each definite mental state, conceived as a "moment," should correspond to a resonant cell assembly, i.e.

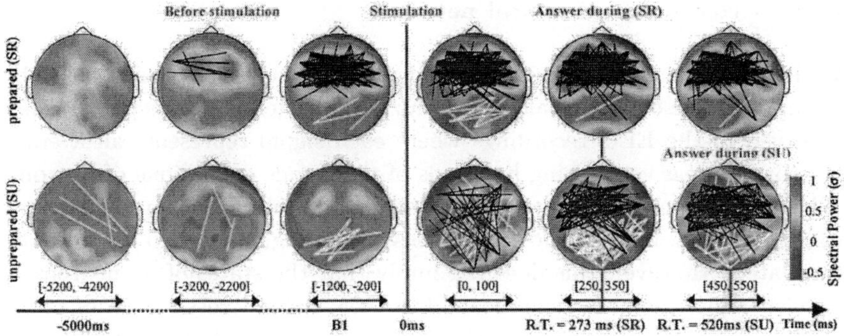

Fig. 12.9 Gamma power distributions and patterns of synchrony in the course of time for the states of "steady readiness" (prepared, upper row) and of "spontaneous unreadiness" (unprepared, bottom row). Black–and–white reproduction of the original color figure. From [Lutz *et al.* (2002)].

to a transient synchronization pattern in a large group of neurons which are broadly distributed in the brain [Varela (1995)]. A transition from one distinct experienced moment to another goes through an active desynchronization episode. Thus, the mental process is organized as a sequence of synchronization flashes alternating with brief desynchronization periods [Rudrauf *et al.* (2003)].

The analysis of human consciousness has lead to the *Dynamic Core Hypothesis* [Tononi and Edelman (1998)]. The dynamic core is a large cluster of neuronal groups that together constitute, on a time scale of hundreds of milliseconds, a unified neural process of high complexity. The neuronal groups in this functional cluster are interacting much strongly among themselves than with the rest of the brain. The neurons in the dynamical core are not directly involved in the automated brain routines, which go on unconsciously. Instead, their primary role is to integrate various perceptions and mental experiences, which is essential for the brain as a whole.

Some neural diseases, such as epilepsy and the Parkinson disease, involve pathological synchronization (see [Tass (1999); Milton and Jung (2003)]). The treatment of these illnesses require termination of persistent disfunctional synchronies which can be achieved by a purposeful phase resetting in some neural oscillating groups. Here, correct anatomical localization of synchronization processes in three dimensions is required. This can be done by the especially developed method of synchronization tomography based on magnetoencephalography [Tass *et al.* (2003)].

12.3 Cross-coupled neural networks

We have seen in the previous section that brain activity is characterized by the appearance of large-scale synchronization patterns. These patterns are observed in the EEG recordings where each signal represents an average over brain areas comprising hundreds of thousands or millions of neurons. When synchronies between the EEG signals originating from different areas are present, neural processes in these brain regions should be mutually correlated. However, this does not imply that the states of all neurons in a given region are then identical. In other words, cross-synchronization cannot simply result from synchronization of the individual states of neurons. If this were the case, large brain regions would have acted as a single neuron—which is not compatible with their complex functions.

Apparently, the elementary entities in such synchronization phenomena are not single neurons, but their relatively large groups. These groups should represent separate neural networks with high internal organization, able to perform complex functions of information processing. In addition to connections inside a network, different networks are connected, or *cross-coupled*. This additional coupling may lead to correlations and synchronization in the activity states of individual networks in their ensemble.

Synchronization phenomena in ensembles of networks are understood much less than synchronization and clustering in populations of individual neurons. Below, we present a study of such phenomena in a neural ensemble with a simple organization [Zanette and Mikhailov (1998b)]. We assume here that the ensemble consists of many layers, each occupied by an identical copy of the same network. The connections are spanning all layers, so that each network is cross-coupled in the same way with all other networks in the ensemble. This means that the considered system essentially represents a globally coupled ensemble of identical neural networks.

Additional simplification consists in the use of the McCulloch–Pitts model of neural networks. To introduce this model, we consider the integrate-and-fire neurons discussed in Section 12.1 and construct a variant of the mean-field approximation for such dynamical elements. Instead of specifying individual instantaneous states of neurons, their activity will be characterized by a variable $v_i(t)$ that is the average spiking rate of neuron i at time t. To obtain the evolution equation for such variables, we start from the equations (12.8), i.e.

$$\dot{\varphi}_i = (1 - \cos\varphi_i) + (1 + \cos\varphi_i)\left(r + \sum_{j=1}^{N} s_{ij}\delta\left(\varphi_j - \pi\right)\right). \tag{12.14}$$

Supposing that a given neuron receives a large number of non-correlated weak inputs, we can replace $\sum_{j=1}^{N} s_{ij}\delta\left(\varphi_j - \pi\right)$ by its mean $q_i = \sum_{j=1}^{N} s_{ij}v_j$, so that the evolution equation for the neuron i becomes

$$\dot{\varphi}_i = (1 - \cos\varphi_i) + (1 + \cos\varphi_i)\left(r + q_i\right). \tag{12.15}$$

This equation describes a single integrate-and-fire neuron with a modified control parameter $\widetilde{r}_i = r + q_i$. Assuming that the mean q_i changes only slowly with time, we treat it as a constant in this equation. If $\widetilde{r}_i < 0$ the neuron resides in its rest state and no spikes are generated ($v_i = 0$). When $\widetilde{r}_i > 0$, the neuron performs oscillations (see Section 12.1) at the frequency $\omega_i = 2\sqrt{\widetilde{r}_i}$ and the spiking rate $v_i = (2\pi)^{-1}\omega_i$. Therefore, the spiking rate $v_i(t)$ of neuron i at time t is given by

$$v_i(t) = H\left(r + \sum_{j=1}^{N} s_{ij}v_j(t)\right), \tag{12.16}$$

where the function $H(x)$ is defined as

$$H(x) = \begin{cases} 0 & \text{for} \quad x \leq 0, \\ (1/\pi)\sqrt{x} & \text{for} \quad x > 0. \end{cases} \tag{12.17}$$

So far, no propagation delays were taken into account. In reality, the propagation velocity of electrical pulses is finite and some delays are always present. Suppose for simplicity that the delays are the same for any pair of neurons and are equal to τ. If we take such delays into account and measure time in units of the delay τ, equation (12.16) should be replaced by a map

$$v_i(t+1) = H\left(r + \sum_{j=1}^{N} s_{ij}v_j(t)\right). \tag{12.18}$$

Note that the particular form (12.17) of the function $H(x)$ results from the reduced integrate-and-fire model (12.8) which is valid only under certain conditions. Generally, the firing rate should undergo saturation for strong applied signals. Therefore, we can phenomenologically choose $H(x)$ as a sharp step function,

$$H(x) = \begin{cases} 0 & \text{for} \quad x \leq 0, \\ 1 & \text{for} \quad x > 0 \end{cases} \tag{12.19}$$

or use a smooth step function defined as

$$H(x) = [1 + \tanh(\beta x)]/2 \tag{12.20}$$

with the parameter β specifying the steepness of the step.

The McCulloch–Pitts model, given by a map (12.18) with a step function $H(x)$, was introduced a long time ago [McCulloch and Pitts (1943)]. For many years, it remained a favorite system for mathematical investigations of neural networks. If the connections are symmetric ($s_{ij} = s_{ji}$), such networks evolve to a steady state characterized by constant firing rates v_i. The system can possess a large number of such steady states, so that multistability is observed. This property of the McCulloch–Pitts model is used to design systems with associative memory (see, e.g., [Mikhailov (1994)]).

On the other hand, these networks can also show persistent regular or chaotic oscillations if the connections between neurons are asymmetric ($s_{ij} \neq s_{ji}$). It is known that such chaotic networks with balanced activatory and inhibitory connections can show rapid response to external stimuli [van Vreeswijk and Sompolinsky (1996)]. Therefore, it is interesting to see what would be the synchronization behavior in an ensemble of such cross-coupled networks.

Following [Zanette and Mikhailov (1998b)], we consider ensembles made of N identical neural networks (or layers), each consisting of K neurons. The collective dynamics of the ensemble is described by a coupled map

$$v_i^n(t+1) = (1-\varepsilon)\, H(q_i^n) + \varepsilon H\left(\sum_{m=1}^{N} q_i^m\right), \tag{12.21}$$

where

$$q_i^n = \sum_{j=1}^{K} s_{ij} v_j^n(t) \tag{12.22}$$

is the signal arriving at neuron i in network n at time t from all other neurons in the same network, s_{ij} are connection weights (the same for all networks, see the description below), and $H(x)$ is the step function (12.20).

The parameter r is chosen equal to zero, so that in absence of external signals a neuron is exactly at a transition from excitable to oscillatory dynamics.

Each of the connection weights s_{ij} between neurons is chosen at random with equal probability from the interval between -1 and 1. On the average, the numbers of activatory and inhibitory connections are equal and the network is balanced. The weights of forward and reverse connections are independently selected, and therefore the connection matrix is asymmetric ($s_{ij} \neq s_{ji}$).

The first of the terms on the right-hand side in equation (12.21) represents an individual response of a neuron to the signals received by it from all other neurons in its own network. The second term takes into account cross-coupling between different networks in the ensemble. It is obtained by summation of individual signals received by neurons occupying the same positions in all N networks. Note that, for a given neuron i, this additional signal is the same for all networks and therefore the cross-coupling is global.

The parameter ε specifies the strength of global coupling. When $\varepsilon = 0$, this coupling is absent and all networks in the ensemble are independent. On the other hand, at $\varepsilon = 1$ the first term vanishes and the activity states of all networks are identical, because they are determined only by the global signals. In the interval $0 < \varepsilon < 1$, the ensemble dynamics is governed by an interplay between internal coupling inside the networks and global coupling across them.

Numerical simulations of this model were performed [Zanette and Mikhailov (1998b)] for ensembles of $N = 100$ identical networks, each with $K = 50$ neurons. The smooth step function (12.20) with $\beta = 10$ was used. The connection weights remained fixed within the entire series of simulations with varying global coupling intensity. It was checked that, for the used patterns of connection weights, the network dynamics was always chaotic. The initial conditions for all neurons in all networks in each simulation have been randomly chosen.

Since subsequent states of all neurons in all networks were recorded, each simulation yielded a large volume of data. To detect patterns of synchrony in these data, several methods were employed.

An important property is the integral time-dependent activity

$$V_n(t) = \sum_{i=1}^{K} v_i^n(t) \tag{12.23}$$

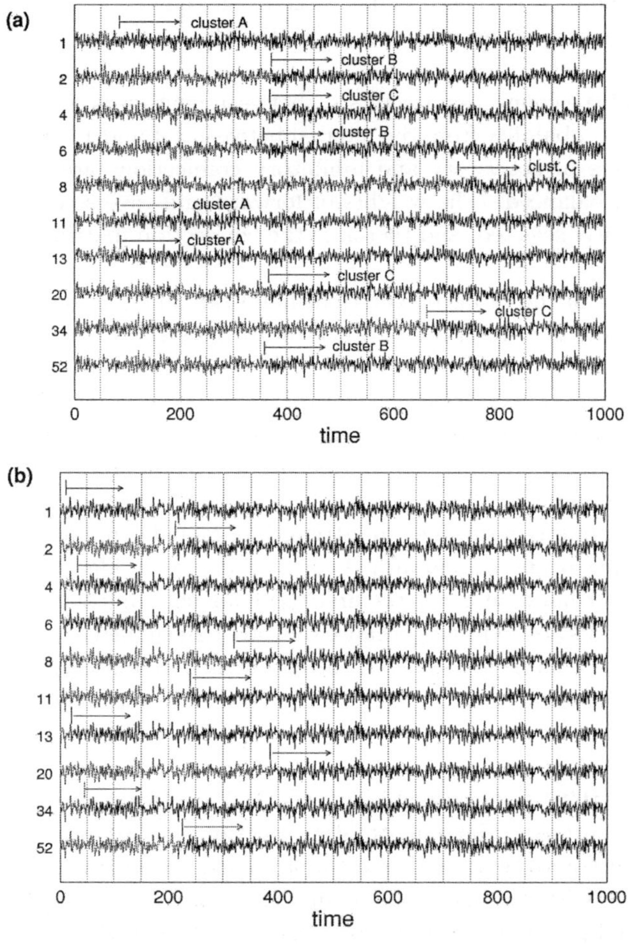

Fig. 12.10 Time-dependent integral activities of ten arbitrarily selected networks in an ensemble of 100 networks for two different intensities of cross-network coupling (a) $\varepsilon = 0.35$ and (b) $\varepsilon = 0.5$. Synchronization of each signal begins at the corresponding bar. From [Zanette and Mikhailov (1998b)].

which can be computed for each network $n = 1, \ldots, N$ in the ensemble. Note that such integral activities, averaged over a whole network, are to a certain extent similar to the EEG signals which also represent averages over large neuronal populations.

Figure 12.10 shows temporal activity patterns in ten arbitrarily chosen networks in the ensemble at two different strengths of cross-coupling. When $\varepsilon = 0.35$ (Fig. 12.10a), the ensemble becomes divided into three synchronous clusters (A, B and C). The integral signals $V_n(t)$ of networks belonging to the same cluster are identical, and the dynamics remains chaotic. For a higher strength $\varepsilon = 0.5$ of cross-coupling (Fig. 12.10b), the integral signals generated by all networks in the ensemble eventually become identical.

The degree of synchronization in the collective ensemble dynamics can be characterized by the dispersion $D(t)$ of activity pattern of individual networks at time t, which is defined as

$$D(t) = \frac{1}{N} \sum_{n=1}^{N} \sum_{i=1}^{K} \left[v_i^n(t) - \overline{v}_i(t)\right]^2 \tag{12.24}$$

where

$$\overline{v}_i(t) = \frac{1}{N} \sum_{n=1}^{N} v_i^n(t) \tag{12.25}$$

is the ensemble-averaged firing rate of a neuron occupying position i in all networks in the ensemble.

The time dependence of the dispersion $D(t)$ at $\varepsilon = 0.5$ is displayed in Fig. 12.11. As evidenced by Fig. 12.10b, synchronization begins with the formation of a coherent nucleus consisting of a few networks. This nucleus grows by aggregation of further networks which become entrained. While some networks remain nonentrained, the dispersion is still relatively large, though it gradually decreases with time. When the last network has approached the coherent cluster, exponential decrease of $D(t)$ is observed. Thus, in the final synchronous regime the states of all neurons occupying the same positions in the networks become identical.

To analyze dynamical clustering in this system, pair distances

$$d_{nm} = \left[\sum_{i=1}^{K} \left(v_i^n - v_i^m\right)^2\right]^{1/2} \tag{12.26}$$

between instantaneous activity patterns of different networks m and n were computed. Figure 12.12 shows normalized histograms of such pair distances for several strengths of coupling between the networks.

When the coupling is weak [Fig. 12.12(a)], the histogram has a single smooth maximum at a typical pair distance between the activity pat-

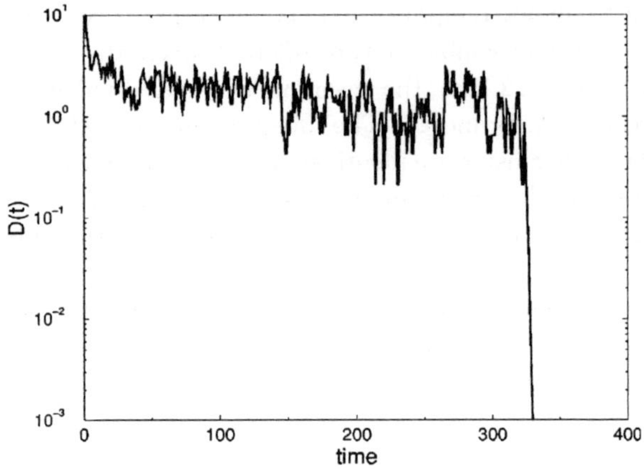

Fig. 12.11 Dispersion $D(t)$ of activity patterns as a function of time for $\varepsilon = 0.5$. From [Zanette and Mikhailov (1998b)].

terns in effectively independent networks. At a larger coupling intensity (Fig. 12.12b), some networks already have exactly the same activity patterns, so that the pair distance between them is zero (as revealed by the presence of a peak at $d = 0$ in the histogram in Fig. 12.12b). However, the activity patterns of most of the networks are still asynchronous. If global coupling is further increased, the number of identical pairs grows (Fig. 12.12c). Now, the histogram exhibits several peaks that correspond to particular network clusters. Nonetheless, there also some networks which do not belong to any synchronous cluster. After a slight increase in the coupling strength, all networks in the ensemble belong to the three synchronous clusters (Fig. 12.12d). Continuing to increase the coupling strength, a transition to full synchronization of activity patterns of networks is found at $\varepsilon \approx 0.4$.

Note that the regime of dynamical clustering and the final transition to full synchronization in the considered ensemble of cross-coupled neural networks are similar to the respective phenomena in a population of globally coupled chaotic Rössler oscillators, earlier described in Section 8.4.

The simulations and the statistical analysis were repeated for different random choices of connection weights in the networks and have revealed the same sequence of changes leading to clustering and full synchronization. However, the critical strengths of cross-coupling necessary to impose synchronization were found to depend on a particular choice of the connection

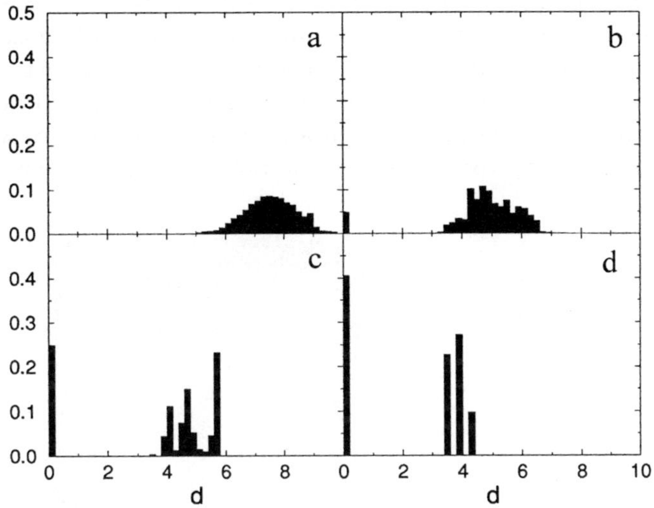

Fig. 12.12 Normalized histograms of distributions over pair distances between networks at different intensities of cross-network coupling (a) $\varepsilon = 0.15$, (b) $\varepsilon = 0.28$, (c) $\varepsilon = 0.34$, and (d) $\varepsilon = 0.35$. From [Zanette and Mikhailov (1998b)].

weights. Moreover, essentially the same results were obtained when ensembles consisting of larger networks of 100 neurons were considered and when other step-like functions $H(x)$ were used [Zanette and Mikhailov (1998b)].

In the model (12.21) we assumed that each neuron in each network in the ensemble is participating in global cross-network interactions. However, similar results are also obtained if only a randomly chosen fraction of all neurons is involved in such interactions. Suppose that the collective dynamics of an ensemble is described by the equations

$$v_i^n(t+1) = (1 - \varepsilon\xi_i)\, H\left(q_i^n\right) + \varepsilon\xi_i H\left(\sum_{m=1}^{N} q_i^m\right) \tag{12.27}$$

where ξ_i are random variables, taking value 1 or 0 with probability p or $1 - p$, respectively. Here, only those neurons i, which have $\xi_i = 1$, are participating in global cross-network interactions. This means that only a fraction pK of all neurons in any given network are sensitive to global signals.

Numerical simulations of the modified model (12.27) are summarized in

Fig. 12.13. Full synchronization of all networks in the ensemble is found inside the dark-gray region. The regimes of partial synchronization and dynamical clustering (defined by the presence of at least two networks with identical activity patterns) are observed inside the light-gray region in this figure. Remarkably, partial synchronization can still be observed when only about 10% of global cross-network connections are left in the ensemble.

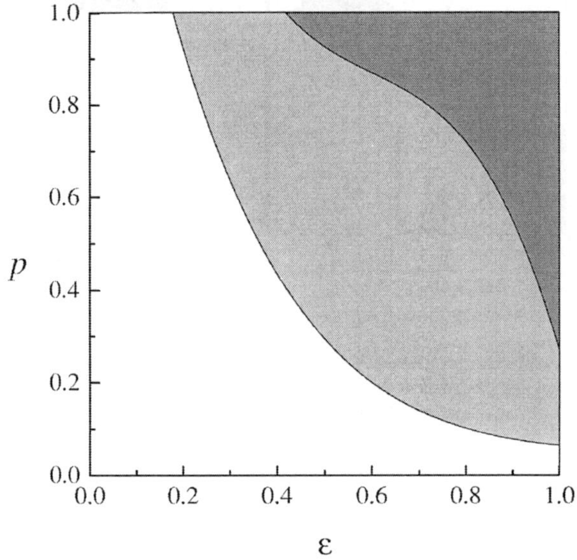

Fig. 12.13 Synchronization diagram. From [Zanette and Mikhailov (1998b)].

Thus, our analysis indicates that both full synchronization and dynamical clustering are possible not only at the level of individual oscillators, but also for systems where each element already represents a whole neural network. Though this analysis has been performed only for a greatly simplified McCullogh–Pitts model, similar results would probably be also found for the cross-coupled networks formed by integrate-and-fire neurons. While not intended to explain any particular neuronal behavior, our study shows that the formation of synchronous clusters is a generic feature of neuronal assemblies, as it is assumed in modern neurophysiology.

Bibliography

Abarbanel, H. D. I., Rabinovich, M. I., Selverston, A., Bazhenov, M. V., Huerta, R., Sushchik, M. M. and Rubchinskii, L. L. (1996), Synchronization in neural networks. *Phys. Usp.* **39**, 4, pp. 337–362.

Abramson, G. (2000). Long transients and cluster size in globally coupled maps, *Europhys. Lett.* **52**, 6, pp. 615–620.

Abramson, G. and Zanette, D. H. (1998). Globally coupled maps with asynchronous updating, *Phys. Rev. E* **58**, 4, pp. 4454–4460.

Alberts, B., Bray, D., Lewis, J., Raff, M., Roberts, K. and Watson, J. D. (1994). The Molecular Biology of the Cell, Garland, New York, 653 p.

Anishchenko, V. S., Vadivasova, T. E., Postnov, D. E. and Safonova, M. A. (1992). Synchronization of chaos, *Int. J. Bif. Chaos* **2**, 3, pp. 633–644.

Ariaratnam, J. T. and Strogatz, S. H. (2001). Phase diagram for the Winfree model of coupled nonlinear oscillators, *Phys. Rev. Lett.* **86**, 19, pp. 4278–4281.

Aronson, D. G., Ermentrout, G. B. and Kopell, N. (1990). Amplitude response of coupled oscillators, *Physica D* **41**, 3, pp. 403–449.

Balmforth, N. J., Jacobson, A. and Provenzale, A. (1999). Synchronized family dynamics in globally coupled maps, *Chaos* **9**, 3, pp. 738–754.

Bar-Eli K. (1985). On the stability of coupled chemical oscillators, *Physica D* **14**, 2, pp. 242–252.

Barahona, M. and Pecora, L. M. (2002). Synchronization in small-world systems, *Phys. Rev. Lett.* **89**, 5, 054101.

Battogtokh, D. and Kuramoto, Y. (2000). Turbulence of nonlocally coupled oscillators in the Benjamin-Feir stable regime, *Phys. Rev. E* **61**, 3, pp. 3227–3229.

Battogtokh, D., Preusser, A. and Mikhailov, A. S. (1997). Controlling turbulence in the complex Ginzburg-Landau equation II. Two-dimensional systems, *Physica D*, **106**, 3–4, pp. 327–362.

Belair, J. (1986). Periodic pulsatile stimulation of a nonlinear oscillator, *J. Math. Biol.* **24**, 2, pp. 217–232.

Belykh, V. N., Belykh, I. V. and Hasler, M. (2000). Hierarchy and stability of partially synchronous oscillations of diffusively coupled dynamical systems,

Phys. Rev. E **62**, 5, pp. 6332–6345.

Belykh, I., Belykh, V., Nevidin, K. and Hasler, M. (2003). Persistent clusters in lattices of coupled nonidentical chaotic systems, *Chaos* **13**, 1, pp. 165–178.

Bertram, M., Beta, C., Pollmann, M., Mikhailov, A. S., Rotermund, H. H. and Ertl, G. (2003). Pattern formation on the edge of chaos: experiments with CO oxidation on a Pt(110) surface under global delayed feedback, *Phys. Rev. E*, **67**, 3, 036208.

Bertram, M. and Mikhailov, A. S. (2003). Pattern formation on the edge of chaos: mathematical modeling of CO oxidation on a Pt(110) surface under global delayed feedback, *Phys. Rev. E*, **67**, 3, 0036207.

Beta, C., Bertram, M., Mikhailov, A. S., Rotermund, H. H. and Ertl, G. (2003). Controlling turbulence in a surface chemical reaction by time-delay autosynchronization, *Phys. Rev. E*, **67**, 4, 046224.

Blasius, B., Huppert, A. and Stone, L. (1999). Complex dynamics and phase synchronization in spatially extended ecological systems, *Nature* **399**, pp. 355–358.

Blasius, B., Montbrió, E. and Kurths, J. (2003). Anomalous phase synchronization in populations of nonidentical oscillators, *Phys. Rev. E* **67**, 035204R.

Blumenfeld, L. A. and Tikhonov, A. N. (1994). Biophysical Thermodynamics of Intracellular Processes: Molecular Machines of the Living Cell, Springer, Berlin, 175 p.

Boccaletti, S., Grebogi, C., Lai, Y. C., Mancini, H., and Maza D. (2000). The control of chaos: theory and applications, *Phys. Rep.* **329**, 3, pp. 103–197.

Boccaletti, S., Kurths, J., Osipov, G., Valladares, D. L. and Zhou, C. S. (2002). The synchronization of chaotic systems, *Phys. Rep.* **366**, 1–2, pp. 1–101.

Buck, J. and Buck, E. (1976). Synchronous fireflies, *Sci. Am.* **234**, 5, pp. 74–85.

Cencini, M., Falcioni, M., Vergni, D. and Vulpiani, A. (1999). Macroscopic chaos in globally coupled maps, *Physica D* **130**, 1–2, pp. 58–72.

Chabanol, M. L., Hakim, V. and Rappel, W. J. (1997). Collective chaos and noise in the globally coupled complex Ginzburg-Landau equation, *Physica D* **103**, 1–4, pp. 273–293.

Chaté, H. and Manneville, P. (1992). Collective behaviors in spatially extended systems with local interactions and synchronous updating, *Prog. Theor. Phys.* **87**, 1, pp. 1–60.

Chen, Y., Rangarajan, G. and Ding, M. (2003). General stability analysis of synchronized dynamics in coupled systems, *Phys. Rev. E* **67**, 026209.

Choi, M. Y., Kim, H. J., Kim, D. and Hong, H. (2000). Synchronization in a system of globally coupled oscillators with time delay, *Phys. Rev. E* **61**, 1, pp. 371–381.

Cosenza, M. G. and González, J. (1998). Synchronization and collective behavior in globally coupled logarithmic maps, *Prog. Theor. Phys.* **100**, 1, pp. 21–38.

Cosenza, M. G. and Parravano, A. (2003). Dynamics of coupling functions in globally coupled maps: Size, periodicity, and stability of clusters, *Phys. Rev. E* **64**, 036224.

Coullet, P. and Emilsson, K. (1992a). Strong resonances of spatially distributed oscillators. A laboratory to study patterns and defects, *Physica D* **61**, 1–4, pp. 119–131.

Coullet, P. and Emilsson, K. (1992b). Pattern formation in the strong resonant forcing of spatially distributed oscillators, *Physica A* **188**, 1–3, pp. 190–200.

Crisanti, A., Falcioni, M. and Vulpiani, A. (1996). Broken ergodicity and glassy behavior in a deterministic chaotic map, *Phys. Rev. Lett.* **76**, 4, pp. 612–615.

Crutchfield, J., Farmer, D., Packard, N., Shaw, R., Jones, G. and Donnelly, R. J. (1980). Power spectral analysis of a dynamical system, *Phys. Lett. A* **76**, 1, pp. 1-4.

Daido, H. (1986). Discrete-time population dynamics of interacting self oscillators, *Prog. Theor. Phys.* **75**, 6, pp. 1460–1463.

Daido, H. (1987). Population dynamics of randomly interacting self-oscillators, *Prog. Theor. Phys* **77**, 3, pp. 622–634.

Daido, H. (1988). Lower critical dimension for populations of oscillators with randomly distributed frequencies: A renormalization-group analysis, *Phys. Rev. Lett.* **61**, 2, pp. 231–234.

Daido, H. (1989). Intrinsic fluctuation and its critical scaling in a class of populations of oscillators with distributed frequencies, *Progr. Theor. Phys.* **18**, 4, pp. 727–731.

Daido, H. (1992). Quasientrainment and slow relaxation in a popultion of oscillators with random and frustrated interactions, *Phys. Rev. Lett.* **68**, 7, pp. 1073–1076.

Daido, H. (1996). Multibranch entrainment and scaling in large populations of coupled oscillators, *Phys. Rev. Lett.* **77**, 7, pp. 1406–1409.

Daido, H. (1997). Strange waves in coupled-oscillator arrays: Mapping approach, *Phys. Rev. Lett.* **78**, 9, pp. 1683–1686.

Daido, H. (2000). Algebraic relaxation of an order parameter in randomply coupled limit-cycle oscillators, *Phys. Rev. E* **61**, 2, pp. 2145–2147.

Danø, S., Sørensen, P. G. and Hynne, F. (1999). Sustained oscillations in living cells, *Nature* **402**, 6759, pp. 320–322.

Davidsen, J. and Kapral, R. (2002). Phase synchronization and topological defects in inhomogeneous media, *Phys. Rev. E* **66**, 055202(R).

Domínguez, D. and Cerdeira, H. (1993). Order and turbulence in rf-driven Josephson junction series array, *Phys. Rev. Lett.* **71**, 20, pp. 3359–3362.

Earl, M. G. and Strogatz, S. H. (2003). Synchronization in oscillator networks with delayed coupling: A stability criterion, *Phys. Rev. E* **67**, 3, 036204, 4p.

Eckmann, J. P. and Ruelle, D. (1985). Ergodic theory of chaos and strange attractors, *Rev. Mod. Phys.* **57**, 3, pp. 617–656.

Ermentrout, G. B. (1990). Oscillator death in populations of "all to all" coupled nonlinear oscillators, *Physica D* **41**, 2, pp. 219–231.

Ermentrout, G. B. and Kopell, N. (1984). Frequency plateaus in a chain of weakly coupled oscillators, *SIAM J. Math. Anal.* **15**, 2, pp. 215–237.

Ernst, U., Pawelzik, K., and Geisel, T. (1995). Synchronization induced by temporal delays in pulse-coupled oscillators, *Phys. Rev. Lett.* **74**, 9, pp. 1570–1573.

Falcke, M. and Engel H. (1994). Pattern formation during the CO oxidation on Pt(110) surfaces under global coupling, *J. Chem. Phys.* **101**, 7, pp. 6225-6263.

Fink, K. S., Johnson, G., Carroll, T., Mar, D. and Pecora, L. (2000). Three coupled oscillators as a universal probe of synchronization stability in coupled oscillators arrays, *Phys. Rev. E* **61**, 5, pp. 5080–5090.

Fujisaka, H. and Yamada, T. (1983). Stability theory of synchronized motion in coupled-oscillator systems. I, *Prog. Theor. Phys.* **69**, 1, pp 32–47.

Fujisaka, H. and Yamada, T. (1986). Stability theory of synchronized motion in coupled-oscillator systems. IV. Instability of synchronized chaos and new intermittency, *Progr. Theor. Phys.* **75**, 5, pp. 1087–1104.

Furusawa, Ch. and Kaneko, K. (2000). Origin of complexity in multicellular organisms, *Phys. Rev. Lett.* **84**, 26, pp. 6130–6133.

Furusawa, Ch. and Kaneko, K. (2001). Theory of robustness of irreversible differentiation in a stem cell system: chaos hypothesis, *J. Theor. Biol.* **209**, 4, pp. 395–416.

Gerstner, W. (1995). Time structure of the activity in neural network models, *Phys. Rev. E* **51**, 1, pp. 738–758.

Gerstner, W. (2000). Population dynamics of spiking neurons: fast transients, asynchronous states, and locking, *Neural Computation* **12**, 1, pp. 43–89.

Gerstner, W. (2001). What is different with spiking neurons?, in Plausible Neural Networks for Biological Modelling, eds. Masterbroek, H. and Vos, J. E., Kluwer, Dordrecht, pp. 23–8.

Ghosh, A. and Chance, B. (1964). Oscillations of glycolytic intermediates in yeast cells, *Biochem. Biophys. Res. Commun.* **16**, 2, pp. 174–181.

Ghosh, A. K., Chance, B. and Pye, E. K. (1971). Metabolic coupling and synchronization of NADH oscillations in yeast cell populations, *Arch. Biochem. Biophys.* **145**, 1, pp. 319–331.

Ginzburg, V. L. and Andryushin, E. A. (1994). Superconductivity, World Scientific, Singapore, 104p.

Glass, L. and Mackey, M. C. (1979). Simple model for phase locking of biological oscillators, *J. Math. Biol.* **7**, 4, pp. 339–352.

Goldberg, K., Newman, M. and Haynsworth E., in *Handbook of mathematical functions*, edited by M. Abramowitz and I. A. Stegun, Dover Publications, New York, 1970.

Golomb, D., Hansel, D., Shraiman, B. and Sompolinsky, H. (1992). Clustering in globally coupled phase oscillators, *Phys. Rev. A* **45**, 6, pp. 3516–3530.

Golomb, D. and Rinzel, J. (1994). Clustering in globally coupled inhibitory neurons, *Physica D* **72**, 3, pp. 259–282.

Gray, C. M., König, P., Engel, A. K. and Singer, W. (1989). Oscillatory responses in cat visual cortex exhibit inter-columnar synchronization which reflects global stimulus properties, *Nature* **338**, 6213, pp. 334–337.

Gruler, H. and Müller-Enoch, D. (1991). Slaving the cytochrome P-450 dependent monooxygenase system by periodically applied light-pulses, *Eur. Biophys. J.* **19**, 4, pp. 217–219.

Guckenheimer, J. and Holmes, P. J. (1983). Nonlinear Oscillations, Dynamical

Systems, and Bifurcations of Vector Fields, Springer-Verlag, Berlin, 480p.

Häberle, W., Gruler, H., Dutkowski, Ph. and Müller-Enoch, D. (1990). Light-induced activation and synchronization of the cytochrome P-450 dependent monooxygenase system, *Z. Naturforsch. C* **45**, 3–4, pp. 273–279.

Haim, D., Lev, O., Pismen, L. M. and Sheintuch, M. (1992). Modeling periodic and chaotic dynamics in anodic nickel dissolution, *J. Phys. Chem.* **96**, 6, pp. 2676–2681.

Hakim, V. and Rappel W. J. (1992). Dynamics of globally coupled complex Ginzburg-Landau equation, *Phys. Rev. A* **46**, 12, pp. R7347–R7350.

Hansel, D., Mato, G. and Meunier, C. (1993). Clustering and slow switching in globally coupled phase oscillators, *Phys. Rev. E* **48**, 5, pp. 3470–3477.

Harwell, L. H., Hopfield, J. J., Leibler, S., and Murray, A. W. (1999). From molecular to modular cell biology, *Nature* **402**, supplement, pp. C47–C52.

Heagy, J. F., Carroll, T. L. and Pecora, L. M. (1994a). Synchronous chaos in coupled oscillator systems, *Phys. Rev. E* **50**, 3, pp. 1874–1885.

Heagy, J. F., Platt, N. and Hammel, S. M. (1994b). Characterization of on-off intermittency, *Phys. Rev. E* **49**, 2, pp. 1140–1150.

Heagy, J. F., Pecora, L. M. and Carroll, T. L. (1995). Short wavelength bifurcations and size instabilities in coupled oscillator systems, *Phys. Rev. Lett.* **74**, 21, pp. 4185-4188.

Hess, B. (1973). Organization of glycolysis: oscillatory and stationary control, *Symp. Soc. Exp. Biol.* **27**, 1, pp. 105–131.

Hess, B. (1997). Periodic patterns in biochemical reactions, *Quart. Rev. Biophys.* **30**, 2, pp. 121–176.

Hess, B. and Markus, M. (1985). The diversity of biochemical time patterns, *Ber. Bunsenges. Phys. Chem.* **89**, 6, pp. 642–651.

Hess, B. and Mikhailov, A. S. (1994). Self-organization in living cells, *Science* **264**, 5156, pp. 223–224.

Hess, B. and Mikhailov, A. S. (1995). Microscopic self-organization in living cells: a study of time matching, *J. Theor. Biol.* **176**, 1, pp. 181–184.

Hess, B. and Mikhailov, A. S. (1996). Transition from molecular chaos to coherent spiking of enzymic reactions in small spatial volumes, *Biophys. Chem.* **58**, 3, pp. 365–368.

Hodgkin, A. L. (1948). The local electric changes associated with repetitive action in a non-medulated axon, *J. Physiol.* **107**, 2, pp. 165–181.

Hoppensteadt, F. C. and Izhikevich, E. M. (1997). Weakly Connected Neural Networks, Springer, New York, 400p.

Hoppensteadt, F.C. and Izhikevich E.M. (1998). Thalamo-cortical interactions modeled by weakly connected oscillators: Could the brain use FM radio principles? *Biosystems* **48**, 1–3, pp. 85–94.

Izhikevich, E. M. (1998). Phase models with explicit time delays, *Phys. Rev. E* **58**, 1, pp. 905–908.

Izhikevich, E. M. (2000). Neural excitability, spiking, and bursting, *Int. J. Bif. Chaos* **10**, 6, pp. 1171–1266.

Jakubith, S., Rotermund, H. H., Engel, W, von Oertzen, A. and Ertl. G. (1990). Spatiotemporal concentration patterns in a surface reaction: propagating and standing waves, rotating spirals, and turbulence, *Phys. Rev. Lett.* **65**, 24, pp. 3013–3016.

Kaneko, K. (1984). Period-doubling of kink-antikink patterns, quasiperiodicity in antiferro-like structures and spatial intermittency in coupled logistic lattice, *Prog. Theor. Phys.* **72**, 3, pp. 480–486.

Kaneko, K. (1985). Spatiotemporal intermittency in coupled map lattices, *Prog. Theor. Phys.* **74**, 5, pp. 1033–1044.

Kaneko, K. (1989). Chaotic but regular posi-nega switch among coded attractors by cluster-size variation, *Phys. Rev. Lett.* **63**, 3, pp. 219–223.

Kaneko, K. (1990a). Clustering, coding, switching, hierarchical ordering, and control in a network of chaotic elements, *Physica D* **41**, 2, pp. 137–172.

Kaneko, K. (1990b). Globally coupled chaos violates the law of large numbers but not the central-limit theorem, *Phys. Rev. Lett.* **65**, 12, pp. 1391–1394.

Kaneko, K. (1991a). Globally coupled circle maps, *Physica D* **54**, 1, pp. 5–19.

Kaneko, K. (1991b). Partition complexity in a network of chaotic elements, *J. Phys. A: Math. Gen.* **24**, 9, pp. 2107–2119.

Kaneko, K. (1992). Mean field fluctuation of a network of chaotic elements, *Phys. D* **55**, 3–4, pp. 368–384.

Kaneko, K. (1994a). Relevance of dynamical clustering to biological networks, *Physica D* **75**, 1-3, pp. 55-73.

Kaneko, K. (1994b). Information cascade with marginal stability in a network of chaotic elements, *Physica D* **77**, 4, pp. 456–472.

Kaneko, K. (1995). Remarks on the mean field dynamics of networks of chaotic elements, *Physica D* **86**, 2, pp. 158–170.

Kaneko, K. and Tsuda, I. (2000). Complex Systems: Chaos and Beyond, A Constructive Approach with Applications in Life Sciences, Springer Verlag, Berlin, 273p.

Kaneko, K. and Yomo, T. (1994). Cell division, differentiation, and dynamic clustering, *Physica D* **75**, 1–3, pp. 89–102.

Kaneko, K. and Yomo, T. (1997). Isologous diversification: a theory of cell differentiation, *Bull. Math. Biol.***59**, 1, pp. 139–196.

Kaneko, K. and Yomo, T. (1999). Isologous diversification for robust development of cell society, *J. Theor. Biol.***199**, 3, pp. 243–256.

Keener, J. P., Hoppensteadt, F. C., and Rinzel, J. (1981). Integrate-and-fire models of nerve membrane response to oscillatory input, *SIAM J. Appl. Math.* **41**, 3, pp. 503–517.

Kim, M., Bertram, M., Pollmann, M., von Oertzen, A., Mikhailov, A. S., Rotermund, H. H. and Ertl, G. (2001). Controlling chemical turbulence by global delayed feedback: pattern formation in catalytic CO oxidation on Pt(110), *Science*, **292**, 5520, pp. 1357–1360.

Kiss, I. Z. and Hudson, J. L. (2003). Chaotic cluster itinerancy and hierarchical cluster trees in electrochemical experiments, *Chaos* **13**, 3, pp. 999–1009.

Kiss, I. Z., Wang, W. and Hudson, J. L. (2002a). Populations of coupled electrochemical oscillators, *Chaos* **12**, 1, pp. 252–263.

Kiss, I. Z., Zhai, Y. and Hudson, J. L. (2002b). Emerging coherence in a population of chemical oscillators, *Science* **296**, 5573, pp. 1676–1678.

Kiss, I. Z., Zhai, Y. and Hudson, J. L. (2002c). Collective dynamics of chaotic chemical oscillators and the law of large numbers, *Phys. Rev. Lett.* **88**, 23, 238301.

Knight, B. W. (1972). Dynamics of encoding in a population of neurons, *J. Gen. Physiol.* **59**, 6, pp. 734–751.

Krischer, K., Eiswirth, M. and Ertl, G. (1992). Oscillatory CO oxidation on Pt(110). Modeling of temporal self-organization, *J. Chem. Phys.* **96**, 12, pp. 9161–9172.

Kuang, Y. (1993). Delay Differential Equations with Applications to Population Dynamics, Academic Press, New York, 398p.

Kuramoto, Y. (1984). Chemical Oscillations, Waves, and Turbulence, Springer, Berlin, 156p.

Kuramoto, Y. (1991). Collective synchronization of pulse-coupled oscillators and excitable units, *Physica D* **50**, 1, pp. 15–30.

Kuramoto, Y. (1995). Scaling behavior of turbulent oscillators with nonlocal interaction, *Prog. Theor. Phys.* **94**, 3, pp. 321–330.

Kuramoto, Y. and Nakao, H. (1996). Origin of power-law spatial correlations in distributed oscillators and maps with nonlocal coupling, *Phys. Rev. Lett.* **76**, 23, pp. 4352–4355.

Kuramoto, Y. and Nakao, H. (1997). Power-law spatial correlations and the onset of individual motions in self-oscillatory media with non-local coupling, *Physica D* **103**, 1–4, pp. 294–313.

Latora, V. and Marchiori, M. (2001). Efficient behavior of small-world networks, *Phys. Rev. Lett.* **87**, 3, 198701.

Lee, K. J., Kwak, Y. and Lim, T. K. (1998). Phase jumps near a phase synchronization transition in systems of two coupled chaotic oscillators, *Phys. Rev. Lett.* **81**, 2, pp. 321–324.

Lerch, H.-Ph., Stange, P., Mikhailov, A. S. and Hess, B. (2002). Mutual synchronization of molecular turnover cycles in allosteric enzymes III. Intramolecular cooperativity, *J. Phys. Chem. B* **106**, 12, pp. 3237–3247.

Lima, D., Battogtokh, D., Mikhailov, A., Borkmans, P. and Dewel, G. (1998). Pattern selection in oscillatory media with global coupling, *Europhys. Lett.* **42**, 6, pp. 631–636.

Liu, Z., Lai Y. C. and Hoppensteadt, F.C. (2001). Phase clustering and transition to phase synchronization in a large number of coupled nonlinear oscillators, *Phys. Rev. E* **63**, 5, 055201(R).

Lorenz, E. N. (1963). Deterministic nonperiodic flow, *J. Atmos. Sci.* **20**, 2, pp. 130–141.

Lutz, A., Lachaux, J.-Ph., Martinerie, J., and Varela, F. J. (2002). Guiding the study of brain dynamics by using first-person data: synchrony patterns correlate with ongoing conscious states during a simple visual task, *Proc. Natl. Acad. Sci. U. S. A.* **99**, 3, pp. 1586–1591.

Mair, T. and Müller, S. C. (1996). Traveling NADH and proton waves during oscillatory glycolysis *in vitro*, *J. Biol. Chem.* **271**, 2, pp. 627–530.

Mandelbrot, B. (1982). The Fractal Geometry of Nature, Freeman, New York, 468p.

Manrubia, S. C., Bastolla, U. and Mikhailov, A. S. (2001). Replica symmetry breaking in dynamical glasses, *Europ. Phys. J. B* **23**, 4, pp. 497–506.

Manrubia, S. C. and Mikhailov, A. S. (1999). Mutual synchronization and clustering in randomly coupled chaotic dynamical networks, *Phys. Rev. E* **60**, 2, pp. 1579–1589.

Manrubia, S. C. and Mikhailov, A. S. (2000a). Synchonization and clustering in coupled saw-tooth maps, *Int. J. Bif. Chaos* **10**, 10, pp. 2465-2478.

Manrubia, S. C. and Mikhailov, A. S. (2000b). Very long transients in globally coupled maps, *Europhys. Lett.* **50** 5, pp. 580–586.

Manrubia, S. C. and Mikhailov, A. S. (2001). Globally coupled logistic maps as dynamical glasses, *Europhys. Lett.* **53**, 4, pp. 451–457.

Marinari, E., Parisi, G. and Ritort, F. (1994). Replica field theory for deterministic models. II. A non-random spin glass with glassy behaviour, *J. Phys. A* **27**, 23, pp. 7647–7668.

Markus, M., Kuschmitz, D. and Hess, B. (1984). Chaotic dynamics in yeast glycolysis under periodic substrate input flux, *FEBS Lett.* **172**, 2, pp. 235–238.

Markus, M., Kuschmitz, D. and Hess, B. (1985). Properties of strange attractors in yeast glycolysis, *Biophys. Chem.* **22**, 1–2, pp. 95–108.

Matthews, P. C., Mirollo, R. E. and Strogatz, S. H. (1991). Dynamics of a large system of coupled nonlinear oscillators, *Physica D* **52**, 2–3, pp. 293–331.

Matthews, E. K. and O'Connor, M. D. L. (1979). Dynamic oscillations in the membrane potential of pancreatic islet cells, in: Cellular Oscillations, ed. M. J. Berridge *et al.*, London, Academic Press, pp. 75–91.

Matthews, P. C. and Strogatz, S. H. (1990). Phase diagram for the collective behavior of limit-cycle oscillators, *Phys. Rev. Lett.* **65**, 14, pp. 1701–1704.

May, R. M. (1974). Biological populations with nonoverlapping generations: Stable points, stable cycles, and chaos, *Science* **186**, 4164, pp. 645–647.

McCulloch, W. C. and Pitts, W. (1943). A logical calculus of the ideas immanent in neural networks, *Bull. Math. Biophys.* **5**, 1, pp. 115–133.

Mézard, M., Parisi, G. and Virasoro, M. (1987). Spin Glass Theory and Beyond. An Introduction to the Replica Method and Its Applications, World Scientific, Singapore, 476p.

Mézard, M. and Virasoro, M. A. (1985). The microstructure of ultrametricity, *J. Phys.* **46**, 8, pp. 1293–1307.

Mikhailov, A. S. (1994). Foundations of Synergetics I. Distributed Active Systems, 2nd edition, Springer, Berlin, 214p.

Mikhailov, A. S. and Battogtokh, D. (1996). Controlling turbulence in the complex Ginzburg-Landau equation, *Physica D*, 1–2, **90**, pp. 84–95.

Mikhailov, A. S. and Calenbuhr, V. (2002). From Cells to Societies. Models of Complex Coherent Action, Springer, Berlin, 299p.

Mikhailov, A. S. and Hess, B. (1996). Microscopic self-organization of enzymic reactions in small volumes, *J. Phys. Chem.* **100**, 49, pp. 19059–19065.

Mikhailov, A. S. and Loskutov, A. Yu. (1996). Foundations of Synergetics II. Chaos and Noise, 2nd enlarged edition, Springer, Berlin, 277p.

Milton, J. and Jung, P. (2003). Epilepsy as a Dynamic Disease, Springer, Berlin, 417p.

Mirollo, R. E. and Strogatz, S. H. (1990a). Amplitude death in an array of limit-cycle oscillators, *J. Stat. Phys.* **60**, 1–2, pp. 245–262.

Mirollo, R. E. and Strogatz, S. H. (1990b). Synchronization of pulse-coupled biological oscillators, *SIAM J. Appl. Math.* **50**, 6, pp. 1645–1662.

Montbrió, E. and Blasius, B. (2003). Using nonisochronicity to control synchronization in ensembles of nonidentical oscillators, *Chaos* **13**, 1, pp 291–308.

Morita, S. and Chawanya, T. (2002). Collective motions in globally coupled tent maps with stochastic updating, *Phys. Rev. E* **65**, 046201.

Mosekilde, E., Maistrenko, Y., Postnov, D. and Maistrenko, Iu. L. (2002). Chaotic Synchronization: Applications to Living Systems, World Scientific, Singapore, 440p.

Muñoz, M. A. and Pastor-Satorras, R. (2003). Stochastic theory of synchronization transitions in extended systems, *Phys. Rev. Lett.* **90**, 204101.

Nakagawa, N. and Komatsu, T. S. (1998). Collective motion occurs inevitably in a class of populations of globally coupled chaotic elements, *Phys. Rev. E* **57**, 2, pp. 1570–1575.

Nakagawa, N. and Kuramoto, Y. (1993). Collective chaos in a population of globally coupled oscillators, *Prog. Theor. Phys.* **89**, 2, pp. 313–323.

Nakagawa, N. and Kuramoto, Y. (1994). From collective oscillations to collective chaos in a globally coupled oscillator system, *Physica D* **75**, 1–3, pp. 74–80.

Nakagawa, N. and Kuramoto, Y. (1995). Anomalous Lyapunov spectrum in globally coupled oscillators, *Physica D* **80**, 3, pp. 307–316.

Niebur, E., Schuster, H. G., Kammen, D. M and Koch, C. (1991). Oscillator-phase coupling for different two-dimensional network connectivities, *Phys. Rev. A* **44**, 10, pp. 6895–6904.

Okuda, K. (1993). Variety and generality of clustering in globally coupled oscillators, *Physica D* **63**, 3–4, pp. 422–436.

O'Rourke, B., Ramza, B. M. and Marban, E. (1994). Oscillations of membrane current and excitability driven by metabolic oscillations in heart cells, *Science* **265**, 5174, pp. 962–966.

Osipov, G. V., Pikovsky, A. S., Rosenblum, M. G. and Kurths, J. (1997). Phase synchronization effects in a lattice of nonidentical Rössler oscillators, *Phys. Rev. E* **55**, 3, pp. 2353–2361.

Ott, E., So, P., Barreto, E. and Antonsen, T. (2002). The onset of synchronization in systems of globally coupled chaotic and periodic oscillators, *Physica D* **173**, 1-2, pp. 29-51.

Ott, E., Sommerer, J. C., Alexander, J. C., Kan, I. and Yorke, J. A. (1993). Scaling behavior of chaotic systems with riddled basins, *Phys. Rev. Lett.* **71**, 25, pp. 4134–4137.

Panter, P. (1965) Modulation, Noise and Spectral Analysis, McGraw-Hill, New York, 311 p.

Park, K., Rhee, S. W. and Choi, M. Y. (1998). Glass synchronization in the network of oscillators with random phase shifts, *Phys. Rev. E* **57**, 5, pp. 5030–5035.

Parravano, A. and Cosenza, M. G. (1998). Driven maps and the emergence of ordered collective behavior in globally coupled maps, *Phys. Rev. E* **58**, 2, pp. 1665–1671.

Pecora, L. M. (1998a). Chaos starts to communicate, *Phys. World* **11**, 4, pp. 25–26.

Pecora, L. M. (1998b). Synchronization conditions and desynchronizing patterns in coupled limit-cycle and chaotic systems, *Phys. Rev. E* **58**, 1, pp. 347–360.

Pecora, L. M. and Carroll, T. L. (1998). Master stability functions for synchronized coupled systems, *Phys. Rev. Lett.* **80**, 10, pp. 2109-2112.

Pecora, L. M., Carroll, T. L., Johnson, G. A., Mar, D. J. and Heagy, J. F. (1997). Fundamentals of synchronization in chaotic systems, concepts, and applications, *Chaos* **7**, 4, pp. 520–543.

Pérez, G. and Cerdeira, H. (1992). Instabilities and nonstatistical behavior in globally coupled systems, *Phys. Rev. A* **46**, 12, pp. 7492–7497.

Peskin, C. S. (1975). Mathematical Aspects of Heart Physiology, Courant Institute of Mathematical Sciences, New York, 321p.

Pikovsky, A. S. and Kurths, J. (1994). Do globally coupled maps really violate the law of large numbers? *Phys. Rev. Lett.* **72**, 11, pp. 1644–1646.

Pikovsky, A. S., Rosenblum, M. G. and Kurths, J. (1996). Synchronization in a population of globally coupled chaotic oscillators, *Europhys. Lett.* **34**, 3, pp. 165–170.

Pikovsky, A. S., Rosenblum, M. G. and Kurths, J. (2000). Phase synchronization in regular and chaotic systems, *Int. J. Bif. Chaos* **10**, 10, pp. 2291–2305.

Pikovsky, A., Rosenblum, M. and Kurths, J. (2001). Synchronization. A universal concept in non-linear sciences, Cambridge Nonlinear Science Series 12, Cambridge University Press, Cambridge, 500p.

Pikovsky, A. S., Rosenblum, M., Osipov, G. V. and Kurths, J. (1997). Phase synchronization of chaotic oscillators by external driving, *Physica D* **104**, 3–4, pp. 219–238.

Popovych, O., Maistrenko, Yu. and Mosekilde, E. (2001). Loss of coherence in a system of globally coupled maps, *Phys. Rev. E* **64**, 026205.

Ramana Reddy, D. V., Sen, A. and Johnston, G. L. (1998). Time delay induced death in coupled limit-cycle oscillators, *Phys. Rev. Lett.* **80**, 23, pp. 5109–5112.

Ramana Reddy, D. V., Sen, A. and Johnston, G. L. (1999). Time delay effects on coupled limit cycle oscillators at Hopf bifurcation, *Physica D* **129**, 1–2, pp. 15–34.

Richard, P., Teussink, B., Van Dam, K. and Westerhoff, H. V. (1996). Sustained oscillations in free-enery state and hexose phoshate in yeast, *Yeast* **12**, 8, pp. 731–740.

Rim, S., Kim, I., Kang, P., Park, Y. J. and Kim C. M. (2002). Routes to complete synchronization via phase synchronization in coupled nonidentical chaotic oscillators, *Phys. Rev. E* **66**, 015205(R).

Rodriguez, E., George, N., Lachaux, J.-Ph., Martinerie, J., Renault, B., and Varela, F. J. (1999). Perception's shadow: long-distance synchronization of human brain activity, *Nature* **397**, 6718, pp. 430–33.

Roelfsema, P. R., Engel, A. K., König, P., and Singer, W. (1997). Visuomotor integration is associated with zero time-lag synchronization among cortical areas, *Nature* **385**, 6612, pp. 157–161.

Rosenblum, M. G., Pikovsky, A. S. and Kurths, J. (1996). Phase synchronization of chaotic oscillators, *Phys. Rev. Lett.* **76**, 11, pp. 1804–1807.

Rosenblum, M. G., Pikovsky, A. S. and Kurths, J. (1997). From phase to lag synchronization in coupled chaotic oscillators, *Phys. Rev. Lett.* **78**, 22, pp. 4193–4196.

Rössler, O. E. (1976). Equation for continuous chaos, *Phys. Lett. A* **57**, 5, pp. 397–398.

Rudrauf, D., Lutz, A., Cosmelli, D., Lachaux, J.-Ph., and Le Van Quyen, M. (2003). From autopoiesis to neurophenomenology: Fransisco Varela's exploration of the biophysics of being, *Biol. Res.* **36**, 1, pp. 27–65.

Sakaguchi, H. (1988). Cooperative phenomena in coupled oscillator systems under external fields, *Prog. Theor. Phys.* **79**, 1, pp. 39–46.

Sakaguchi, H. and Kuramoto, Y. (1986). A soluble active rotator model showing phase transitions via mutual entrainment, *Prog. Theor. Phys.* **76**, 3, pp. 576–581.

Sakaguchi, H., Shinimoto, S. and Kuramoto, Y. (1987). Local and global self-entrainments in oscillator lattices, *Prog. Theor. Phys.* **77**, 5, pp. 1005–1010.

Sameshima, T., Fukushima, K. and Yamada,T. (2001). Chaotic transition in a five-coupled ϕ^4-field soliton system, *Physica D* **150**, 1–2, pp. 104–117.

Schienbein, M. and Gruler, H. (1997). Enzymic kinetics, self-organized molecular machines, and parametric resonance, *Phys. Rev. E* **56**, 6, pp. 7116–7127.

Schuster, H. G. and Wagner, P. (1989). Mutual entrainment of two limit cycle oscillators with time delayed coupling, *Prog. Theor. Phys.* **81**, 5, pp. 939–945.

Sharma, A. and Gupte, N. (2002). Spatiotemporal intermittency and scaling laws in inhomogeneous coupled map lattices, *Phys. Rev. E* **66**, 036210.

Shibata, T., Chawanya, T. and Kaneko, K. (1999). Noiseless collective motion out of noisy chaos, *Phys. Rev. Lett.* **82**, 22, pp. 4424–4427.

Shibata, T. and Kaneko, K. (1997). Heterogeneity induced order in globally coupled chaotic systems, *Europhys. Lett.* **38**, 6, pp. 417–422.

Shiino, M. and Francowicz, M. (1989). Synchronization of infinitely many coupled limit-cycle type oscillators, *Phys. Lett. A* **136**, 2, pp. 103–108.

Shinbrot, T. (1994). Synchronization of coupled maps and stable windows, *Phys. Rev. E* **50**, 4, pp. 3230–3233.

Shinomoto, S. and Kuramoto, Y. (1986a). Phase-transitions in active rotator systems, *Prog. Theor. Phys.* **75**, 5, pp. 1105–1110.

Shinomoto, S. and Kuramoto, Y. (1986b). Cooperative phenomena in two-dimensional active rotator systems, *Prog. Theor. Phys.* **75**, 6, pp. 1319–1327.

Shinomoto, S. and Kuramoto, Y. (1988). Phase transitions and their bifurcation analysis in a large population of active rotators with mean-field coupling, *Prog. Theor. Phys.* **79**, 3, pp. 600–607.

Sismondo, E. (1990). Synchronous, alternating, and phase-locked stridulation by a tropical katydid, *Science* **249**, 4964, pp. 55–58.

Sosnovtseva, O. V., Balanov, A. G., Vadivasova, T. E., Astakhov, V. V. and

Mosekilde, E. (1999). Loss of lag synchronization in coupled chaotic systems, *Phys. Rev. E* **60**, 6, p. 6560–6565.

Stange, P., Mikhailov, A. S. and Hess, B. (1998). Mutual synchronization of molecular turnover cycles in allosteric enzymes, *J. Phys. Chem. B* **102**, 32, pp. 6273–6289.

Stange, P., Mikhailov, A. S. and Hess, B. (1999). Mutual synchronization of molecular turnover cycles in allosteric enzymes. II. Product inhibition, *J. Phys. Chem. B* **103**, 29, pp. 6111–6120.

Stange, P., Mikhailov, A. S. and Hess, B. (2000). Coherent intramolecular dynamics of enzymic reaction loops in small volumes, *J. Phys. Chem. B* **104**, 8, pp. 1844–1853.

Stange, P., Zanette, D. H., Mikhailov, A. S. and Hess, B. (1998a). Self-organizing molecular networks, *Biophys. Chem.* **72**, 1–2, pp. 73–85.

Stiller, J. C. and Radons, G. (1998). Dynamics of nonlinear oscillators with random interactions, *Phys. Rev. E* **58**, 2, p. 1789–1799.

Strang, G. (1986). Introduction to Applied Mathematics, Wellesley-Cambridge Press, Wellesley, 758p.

Strogatz, S. H. (1994). Nonlinear Dynamics and Chaos: With Applications in Physics, Biology, Chemistry, and Engineering, Perseus Publishing, New York, 498p.

Strogatz, S. H. (1998). Death by delay, *Nature* **394**, 6691, pp. 316–317.

Strogatz, S. H. (2000). From Kuramoto to Crawford: Exploring the onset of synchronization in populations of coupled oscillators, *Physica D* **143**, 1–4, pp. 1–20.

Strogatz, S. H. (2003). Sync: The Emerging Science of Spontaneous Order, Hyperion Press, New York, 338p.

Strogatz, S. H., Mirollo, R. E. and Matthews, P. C. (1992). Coupled nonlinear oscillators below the synchronization threshold: Relaxation be generalized Landau damping, *Phys. Rev. Lett.* **68**, 18, pp. 2730–2733.

Tass, P. (1997). Phase and frequency in a population of phase oscillators, *Phys. Rev. E* **56**, 2, pp. 2043–2060.

Tass, P. (1999). Phase Resetting in Medicine and Biology, Springer, Berlin, 329p.

Tass, P. A., Fieseler, T., Dammers, J., Dolan, K., Morosan, P., Majtanik, M., Boers, F., Muren, A., Zilles, K., and Fink, G. R. (2003). Synchronization tomography: a method for three-dimensional localization of phase synchronized neuronal populations in the human brain using magnetoencephalography, *Phys. Rev. Lett.* **90**, 8, 088101.

Tononi, G. and Edelman, G. M. (1998). Neuroscience. Consciousness and complexity, *Science* **282**, 5395, pp. 1846–1851.

Tononi, G., Srinivasan, R., Russell, D. P., Edelman, G. M. (1998). Investigating neural correlates of conscious perception by frequency-tagged neuromagnetic responses, *Proc. Natl. Acad. Sci. U. S. A.* **95**, 6, pp. 3198–3203.

Tornheim, K. (1997). Are metabolic oscillations responsible for normal oscillatory insulin secretion?, *Diabetes* **46**, 9, pp. 1375–1380.

Tsang, K. Y., Mirollo, R. E., Strogatz, S. H. and Wiesenfeld K. (1991). Dynamics of a globally coupled oscillator array, *Physica D* **48**, 1, pp. 102–112.

Tsodyks, M., Mitkov, I., and Sompolinsky, H. (1993). Pattern of synchrony in inhomogeneous networks of oscillators with pulse interactions, *Phys. Rev. Lett.* **71**, 8, pp. 1280–1283.

Turing, A. M. (1952). The chemical basis of morphogenesis, *Philos. Trans. Roy. Soc. (London) B* **237**, 641, pp. 37–72.

van Vreeswijk, C. and Sompolinsky, H. (1996). Chaos in neuronal networks with balanced excitatory and inhibitory activity, *Science* **274**, 5293, pp. 1724–1726.

Varela, F. J. (1995). Resonant cell assemblies: a new approach to cognitive functions and neural synchrony, *Biol. Res.* **28**, 1, pp. 81–95.

Venkataramani, S. C., Antonsen, T. M., Ott, E. and Sommerer, J. C. (1996). On-off intermittency: power spectrum and fractal properties of time series, *Physica D* **96**, 1–4, pp. 66–99.

Veser, G., Mertens, F., Mikhailov, A. S. and Imbihl, R. (1993). Global coupling in the presence of defects: synchronization in an oscillatory surface reaction, *Phys. Rev. Lett.* **71**, 6, pp. 935–939.

von der Malsburg, C. and Schneider, W. (1986). A neural cocktail-party processor, *Biol. Cybern.* **54**, 1, pp. 29–40.

von der Malsburg, C., and Singer, W. (1988). Principles of cortical network organization, in Neurology of Neocortex, eds. Rakic, P. and Singer, W., Wiley, Chichester, pp. 69–99.

Walker, T. J. (1969). Acoustic synchrony: Two mechanisms in the snowy tree cricket, *Science* **166**, 3907, pp. 891–894.

Wang, W., Kiss, I. Z., Hudson, J. L. (2000). Experiments on arrays of globally coupled chaotic electrochemical oscillators: synchronization and clustering, *Chaos* **10**, 1, pp. 248–256.

Wang, W., Kiss, I. Z., Hudson, J. L. (2001). Clustering of arrays of chaotic chemical oscillators by feedback and forcing, *Phys. Rev. Lett.* **86**, 21, pp. 4954–4957.

Watts, D. J. and Strogatz, S. H. (1998). Collective dynamics of 'small-world' networks, *Nature* **393**, 6684, pp. 440-442.

Wiener N. (1948). Cybernetics, or Control and Communication in the Animal and the Machine, MIT Press, Cambridge, 212p.

Wiesenfeld, K. and Hadley, P. (1989). Attractor crowding in oscillator arrays, *Phys. Rev. Lett.* **62**, 12, pp. 1335–1338.

Wiggins, S. (1990). Introduction to Applied Nonlinear Dynamical Systems and Chaos, Springer-Verlag, New York, 672p.

Winfree, A. T. (1967). Biological rhythms and the behavior of populations of coupled oscillators, *J. Theor. Biol.* **16**, 1, pp. 15–42.

Winfree, A. T. (1972). Oscillatory glycolysis in yeast: the pattern of phase resetting by oxygen, *Arch. Biochem. Biophys.* **149**, 2, pp. 388–401.

Winfree, A. T. (2001). The Geometry of Biological Time, Springer, New York, 777p.

Wolfram, S. (1983). Statistical mechanics of cellular automata, *Rev. Mod. Phys.* **55**, 3, pp. 601-644.

Xie, F. and Hu, G. (1997). Clustering dynamics in globally coupled map lattices,

Phys. Rev. E **56**, 2, pp. 1567–1570.

Yamada, T. and Fujisaka, H. (1983). Stability theory of synchronized motion in coupled-oscillator systems. II, *Prog. Theor. Phys.* **70**, 5, pp. 1240–1248.

Yamada, T. and Fujisaka, H. (1984). Stability theory of synchronized motion in coupled-oscillator systems. III, *Prog. Theor. Phys.* **72**, 5, pp. 885–894.

Yeung, M. K. S. and Strogatz, S. H. (1999). Time delay in the Kuramoto model of coupled oscillators, *Phys. Rev. Lett.* **82**, 3, pp. 648–651.

Zanette, D. H. (1999). Dynamics of chaotic maps with global inhomogeneous coupling, *Europhys. Lett.* **45**, 4, pp. 424–430.

Zanette, D. H. (2000). Propagating structures in globally coupled systems with time delays, *Phys. Rev. E* **62**, 3, pp. 3167–3172.

Zanette, D. H. and Mikhailov, A. S. (1998a). Condensation in globally coupled populations of chaotic dynamical systems, *Phys. Rev. E* **57**, 1, pp. 276–281.

Zanette, D. H. and Mikhailov, A. S. (1998b). Mutual synchronization in ensembles of globally coupled neural networks, *Phys. Rev. E* **58**, 1, pp. 872–875.

Zanette, D. H. and Mikhailov, A. S. (2000). Dynamical clustering in large populations of Rössler oscillators under the action of noise, *Phys. Rev. E* **62**, 6, pp. R7571–R7574.

Zhai, Y., Kiss, I. Z. and Hudson, J. L. (2003). Emerging coherence of oscillating chemical reactions on arrays: experiments and simulations, *Ind. Eng. Chem. Res.*, in press.

Zhou, C. and Kurths, J. (2001). Spatiotemporal coherence resonance of phase synchronization in weakly coupled chaotic oscillators, *Phys. Rev. E* **65**, 040101(R).

Index

amplitude, 252
  in chaotic systems, 116
analytic signal, 252
Arnol'd tongue, 18
array
  one-dimensional, 62, 77, 156
  regular, 200
  two-dimensional, 156
asynchronous updating, 184, 190
attraction basin, 179, 180
  globally riddled, 179
  weight of an, 220
attractor coding, 211
attractors
  chaotic, 109, 275
  coexistence of, 174, 179, 193
  number of, 181
  reconstruction of, 234
  weight of, 210

bifurcation
  Andronov–Hopf, 83, 89, 97, 228, 230, 265, 269, 277, 311
  period doubling, 110, 234
  pitchfork, 269
  saddle-node, 303, 306
  Turing, 269
biochemical reactions, 273
brain, 53, 312
  synchronization in the, 312, 320

cell, 273
  yeast, 275
cell differentiation, 279, 285
cellular structures, 262
  hexagonal, 270
chaos, 109
chemical turbulence, 257
cluster patterns, 260
clustered state, 30
clustering, 22, 171, 187
  dynamical, 239, 279, 285, 328
  frequency, 43
  fuzzy, 193
  hierarchical, 194, 197, 240
clustering phase, 178, 212
clusters, 43
  frequency-synchronized, 45
collective chaos, 97, 100
connectivity threshold, 144
correlation function, 102, 126
coupling
  global, 19, 136, 229
    complex, 94
  in chaotic systems, 131
  local, 158
  non-local, 99
  shift-invariant, 132, 150
  vector, 139, 197
  weak, 94
coupling field, 175, 182, 191, 217, 223
coupling intensity, 15, 229
  critical, 38, 42, 56, 72, 165, 233
  distribution of, 161

truncated Gaussian, 161
uniform, 161
critical dimension, 69
critical point, 49, 169

delays
global, 248, 255, 266
time, 51, 75, 91, 230, 323
uniform, 81
disorder, 125
dynamical, 47
quenched, 150, 164, 166
distance, 209

eigenvalue, 24, 57, 91
longitudinal, 21, 24, 28, 54, 80
transversal, 21, 25, 80
eigenvalue problem, 21, 53, 82
eigenvector, 21
eigenvector problem, 24
enzymes, 274, 290
ergodic hypothesis, 203
Ermentrout–Kopell theorem, 308

fixed point, 15, 28, 37, 65, 83
force
external, 46, 98, 104
fluctuating, 75
random, 47
fractal dimension, 104
frequency
effective, 36, 66, 152, 232
natural, 14, 36, 152, 232
identical, 52, 53, 62, 75, 85, 86
synchronization, 23, 36, 51, 54, 55, 62, 65, 77, 81, 85, 152
frequency distribution, 41, 44, 74
Gaussian, 41, 67
Lorentzian, 42, 57
power-law, 69
symmetric, 55
frustration, 73, 204, 207

Ginzburg–Landau equation, 84, 99, 265
glycolytic pathway, 274
graph, 61
directed, 81

heterogeneous ensemble, 35, 56, 66, 70, 88, 147, 149, 161

initial condition, 26, 33, 64, 80, 174, 179, 190, 206, 217, 221
sensitivity to the, 109
interaction coefficients, 70
interaction functions, 14, 15, 22, 25, 27, 62, 75, 133
uniform, 19
interaction network
biochemical reactions, 280
complex, 61
eigenvalues of the, 82
enzyme, 274
random, 70, 144
regular, 144
small-world, 145
symmetric, 142
uniform, 77
interactions
frustrated, 72
heterogeneous, 61
local, 62, 81
time-delayed, 50, 56
intermittent behavior, 101, 128, 243
intermittent turbulence, 248, 258
Ising ferromagnet, 204
itinerancy, 194, 196

Jacobian matrix, 113, 122
Josephson junction, 27

Langevin equation, 297
lattice
coupled map, 167
hypercubic, 68
regular, 62, 99
three-dimensional cubic, 68
law of large numbers, 184
linear stability analysis, 20, 23, 28, 29, 53, 54, 56, 63, 79, 86, 90, 117, 133, 175, 268

local field, 73
logistic map, 109
Lorenz oscillator, 128
Lyapunov exponent, 113, 135, 139
  conditional, 123
  spectrum of, 125, 127
  splitting, 176
  transversal, 117, 122, 135, 160, 175

macroscopic state, 203
manifold
  cluster synchronization, 201
  invariant, 121
  synchronization, 114, 121, 134
  transverse, 121
maps
  contracting, 187
  driven, 188
  globally coupled, 172
  identical chaotic, 187
  logarithmic, 190
  logistic, 141
    globally coupled, 159
    one-dimensional array of, 169
    randomly coupled, 164, 193
  one-dimensional, 114
  saw-tooth, 202
    nonlinear, 192
  tent, 188
  two coupled, 116
master stability function, 137, 139
mean-field approximation, 40, 159, 206, 322
microscopic state, 205
molecular machines, 290
motion
  aperiodic, 98
  chaotic, 110
  incoherent, 17, 32, 86
  intermittent, 118, 130, 167
  periodic, 32, 162
  quasiperiodic, 185
multistability, 27, 81, 97, 324

network connectivity, 144, 164
neural networks, 322, 324
neurons, 303, 304
  Class I, 305
  Class II, 311
  excitable, 309
  integrate-and-fire, 303, 310, 311, 323
noise, 47, 56, 150, 166, 185, 197, 227, 244

order
  hidden, 171, 182, 185, 187, 189, 191
  local, 169
order parameter, 39, 48, 50, 55, 68, 71, 74, 148, 155, 165, 206, 220, 233, 236, 301
oscillation death, 86, 91
oscillators
  chemical, 86
  electrochemical, 227–229
    chaotic, 234
    periodic, 230
  entrained, 40, 71
  globally coupled, 84
  harmonic, 84
  identical, 20, 27, 142
  limit-cycle, 83
    identical, 92
  non-entrained, 46
  non-identical, 147
  periodic, 13, 14
    coupled, 13
    locally coupled, 255
    non-identical, 35
  phase, 13, 14, 53, 56, 75
    globally coupled, 51
    identical, 22, 47
    non-identical, 70
  Rössler, 137, 150, 197
    non-identical, 152, 158
  two coupled, 44, 51, 85, 91, 121, 128
  weakly nonlinear, 84, 267
overlap, 208
  dynamical, 215, 216
overlap distribution, 208, 218, 219, 223

oxidation of carbon monoxide, 245
  mathematical model of, 255

pair-distance distribution, 195, 198, 236
parameter space, 18, 25, 33, 57, 86, 193
partition, 180
  probability of a, 213
  two-cluster, 174
partition variety, 212
periodic windows, 172, 184
permutation symmetry, 200
phase, 14, 216, 230, 243, 252
  in chaotic systems, 115
phase density, 31, 36, 48, 56
phase difference, 15, 51, 78
phase oscillator, 14, 296
phase resetting, 277
phase transition, 40, 169, 246
Poincaré section, 98, 115, 126
population
  cellular, 273, 279
  chaotic, 131
  density, 110
power spectrum, 103, 156

Rössler oscillator, 111
replica, 206, 216
replica method, 75
return map, 185
ring, 62, 77, 137, 157

saw-tooth map, 118
self-averaging variable, 211
spatiotemporal pattern, 62, 100, 158, 247, 248, 252
spin, 204
spin glass, 204
  mean-field, 204
stability regions, 25, 33, 54, 58, 86, 92, 95
surface reaction, 227, 245
symmetry breaking, 204
  replica, 208
  spontaneous, 206
synchronization, 292, 312
  $\delta-$, 147
  amplitude, 86
  anti-phase, 16
  chaotic, 146
  enzyme activity, 299
  frequency, 17, 63, 69, 70, 76, 85, 90, 232
  full, 16, 19, 20, 28, 48, 54, 62, 79, 81, 85, 87, 95, 114, 134, 154, 163, 172, 188, 285, 328
  lag, 126
  large-scale, 322
  mutual, 13, 22
  partial, 36, 88, 330
  phase, 123, 152, 244, 279
    partial, 153
    weak, 243
synchronous cluster, 38

thermodynamic limit, 217
time evolution operator, 135
transition
  destabilization, 140
  full synchronization, 128, 238
  phase synchronization, 123, 155
  synchronization, 39, 148, 252
  synchronous phase, 220
turbulent phase, 182

ultrametricity, 204, 210, 224
universality class, 118, 187

waves
  standing, 247, 262, 270
  traveling, 63, 247